Yeast Physiology and Biotechnology

Yeast Physiology and Biotechnology

Edited by Harvey Marshall

SYRAWOOD
PUBLISHING HOUSE
New York

Published by Syrawood Publishing House,
750 Third Avenue, 9th Floor,
New York, NY 10017, USA
www.syrawoodpublishinghouse.com

Yeast Physiology and Biotechnology
Edited by Harvey Marshall

International Standard Book Number: 978-1-64740-397-3 (Hardback)

Cataloging-in-publication Data

Yeast physiology and biotechnology / edited by Harvey Marshall.
p. cm.
Includes bibliographical references and index.
ISBN 978-1-64740-397-3
1. Yeast fungi--Physiology. 2. Yeast fungi--Biotechnology. 3. Microbial biotechnology.
4. Biotechnology. I. Marshall, Harvey.
TP248.27.Y43 Y44 2023
660.62--dc23

TABLE OF CONTENTS

PREFACE

Yeasts are eukaryotic and single-celled microorganisms that belong to the fungus kingdom. The beneficial physiological characteristics of yeasts have led to their application in the field of biotechnology. Sugar fermentation by yeast is the oldest and most widespread application of this technology. The production of different foods involve the use of different type of yeasts, including brewer's yeast for fermenting beer, baker's yeast for production of bread, as well as yeast for making xylitol and wine. They are also utilized as biocontrol agents in agriculture. They can also be used as a source of biocatalysts for the creation of protein medicines and high-value fine chemicals. This book is a valuable compilation of topics, ranging from the basic to the most complex advancements in the field of yeast biotechnology. It will help the readers in keeping pace with the rapid changes in these fields.

The information contained in this book is the result of intensive hard work done by researchers in this field. All due efforts have been made to make this book serve as a complete guiding source for students and researchers. The topics in this book have been comprehensively explained to help readers understand the growing trends in the field.

I would like to thank the entire group of writers who made sincere efforts in this book and my family who supported me in my efforts of working on this book. I take this opportunity to thank all those who have been a guiding force throughout my life.

Editor

1

Exploring the Impact of Lipid-Rich Food Industry Waste Carbon Sources on the Growth of *Candida cylindracea* DSM 2031

Bartłomiej Zieniuk *, Patrycja Mazurczak-Zieniuk and Agata Fabiszewska

Department of Chemistry, Institute of Food Sciences, Warsaw University of Life Sciences-SGGW, 159c Nowoursynowska St., 02-776 Warsaw, Poland; patrycja.mazurczak@op.pl (P.M.-Z.); agata_fabiszewska@sggw.edu.pl (A.F.)

* Correspondence: bartlomiej_zieniuk@sggw.edu.pl

Abstract: The aim of this study was to evaluate the possibility of using several lipid-rich food industry wastes in the culture medium on the growth of *Candida cylindracea* DSM 2031 yeast strain. Four lipid wastes from the food industry: waste fish oil, rancid ghee, waste pork lard, and waste duck processing oil were investigated. It has been shown in the laboratory scale that the above-mentioned wastes can be used to obtain biomass and produce lipolytic enzymes by the tested strain and the *C. cylindracea* extracellular lipase is not constitutive. High yields of biomass (12.84, 12.75, and 12.24 g/dm^3) were obtained in media containing waste duck processing oil, olive oil, and waste pork lard, respectively. The highest lipolytic activity was obtained in the media containing waste fish oil and rancid ghee (0.050 and 0.047 U/cm^3). During 192-h flask cultures the highest extracellular lipase activity and biomass yield were observed in the late logarithmic phase. The study showed that there is a potential for waste management to produce lipolytic enzymes or to produce yeast biomass. The use of waste substrates may contribute to lowering the costs of commercial production, and such a solution is part of the sustainable development strategy.

Keywords: *Candida cylindracea*; lipases; lipid-rich food industry wastes; lipolytic activity

---

1. Introduction

Waste production is an integral feature of human activity and their proper management is an important problem. Wastes are burdensome for the environment, which is mainly associated with water, air and soil pollution. In 2018, 128 million tonnes of waste was generated in Poland, of which almost 10% was municipal waste and about 22% of the waste came from industrial processing [1]. A very interesting approach to waste management is the use of microorganisms that simultaneously produce many valuable metabolites. Biotechnological production of chemical compounds and biocatalysts in the form of enzymes with the use of waste substrates reduce the costs of obtaining valued metabolites. At the same time, enzymes are biodegradable, and their use in the industry contributes to the reduction of the use of chemical catalysts. The microorganisms described in the literature are both bacteria, moulds, and yeasts, and among them varied species from *Candida* genera.

The genus *Candida* is widespread in natural habitats. Up to 2014, three hundred fourteen species belonging to the genus *Candida* have been described [2]. These yeasts are widespread in the natural environment, especially in humid conditions, rich in organic compounds including organic acids and ethanol. The influence of *Candida* yeast is not without significance for the food industry. Various species are involved in both food and feed production, as well as, can be the cause of their spoilage. They are isolated from dairy products, sourdoughs, meats, and sausages, most often from local, fermented products and snacks. Because of their uniqueness, *Candida* yeasts can be useful in miscellaneous biotechnological processes and selected enzymes produced by *Candida* yeasts, like lipases from *C. antarctica*, *C. rugosa*, or *C. cylindracea*, are of commercial and industrial interest [2,3].

The latter one, *C. cylindracea* is worth mentioning, but it should be clarified that yeast of the species *C. cylindracea* are often misled as synonymous with the species of yeast *C. rugosa*. Kurtzman and Robnett [4] tested the ribosomal DNA of both of these species and showed that *C. cylindracea* and *C. rugosa* are separate species representing the Ascomycota phylum. Subsequently, Boontham et al. [5] assigned *C. cylindracea* to the genus *Limtongozyma*. *C. cylindracea* was first isolated from soil in Japan in 1962 by Yamada and Machida [6]. The described yeast species was also isolated from the strelitzia flower in French Guiana, from the fruit of *Sapindus* sp. on one of the Hawaiian islands or from undercooked shrimps. They are non-pathogenic yeasts, which prefer aerobic growth and do not grow at 37 °C, therefore they do not pose a threat to human and animal health [3]. For many years, this species has been of great interest due to its ability to produce lipolytic enzymes and its potential use in the production of food and pharmaceuticals [7–11]. The analysis of the GenBank database reveals the known sequences of five triacylglycerol lipases and products of *LIP1*, *LIP2*, *LIP3*, *LIP4*, and *LIP5* genes.

The aim of the study was the possibility of using selected waste lipid substrates as the main carbon source in the culture of *C. cylindracea* DSM 2031. The work consists of two main stages. First—assessment of biomass yield and extracellular lipolytic activity of *C. cylindracea* DSM 2031 in shake flask cultures in media containing lipid-rich food industry wastes as carbon sources and olive oil as a control carbon source. In the second stage, the growth and activity of extracellular lipolytic enzymes of *C. cylindracea* DSM 2031 were evaluated during 192 h of shake culture in media containing waste fish oil.

2. Materials and Methods

2.1. Microorganism

The following yeast strain was used in the study: *Candida cylindracea* DSM 2031 from the German Collection of Microorganisms and Cell Cultures (Deutsche Sammlung von Mikroorganismen und Zellkulturen GmbH, DSMZ, Braunschweig, Germany). The mentioned strain has been stored in vials containing cryobeads at −20 °C (Protect Select, Technical Service Consultants Ltd., Heywood, UK).

2.2. Materials

Four types of waste oils and fats were used in the experiments: waste fish oil obtained in the process of smoking fish carcasses from the fish processing plant in Poland; ghee, which went rancid and was purchased at retail in a local supermarket in Warsaw (Poland); waste pork lard obtained from smokehouse company located in Podlaskie Voivodeship (Poland); as well as waste duck processing oil from the household poultry roasting process. Moreover, olive oil also acquired in the local supermarket was used as a control carbon source in yeast cultures. Culture media components were purchased from BTL Sp. z o. o. (Łódź, Poland). Chemicals were purchased from Sigma-Aldrich (Poznań, Poland) and Avantor Performance Materials Poland S.A. (Gliwice, Poland), and *p*-nitrophenyl laurate was synthesized in the Department of Chemistry, Institute of Food Sciences, Warsaw University of Life Sciences-SGGW [12].

2.3. Determination of the Fatty Acid Composition of Oils and Fats Used

Fatty acid compositions of oils and fats were determined after derivatization to methyl esters according to Zieniuk et al. [13] and Agilent 7820A (Agilent Technologies, Santa Clara, CA, USA) with flame ionization detector (FID) was used. The fatty acid profiles of oils and fats are shown in Table 1.

Table 1. Profiles of fatty acids in olive oil and lipid-rich food industry wastes [%].

	Fatty Acid Profiles [%]				
	Olive Oil	**Waste Fish Oil**	**Rancid Ghee**	**Waste Pork Lard**	**Waste Duck Processing Oil**
SFA [1]	14.0	24.3	68.2	42.3	36.6
UFA	85.0	75.6	29.8	55.7	59.4
MUFA	78.2	51.0	27.5	41.9	45.4
PUFA	6.8	24.6	2.3	13.8	14.0

[1] SFA—saturated fatty acids; UFA—unsaturated fatty acids; MUFA—monounsaturated fatty acids; PUFA—polyunsaturated fatty acids.

2.4. Media and Culture Conditions

Inoculum cultures of *C. cylindracea* were carried out in flat-bottom flasks in 100 cm^3 of YPG medium (10 g/L yeast extract (Y), 20 g/L peptone (P), 20 g/L glucose (G), pH = 5) for 24 h at 28 °C on a rotary shaker (the temperature conditions were chosen on the basis of preliminary, which were not shown). In experimental cultures, glucose was replaced with lipid wastes or olive oil in control cultures. After 65 h of the culture biomass yield and extracellular lipase activity were determined.

2.5. Determination of Biomass Yield

Yeasts cells were harvested by centrifugation (8000 rpm, 10 min, 4 °C), washed in distilled water and dried at 105 °C until constant weight.

2.6. Determination of Extracellular Lipase Activity

Extracellular lipase activity of *C. cylindracea* was determined by spectrophotometric measurement of *p*-nitrophenyl laurate hydrolysis [14]. Briefly, yeast culture was centrifuged, then 15 cm^3 of supernatant was added to 0.3 mmol of *p*-nitrophenyl laurate dissolved in 2 cm^3 of heptane. After 15 min of stirring at 37 °C, absorbance was measured at 410 nm in UV/Vis spectrophotometer. Results were expressed as U/cm^3, i.e., units of activity (the enzyme quantity that liberated 1 μmol of *p*-nitrophenol per minute under the assay conditions at 37 °C) per 1 cm^3 of supernatant.

2.7. Determination of Selected Culture Parameters

For the comparison of flask cultures of *C. cylindracea* in media with lipid-rich wastes, the following parameters characterizing the growth and production of lipases were determined: product-biomass yield ($Y_{(P/X)}$), biomass-substrate yield ($Y_{(X/S)}$), specific production rate (q_P), and volumetric productivity (v_P).

2.8. Determination of the Growth Curve of C. cylindracea

Based on the first stage of the experiment, waste fish oil was selected for the next stage—determination of the growth curve of *C. cylindracea* in 192 h culture. Flasks with 100 cm^3 of YP medium (10 g/L yeast extract (Y), 20 g/L peptone (P), pH = 5) with the addition of waste fish oil (20 g/L) were prepared and inoculated with 0.1 cm^3 of the inoculum culture. Cultures were carried out in the same way as in the inoculum and in the first stage of the experiment, i.e., at 28 °C and 150 rpm on IKA KS 4000 ic control shaker (IKA, Königswinter, Germany). The following parameters were determined after 24, 41, 48, 65, 72, 89, 96, 113, 120, 144, 168, and 192 h: biomass yield, extracellular lipolytic activity, pH, and optical density (OD_{600}).

2.9. Measurement of the Yeast Optical Density

The optical density of the yeast cultures was determined by spectrophotometry at a wavelength of 600 nm (OD_{600}), using a Rayleigh UV-1601 spectrophotometer (BRAIC, Beijing, China). For this purpose, 1 cm^3 of the culture medium was taken and centrifuged in Eppendorf miniSpin plus centrifuge (Eppendorf AG, Hamburg, Germany) for 2 min at 10,000 rpm. The yeast biomass pellet was then suspended in 1 cm^3 of distilled water and the absorbance was measured. Appropriate dilutions of the yeast suspension were made when measuring the optical density of the culture.

2.10. *Measurement of the pH of the Culture Medium*

The pH values of the cultures were determined using a CP-551 pH meter (Elmetron, Zabrze, Poland).

2.11. *Statistical Analysis*

Statistical analysis was performed using Statistica 13.3 software (TIBCO Software Inc., Palo Alto, CA, USA). The results were analyzed using one-way analysis of variance (ANOVA) and Tukey's post-hoc test. The significance level was $\alpha = 0.05$.

3. Results and Discussion

3.1. *The Use of Lipid-Rich Food Industry Wastes in the Cultivation of C. cylindracea Yeast*

Most of the lipases are inductive enzymes. Microorganisms that produce lipases need selected substances that induce the synthesis of these enzymes. A common example of a lipolytic enzyme inducer reported in the literature is olive oil [15,16], which also served as a control carbon source in the current study. Therefore, before starting the cultivation, the analysis of fatty acid profiles was performed in triacylglycerols of waste lipid substrates and in olive oil. The results of the analyses are presented in Table 1.

It has been shown that the fatty acid compositions in wastes are diversified. Lipid wastes were characterized by higher content of saturated fatty acids (SFA) than olive oil, and the highest content of SFA was found in ghee (68.2%). Waste pork lard, duck oil, and fish oil were composed of 42.3, 36.6, and 24.3% of SFA, respectively. Olive oil contained the highest amount of unsaturated fatty acids (UFA)—85%, and fish oil had a similar content (75.6%), where one third of UFA were polyunsaturated fatty acids (PUFA).

High yields of biomass (12.84, 12.75, and 12.24 g/dm^3) were obtained in media containing waste duck processing oil, olive oil, and waste pork lard, respectively (Figure 1). In the medium where the carbon source was waste fish oil, biomass yield was 10.58 g/dm^3. The lowest biomass yield was obtained in the medium containing ghee (4.93 g/dm^3) and in the YP medium (1.89 g/dm^3).

Growth of yeast in the YP medium, despite the lack of the addition of a carbon source, was possible because peptides contained in peptone and yeast extract were used to build cells and for energy processes. Although peptone and yeast extract are added to the medium to provide amino acids, vitamins and minerals, yeast also uses these substrates for energy purposes and as building components, but it is an insufficient source of carbon, therefore in YP medium the lowest yield of biomass was observed.

Figure 2 shows the dependence of extracellular lipolytic activity on the waste carbon source used in the medium in yeast culture. YP medium (containing no lipid carbon source) served as a control to determine the activity of constitutively produced lipolytic enzymes, regardless of the presence of hydrophobic molecules.

The highest lipolytic activity was obtained in the media containing waste fish oil and rancid ghee (0.050 and 0.047 U/cm^3). Almost two times lower activities were obtained in the media with the other substrates, i.e., 0.028 U/cm^3 (waste pork lard), 0.027 U/cm^3 (olive oil), and 0.024 U/cm^3 (waste duck oil). No lipolytic activity was observed in the YP medium. The obtained results suggest that the lipases produced by *C. cylindracea* are not constitutive enzymes as there was observed a need of lipid carbon source presence in the culture medium. Moreover, no relationship was found between extracellular lipolytic activity and biomass yield. Waste fish oil and rancid ghee were characterized by the most diverse fatty acids profile (Table 1). Waste fish oil contained 24.3% of SFA (almost the lowest amount apart from olive oil) and the highest amount of PUFA (24.6%). On the other hand, rancid ghee showed the highest SFA and the lowest PUFA concentration, but still, both waste lipids influenced the secretion of lipolytic enzymes to the greatest extent.

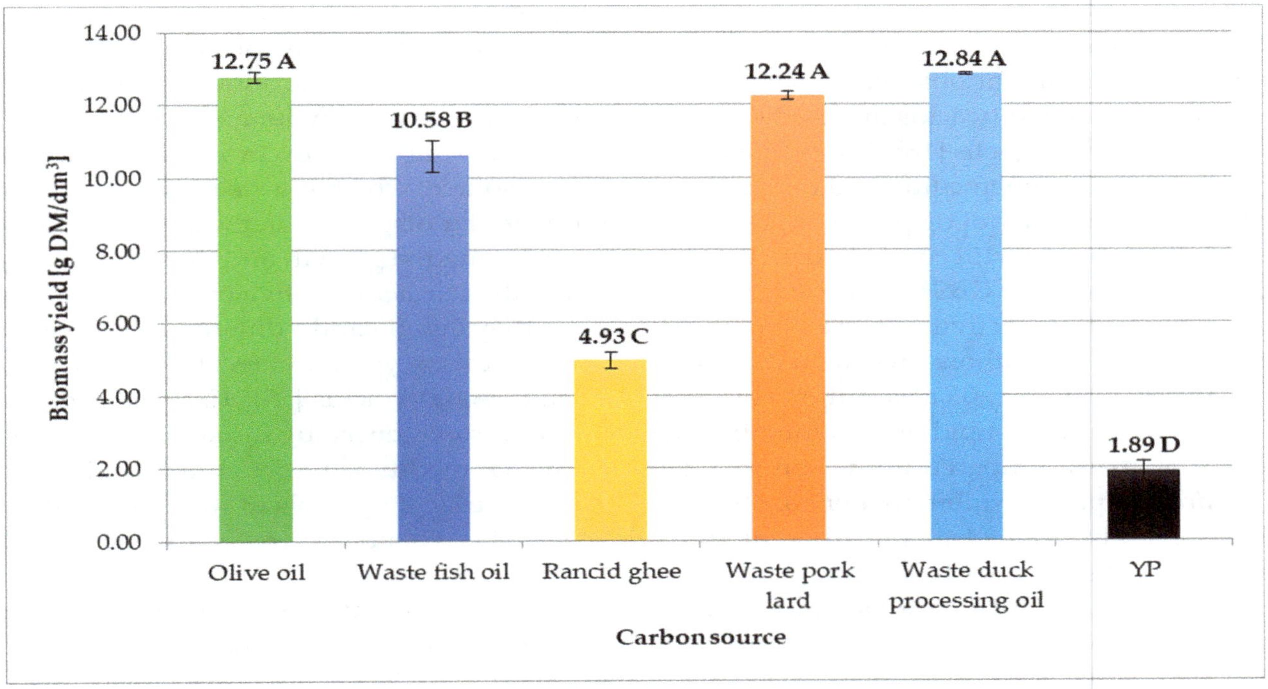

Figure 1. Effect of lipid-rich wastes on the biomass yield of *C. cylindracea* DSM 2031 after 65 h cultures. YP—control medium (10 g/L yeast extract (Y), 20 g/L peptone (P), pH = 5). Means with the same letter did not differ significantly ($\alpha < 0.05$).

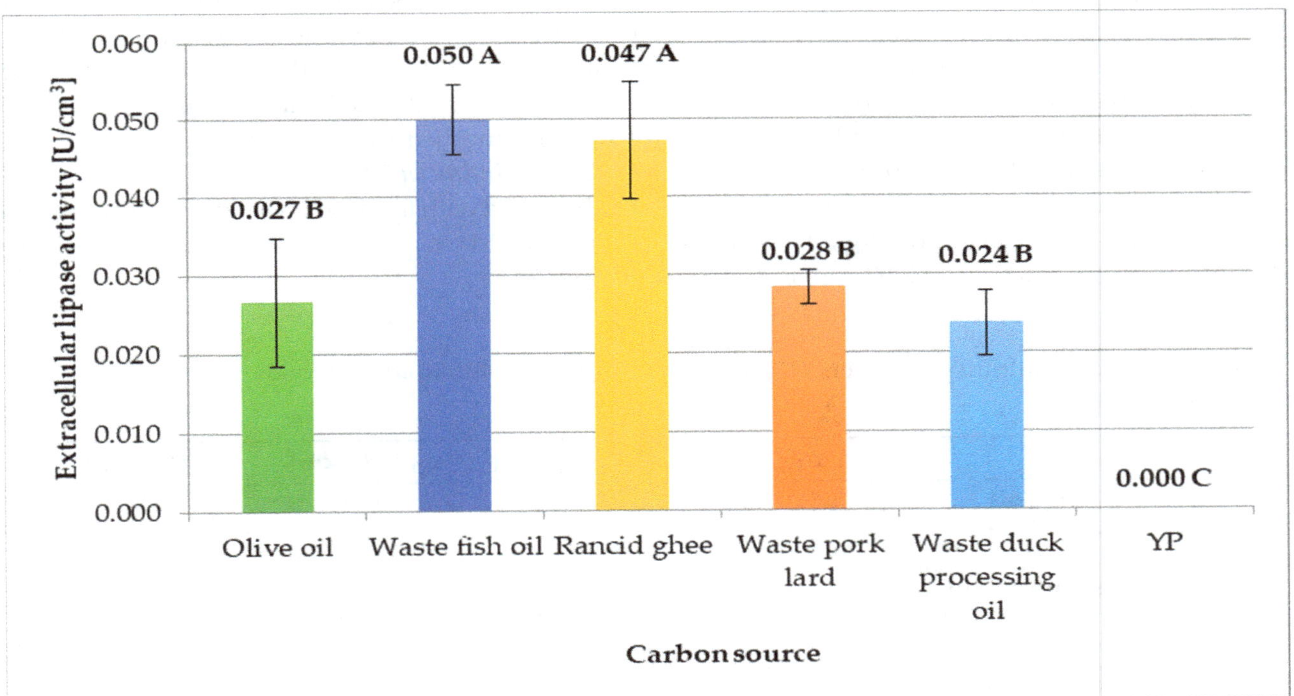

Figure 2. Effect of lipid-rich wastes on the extracellular lipase activity of *C. cylindracea* DSM 2031 after 65 h cultures. YP—control medium (10 g/L yeast extract (Y), 20 g/L peptone (P), pH = 5). Means with the same letter did not differ significantly ($\alpha < 0.05$).

Ghee, pork lard, and waste duck oil contained a relatively high amount of saturated fatty acids, which determined their solid or semi-solid consistency, unlike other fats. The aggregate state of fats may be of significant importance in the cultivation of microorganisms, because their solid form is not conducive to homogeneous distribution in a liquid medium, and thus reduces the contact surface of cells or their enzymes with the substrate, and consequently, fat hydrolysis and assimilation may be difficult. Similar observations were made by Tan and Gill [17] who showed that the higher content of saturated fatty acids in fats, the more of them was left unused by fungi. On the other hand, Papanikolaou and Aggelis [18] showed that *Yarrowia lipolytica* exhibited growth and accumulation of intracellular lipids when cultivated on stearin as a carbon source, which is a stearic acid-rich waste, but emulsifiers and use of disperser were necessary for better fat dispersion in the aqueous phase.

Sokolovska et al. [15] conducted an experiment comparing the growth and lipolytic activity of the yeast *C. cylindracea* CBS 6330 (DSM 2031) in media with oleic acid or olive oil. In both cultures, comparable lipolytic activities were obtained, however, they significantly differed in the obtained biomass yields. The authors presumed that the discussed species of yeast prefers glycerol released by triacylglycerols hydrolysis as a carbon source rather than free fatty acids [15]. *C. cylindracea* yeasts were able to assimilate lipid wastes contained in the media, as evidenced by the significantly higher yields of biomass compared to the control culture in the YP medium. Other studies indicate that *C. cylindracea* lipases exhibit greater substrate specificity for triacylglycerols in which the fatty acid residues contain from 14 to 18 carbon atoms. However, there is no preference for the degree of triacylglycerols saturation [19].

Fabiszewska and Białecka-Florjańczyk [20] revised and compared the results of different authors related to the extracellular lipolytic activity of the *Y. lipolytica* non-conventional yeast species and concluded that the lipase secretion is a resultant of at least several factors and it is most likely mistaken to correlate it only with the oleic acid content of the plant oil used as a carbon source in culture medium [20]. The authors agree with the hypothesis, but further investigation should be taken including *C. cylindracea* species.

The presented results of the experiments showed that it is possible to use lipid wastes in the cultivation of *C. cylindracea*. Goncalves et al. [21] in their research tried to valorize olive mill wastewater. The following yeast strains were selected for the experiment: *Candida rugosa* PYCC 3238, *C. rugosa* CBS 2275, *C. cylindracea* CBS 7869, *Yarrowia lipolytica* W29, *Y. lipolytica* CBS 2073, and *Y. lipolytica* IMUFRJ 50682. All the strains used showed lipolytic activity in the medium with the mentioned substrate, and the most effective lipase producer turned out to be *C. cylindracea* CBS 7869 [21].

For better comparison of flask cultures of *C. cylindracea* in media with lipid-rich wastes, culture parameters such as product-biomass yield ($Y_{(P/X)}$), biomass-substrate yield ($Y_{(X/S)}$), specific production rate (q_P), and volumetric productivity (v_P) were determined (Table 2).

Table 2. Selected culture parameters of *C. cylindracea* in media with lipid-rich food industry wastes as carbon sources.

	$Y_{(P/X)}$ [U/g] [1]	$Y_{(X/S)}$ [g/g]	q_P [U/g × h]	v_P [U/L × h]
Olive oil	2.09	0.64	0.03	0.41
Waste fish oil	4.72	0.53	0.07	0.77
Rancid ghee	9.57	0.25	0.15	0.73
Waste pork lard	2.31	0.61	0.04	0.44
Waste duck processing oil	1.85	0.64	0.03	0.36
YP	0.00	0.09	0.00	0.00

[1] $Y_{(P/X)}$—product-biomass yield; $Y_{(X/S)}$—biomass-substrate yield; q_P—specific production rate; v_P—volumetric productivity.

Obtained results indicated that waste fish oil was the best carbon source for simultaneous production of high biomass yield and high lipolytic activity. In case of product-biomass yield higher value was found for rancid ghee, due to the lower biomass yield. Values of biomass-substrate yield for olive oil, waste fish oil, waste pork lard, and waste duck processing oil were comparable and ranged 0.53–0.64. Specific production rate and volumetric productivity were also highest for cultures with waste fish oil or ghee as carbon sources. Despite similar v_P values in cultures with ghee and fish oil and a higher $Y_{(P/X)}$ value in culture with ghee as a carbon source, the waste fish oil was applied in the next stage of the experiment due to the better use of the waste for the simultaneous production of both biomass and enzymes. An important factor in the microbiological disposal of waste should be its maximized use, i.e., the processing of more lipid waste into yeast biomass and other valuable metabolites that can be used then in different industries.

3.2. Investigation of Changes in Biomass Yield, Lipolytic Activity, and pH Value During the Culture of C. cylindracea in a Medium with Waste Fish Oil

Obtaining a relatively high yield of biomass and enzymatic activity in the medium with waste fish oil allowed for the next part of the experiment, where it was decided to check and determine the optimal time for culturing *C. cylindracea* DSM 2031 to obtain the highest possible extracellular lipolytic activity.

The determination of the growth curve of *C. cylindracea* was based on the determination of biomass yield, extracellular lipolytic activity, pH and optical density. The results of 192-h culture are shown in Figures 3 and 4.

The highest yield of biomass (16.06 g/dm^3), as well as OD_{600} were obtained in 120 h of shake culture and the resulting biomass yield and optical density curves were very comparable. During the first day of culture, the cells adapted to the environmental conditions, and then a logarithmic growth phase started, in which yeast cells were dividing intensively, and the biomass yield increased. After 120 h of cultivation, the yeast clearly slowed down their growth, which was related to entering the stationary phase.

An increase in pH value was also noted during the cultivation (Figure 4). A sudden increase in pH was observed during the first two days from 4.85 to 6.40, and then the medium environment was relatively stable and finally, pH of 7.25 was obtained. According to Vylkova et al. [22] the increase in pH occurs in media with the presence of exogenous amino acids and in the current study peptone and yeast extract as sources of proteins and amino acids were used. Amino acids are metabolized and used for cell growth, and the resulting products of nitrogen metabolism, e.g., in the form of ammonia, alkalize the culture environment [22].

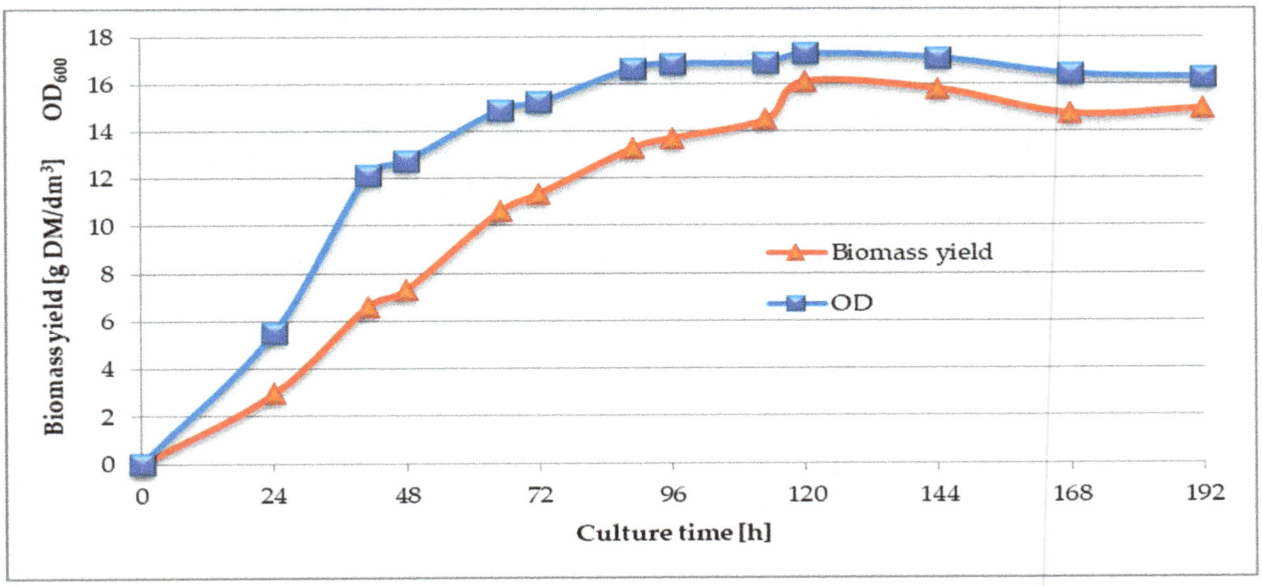

Figure 3. Changes in biomass yield and optical density (OD_{600}) during 192-h culture of *C. cylindracea* DSM 2031 in a medium with waste fish oil.

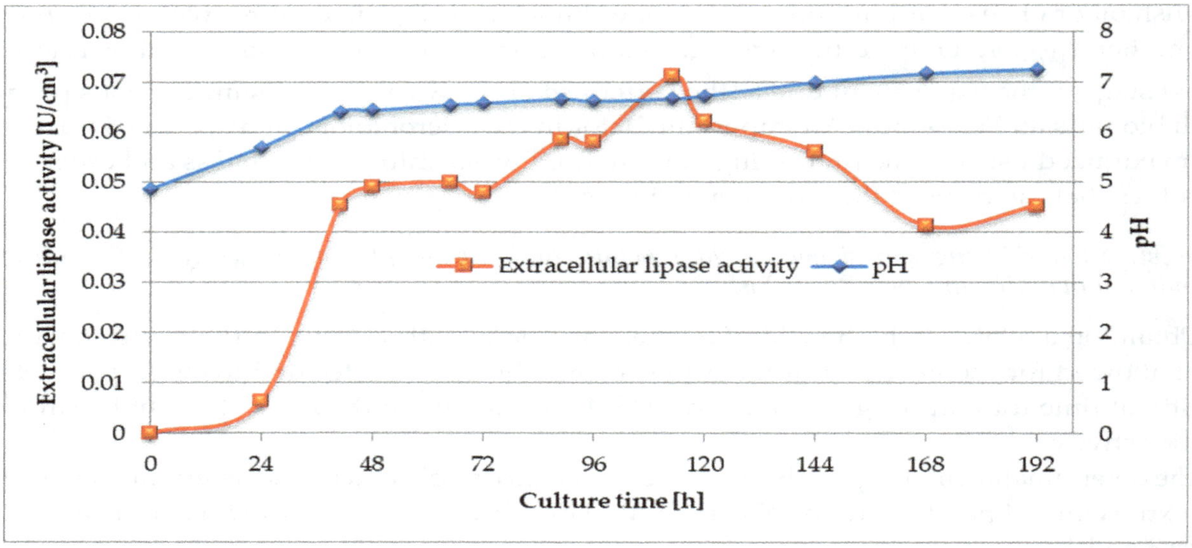

Figure 4. Changes in extracellular lipase activity and pH value during 192-h culture of *C. cylindracea* DSM 2031 in a medium with waste fish oil.

The extension of the cultivation time also had a positive effect on lipolytic activity. The highest lipase activity was obtained at 113 h of cultivation (0.07 U/cm^3) in the late logarithmic phase, which indicates the need for longer cultures in order to obtain more biomass and extracellular enzymes, as well as more efficient waste management. The obtained activity was 1.4-fold higher than reported in the first stage of the experiment (0.05 U/cm^3). Similar observations have been reported in the literature, both Sokolovska et al. [15] and Krastanov et al. [23] observed that the highest lipase activity is obtained near the end of the logarithmic phase and alike to the cited papers, lipolytic activity decreased in the stationary phase. The results presented in the paper are also in accordance with Fabiszewska et al. [24] who observed a different pattern of extracellular and cell-bound lipase production, which was the highest in the early exponential phase of model non-conventional yeast species *Yarrowia lipolytica*. The extracellular lipase activity increased in the late exponential phase due to the lower accumulation of lipase molecules in cell walls [24]. The observations seemed to be similar for both species.

The production of lipases by *Candida cylindracea* has also been investigated by other scientists. Salihu et al. [25] optimized production of lipase by *C. cylindracea* ATCC 14830 in shake flask cultures with palm oil mill effluent (POME) as a basal medium. They suggested that amounts of peptone, Tween-80, and inoculum significantly influenced extracellular lipase production. Scientists provided also that statistical experimental design allowed to utilize POME along with lipase production on a considerable level, and compared to the unoptimized medium, use of 0.45% (*w/v*) peptone, 0.65% (*v/v*) Tween-80, and 2.2% (*v/v*) inoculum resulted in 5.19-fold higher activity of lipolytic enzymes [25].

A statistical approach was also applied by Muralidhar et al. [26]. The central composite design was employed to optimize five components of the medium. Besides yeast extract, malt extract, peptone, and Tween-80, two different carbon sources—glucose and olive oil—were investigated. In optimized media, similarly to the current research, biomass yield and lipase activity increased until 4–5-day of the cultivation. Significantly higher enzymatic activity was achieved in medium with olive oil as a carbon source, what was proved also were results of the current study, where the presence of lipid molecules in the medium affect the lipase production.

4. Conclusions

In the current study, the ability of *C. cylindracea* DSM 2031 to utilize several lipid wastes from the food industry on a laboratory scale was assessed. Lipid-rich wastes can serve as a valuable source of carbon in the media used for the cultivation of lipolytic microorganisms. Thus, there is a potential for their use for the production of lipolytic enzymes or the production of yeast biomass for animal feed. The use of waste substrates can contribute to lowering the costs of commercial production, and such a solution is part of the sustainable development strategy. The presented work is preliminary research and it is justified to conduct further research on the use of waste materials in the microbiological synthesis of enzymes in order to obtain higher yields and possibly the application of such technology on an industrial scale.

Author Contributions: Conceptualization, B.Z., A.F. and P.M.-Z.; methodology, B.Z., P.M.-Z. and A.F.; formal analysis, B.Z. and P.M.-Z.; investigation, B.Z. and P.M.-Z.; writing—original draft preparation, P.M.-Z. and B.Z.; writing—review and editing, B.Z., A.F. and P.M.-Z.; visualization, B.Z.; supervision, A.F. All authors have read and agreed to the published version of the manuscript.

References

1. Domańska, W. *Environment 2019*; Statistics Poland: Warsaw, Poland, 2019; pp. 148–160.
2. Hommel, R.K. CANDIDA Introduction. In *Encyclopedia of Food Microbiology*, 2nd ed.; Batt, C.A., Tortorello, M.L., Eds.; Academic Press: Cambridge, MA, USA, 2014; pp. 367–373.
3. Lachance, M.A.; Boekhout, T.; Scorzetti, G.; Fell, J.W.; Kurtzman, C.P. Candida Berkhout (1923). In *The Yeast, a Taxonomic Study*, 5th ed.; Kurtzman, C.P., Fell, J.W., Boekhout, T., Eds.; Elsevier: Amsterdam, The Netherlands, 2011; pp. 978–1278.
4. Kurtzman, C.P.; Robnett, C.J. Indentification and phylogeny of ascomycetous yeasts from analysis of nuclear large subunit (26S) ribosomal DNA partial sequences. *Anton. Leeuw.* **1998**, *73*, 331–371. [CrossRef] [PubMed]
5. Boontham, W.; Angchuan, J.; Boonmak, C.; Srisuk, N. *Limtongozyma siamensis* gen. nov., sp. nov., a yeast species in the Saccharomycetales and reassignment of *Candida cylindracea* to the genus *Limtongozyma*. *Int. J. Syst. Evol. Microbiol.* **2020**, *70*, 199–203. [CrossRef] [PubMed]
6. Yamada, K.; Machida, H. Studies on the production of lipase by microorganisms. *Nippon Nogeikagaku Kaishi* **1962**, *36*, 858–860. [CrossRef]
7. Kim, B.S.; Hou, C.T. Production of lipase by high density fed-batch culture of *Candida cylindracea*. *Bioprocess Biosyst. Eng.* **2006**, *29*, 59–64. [CrossRef] [PubMed]
8. Zalcain, I.; Zapalena, M.J.; Astiasaran, I.; Bello, J. Dry-fermented Sausages Elaborated with Lipase from *Candida cylindracea*. Comparison with Traditional Formulations. *Meat Sci.* **1995**, *40*, 55–61. [CrossRef]
9. Belkacemi, F.Z.; Merabet-Khelassi, M.; Aribi-Zouioueche, L.; Riant, O. Productionof l-mentyl acetate through kinetic resolution by *Candida cylindracea* lipase: Effect of alkaloids as additives. *Res. Chem. Intermed.* **2018**, *44*, 6847–6860. [CrossRef]
10. Alcântara, T.D.A.P.; Oliveira, J.M.; Evangelista-Barreto, N.S.; Marbac, P.A.S.; Cazetta, M.L. Aerobic decolorization of azo dye orange g by a new yeast isolate *Candida cylindracea* SJL6. *Biosci. J.* **2017**, *33*, 1340–1350. [CrossRef]
11. Tang, A.; Zhang, Y.; Wei, T.; Wu, J.; Li, Q.; Liu, Y. Immobilization of *Candida cylindracea* Lipase by Covalent Attachment on Glu-Modified Bentonite. *Appl. Biochem. Biotechnol.* **2019**, *187*, 870–883. [CrossRef]
12. Zieniuk, B.; Fabiszewska, A.; Białecka-Florjańczyk, E. Screening of solvents for favoring hydrolytic activity of *Candida antarctica* Lipase B. *Bioproc. Biosyst. Eng.* **2020**, *43*, 605–613. [CrossRef]
13. Zieniuk, B.; Wołoszynowska, M.; Białecka-Florjańczyk, E. Enzymatic synthesis of biodiesel by direct transesterification of rapeseed cake. *Int. J. Food Eng.* **2020**, *16*, 3. [CrossRef]
14. Kapturowska, A.; Stolarzewicz, I.; Krzyczkowska, J.; Białecka-Florjańczyk, E. Studies on lipolytic activity of sonicated enzymes from *Yarrowia lipolytica*. *Ultrason. Sonochem.* **2012**, *19*, 186–191. [CrossRef] [PubMed]
15. Sokolovska, I.; Albasi, C.; Riba, J.P.; Bales, V. Production of extracellular lipase by *Candida cylindracea* CBS 6330. *Bioprocess Biosyst. Eng.* **1998**, *19*, 179–186. [CrossRef]
16. Fickers, P.; Marty, A.; Nicaud, J.M. The lipases from *Yarrowia lipolytica*: Genetics, production, regulation, biochemical characterization and biotechnological applications. *Biotechnol. Adv.* **2011**, *29*, 632–644. [CrossRef] [PubMed]
17. Tan, K.H.; Gill, C.O. Batch growth of *Saccharomycopsis lipolytica* on animal fats. *Appl. Microbiol. Biotechnol.* **1985**, *21*, 292–298. [CrossRef]
18. Papanikolaou, S.; Aggelis, G. Selective uptake of fatty acids by the yeast *Yarrowia lipolytica*. *Eur. J. Lipid Sci. Technol.* **2003**, *105*, 651–655. [CrossRef]

19. Lie, O.; Lambertsen, G. Fatty Acid Specificity of *Candida cylindracea* Lipase. *Eur. J. Lipid Sci. Technol.* **1986**, *88*, 365–367. [CrossRef]
20. Fabiszewska, A.U.; Białecka-Florjańczyk, E. Factors influencing synthesis of extracellular lipases by *Yarrowia lipolytica* in medium containing vegetable oils. *J. Microbiol. Biotechnol. Food Sci.* **2014**, *4*, 231–237. [CrossRef]
21. Goncalves, C.; Lopes, M.; Ferreira, J.P.; Belo, I. Biological treatment of olive mill wastewater by non-conventional yeasts. *Bioresour. Technol.* **2009**, *100*, 3759–3763. [CrossRef]
22. Vylkova, S.; Carmar, A.J.; Danhof, H.A.; Collete, J.R.; Zhou, H.; Lorenz, M.C. The fungal pathogen *Candida albicans* autoinduces hyphal morphogenesis by raising extracellular pH. *mBio* **2011**, *2*, e00055-11. [CrossRef]
23. Krastanov, A.; Govindarajan, A.; Daniel, D. Studies on lipase fermentation using *Candida cylindracea* NRRL Y-175061N a stirred tank bioreactor. *Bulg. J. Agric. Sci.* **2008**, *14*, 290–299.
24. Fabiszewska, A.U.; Stolarzewicz, I.; Zamojska, W.; Białecka-Florjańczyk, E. Carbon source impact on *Yarrowia lipolytica* KKP 379 lipase production. *Appl. Biochem. Microbiol.* **2014**, *50*, 404–410. [CrossRef]
25. Salihu, A.; Alam, Z.M.; AbdulKarim, M.I.; Salleh, H.M. Optimization of lipase production by *Candida cylindracea* in palm oil mill effluent based medium using statistical experimental design. *J. Mol. Catal. B Enzym.* **2011**, *69*, 66–73. [CrossRef]
26. Muralidhar, R.V.; Chirumamila, R.R.; Marchant, R.; Nigam, P. A response surface approach for the comparison of lipase production by *Candida cylindracea* using two different carbon sources. *Biochem. Eng. J.* **2001**, *9*, 17–23. [CrossRef]

2

Non-*Saccharomyces* in Winemaking: Source of Mannoproteins, Nitrogen, Enzymes and Antimicrobial Compounds

Ricardo Vejarano

Dirección de Investigación y Desarrollo, Universidad Privada del Norte (UPN), Trujillo 13001, Peru; ricardo.vejarano@upn.edu.pe

Abstract: Traditionally, non-*Saccharomyces* yeasts have been considered contaminants because of their high production of metabolites with negative connotations in wine. This aspect has been changing in recent years due to an increased interest in the use of these yeasts in the winemaking process. The majority of these yeasts have a low fermentation power, being used in mixed fermentations with *Saccharomyces cerevisiae* due to their ability to produce metabolites of enological interest, such as glycerol, fatty acids, organic acids, esters, higher alcohols, stable pigments, among others. Additionally, existing literature reports various compounds derived from the cellular structure of non-*Saccharomyces* yeasts with benefits in the winemaking process, such as polysaccharides, proteins, enzymes, peptides, amino acids, or antimicrobial compounds, some of which, besides contributing to improving the quality of the wine, can be used as a source of nitrogen for the fermentation yeasts. These compounds can be produced exogenously, and later incorporated into the winemaking process, or be uptake directly by *S. cerevisiae* from the fermentation medium after their release via lysis of non-*Saccharomyces* yeasts in sequential fermentations.

Keywords: non-*Saccharomyces*; winemaking; aging-on-lees; yeast assimilable nitrogen

---

1. Introduction

The genus *Saccharomyces* has been the most industrially used in the production of wine. This aspect has been changing in recent years due to an increased interest in the use of non-*Saccharomyces* yeasts [1]. The high potential of these yeasts makes them a useful tool for improving oenological parameters such as nutrients content, stability, aromatic profile, or bioactive profile, in spite of their low fermentative power, and, in some cases, their high production of certain metabolites with negative connotations in the wine [2,3].

Currently, species such as *Torulaspora delbrueckii*, *Lachancea thermotolerans*, *Metschnikowia pulcherrima*, *Schizosaccharomyces pombe*, *Pichia kluyveri*, among others, are sold for wine production [4]. These yeasts are often isolated from the grape, grape-must, and wine, as well as from the soil, winery surfaces, harvesting machinery, and other objects associated with the wine production. In the case of grape-must/wine, the majority are isolated during the first stages of fermentative process given their low fermentation ability and low tolerance to ethanol, with respect to *S. cerevisiae*, so that its use would mainly be associated with the contribution of the specific enological metabolites such as enzymes, aromatic compounds, glycerol, organic acids, fatty acids, proteins, amino acids, polysaccharides, among others [3].

Figure 1 summarizes some of the potential applications of non-*Saccharomyces* yeasts, taking as reference the available commercial preparations for wine production. These preparations are sold in the form of (a) dry yeast, inactivated through thermal treatment and then dried; (b) cell autolysates, which include soluble and insoluble yeast components, partially degraded by the endogenous enzymes; (c) soluble cell extracts, derived from the cytoplasm; and (d) insoluble cell hulls, mainly containing cell walls [5]. These preparations may be the source of compounds of interest such as polysaccharides, proteins, enzymes, or other metabolites with the potential for use in wine production and can be produced exogenously [6].

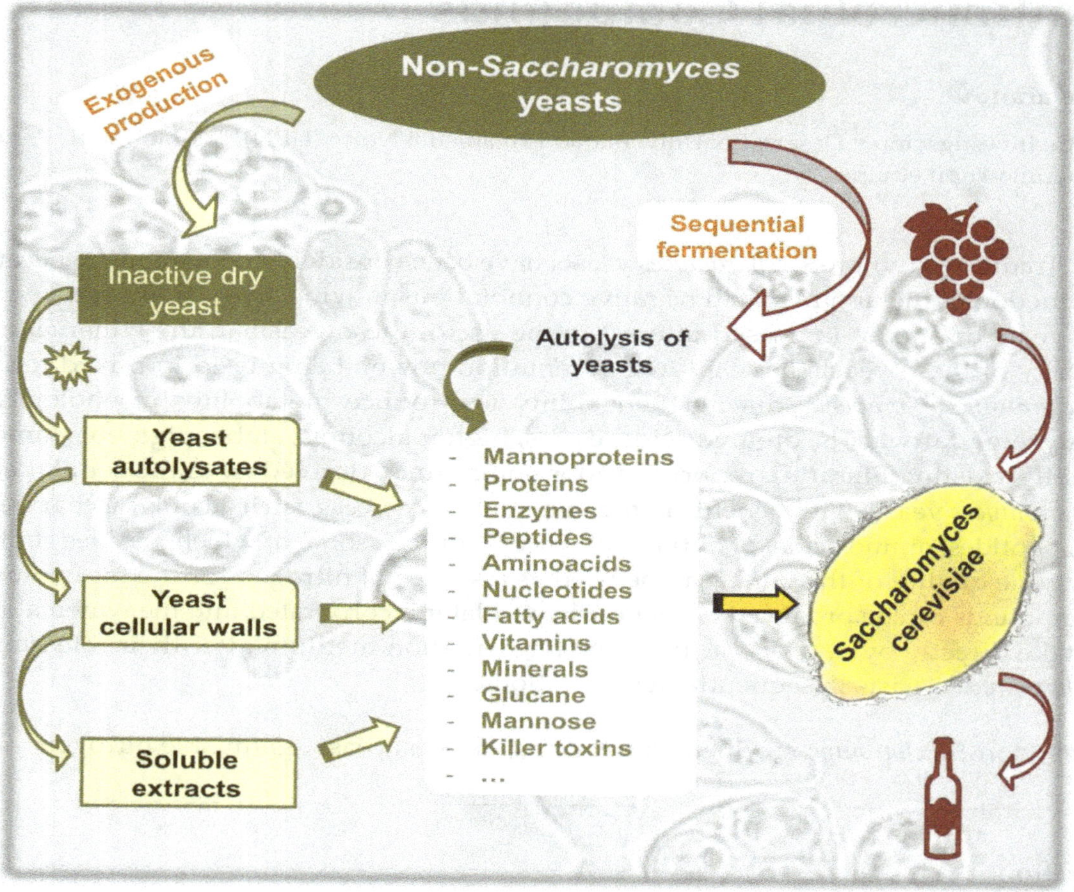

Figure 1. Main cell components of non-*Saccharomyces* yeasts with the potential for use in wine production.

In addition to these preparations, compounds of interest can be obtained directly from the fermentation process as a result of their release from the yeast during cell lysis, for example, in mixed cultures.

2. Non-*Saccharomyces* Yeasts as a Source of Polysaccharides

During aging-on-lees (AOL), a series of changes take place with a direct impact on the wine's properties, due to the interaction between the wine's components, the enzymatic activity, and the polysaccharides released by the lees [7]. The use of non-*Saccharomyces* yeasts for the exogenous production of polysaccharides [6], which can be added to wine, is an interesting alternative. The addition of yeast polysaccharides is a practice authorized by the International Code of Oenological Practice of the International Organization of Vine and Wine (OIV).

These polysaccharides, particularly in red wines, can improve the mouthfullness and body [8], sweetness and roundness [9], aromatic persistence [10], protein and tartaric stability [11,12], and interact with tannins and reduce astringency [13], as well as have an antioxidant effect (due to glutathione) that protects the aromatic compounds and anthocyanins [14], enabling its antioxidant and anti-inflammatory potential to be maintained [15]. They also interact with the tertiary aromatic compounds, reducing the perception of woody aromas in long-aged wines [16], encourage malolactic fermentation [17], and adsorb undesirable and dangerous compounds such as ochratoxin A (OTA) [18]. In sparkling wines, they improve the quality of the foam [19]. However, during AOL, a reduction of aromatic compounds such as terpenes, esters, and aldehydes [20], and, in some cases, anthocyanins in red wines [16,21], has also been reported.

The most important polysaccharides released during AOL are the mannoproteins (Table 1), which are fixed on a three-dimensional network of glucan and chitin in the cell wall [21,22]. The mannoproteins represent between 80% and 100% of the fraction of polysaccharides founded in the wine (molecular mass 100–2000 kDa: 90% mannose and 10% protein), while glucomannoproteins (molecular mass 20–90 kDa: 25% glucose, 25% mannose, and 50% protein) represent between 10% and 20% of the total [23]. Furthermore, α-galactomannose rather than mannose has been found as part of the structure of polysaccharides in *S. pombe* [22].

The enzymes involved in the cell autolysis and subsequent releasing of polysaccharides into the wine are endo β(1→3)-glucanases, endo β-(1→6)-glucanases, exo-(1→6)-α-D-mannose, exo-(1→2)-α-mannose, and α-D-mannosidase [23].

2.1. Non-Saccharomyces Species as a Source of Polysaccharides

Even though *S. cerevisiae* is the most commonly used yeast as a source of polysaccharides, the literature reports different non-*Saccharomyces* species with the potential to produce and release polysaccharides during AOL: *S. pombe*, *Saccharomycodes ludwigii*, *Wickerhamomyces anomalus*, *Hanseniaspora vineae*, *L. thermotolerans*, *M. pulcherrima*, *Starmerella bacillaris*, *T. delbrueckii*, *Zygosaccharomyces rouxii*, and *Zygotorulaspora florentina* (formerly *Zygosaccharomyces florentinus*), among others.

The genus *Schizosaccharomyces* has shown high rates of release of polysaccharides during alcoholic fermentation, of up to seven times higher than *S. cerevisiae* [24]. During AOL, *S. pombe* together with *S'codes ludwigii* are found among the species with high potentials for releasing polysaccharides [21]. *S. pombe* primarily releases galactomannoproteins (Table 1), at concentrations up to 10 times higher than some strains of *Saccharomyces* and *Pichia* [21]. This potential could mean an advantage with regard to accelerating aging processes.

S'codes ludwigii has also shown a high potential for releasing polysaccharides during fermentation and AOL [21,25,26]. Palomero et al. [21] obtained higher rates of polysaccharide release from *S'codes. ludwigii* (110.51 gm/L) and *S. pombe* (103.61 mg/L) with respect to *S. cerevisiae* (36.65 mg/L). Furthermore, the polysaccharides from these non-*Saccharomyces* yeasts show a greater molecular size and may potentially impact the wine's palatability.

However, one aspect to take into account during AOL in red wines is color loss, due to the interaction between the lees and anthocyanins, mainly through adsorption of pigments by the lees [16] and through degradation due to anthocyanin-β-glucosidase activity [27]. The lees from *M. pulcherrima*, *S'codes ludwigii*, or *S. pombe* have shown a low adsorption of anthocyanins with respect to the lees of *S. cerevisiae* and other non-*Saccharomyces* yeasts such as *T. delbrueckii* or *L. thermotolerans* [21,28]. Furthermore, there may be less color loss in wines in the presence of higher pyranoanthocyanins content, which are more stable due to its chemical structure [29], with *S. pombe* being one of the yeasts with the greatest synthesis of these pigments [30].

Additionally, *S. pombe* has also shown the ability to decrease the biogenic amine content in sparkling wines, with better results than *S. cerevisiae* [30], which would be related to the adsorption of the amines on the lees during second fermentation and subsequent aging in the bottle [31]. However, considering the short duration of AOL studies, it is necessary to evaluate the evolution of amines over longer aging periods, such as those carried out in traditional sparkling wine production, given that other studies have reported an increase in the biogenic amines content in wines in contact with lees [32].

In terms of other non-*Saccharomyces* yeasts, *Hanseniaspora vineae* has shown a higher rate of cell lysis with respect to commercial *S. cerevisiae* strains [33]. Other yeasts that have shown a notable polysaccharide contribution are *L. thermotolerans* [34], *Z. rouxii* [26], and *Z. florentina*, particularly in mixed fermentations with *S. cerevisiae* [35].

Table 1. Some non-*Saccharomyces* yeasts with the potential to release mannoproteins.

Yeast	Protein and Monosaccharide Content in Mannoproteins [25]				Nitrogen Requirement
	Protein (%) [a]	Mannose (%) [b]	Glucose (%) [b]	Galactose (%) [b]	
Saccharomyces cerevisiae	24	88	12	-	Slow ammonium uptake [4].
Metschnikowia pulcherrima	25	86	14	-	Weak or no growth in nitrate agar, and unable to develop in YPD agar at 37 °C [28].
Wickerhamomyces anomalus [c]	9	74	26	-	Capable to uptake nitrate [36].
Saccharomycodes ludwigii [d]	12	93	7	-	Unable to uptake nitrate [37]. Capable to uptake cadaverine and ethylamine [37].
Schizosaccharomyces pombe [e]	11	55	22	23	
Starmerella bombicola [f]	14	73	27	-	
Pichia fermentans	15	87	13	-	
Hanseniaspora uvarum [g]	23	81	19	-	
Hanseniaspora valbyensis	20	75	25	-	
Lachancea thermotolerans	16	82	18	-	
Torulaspora delbrueckii	18	85	15	-	
Zygosaccharomyces bailii	29	79	21	-	
Brettanomyces bruxellensis	16	88	12	-	

[a] Percentage of dry matter. [b] Sugars (%) in the polysaccharide fraction. [c] *Wickerhamomyces anomalus* (formerly *Pichia anomala*). [d] *Saccharomycodes ludwigii*: high autolytic activity: polysaccharide-releasing [21,25]. [e] *Schizosaccharomyces pombe*: high autolytic activity: polysaccharide-releasing [21,38]. [f] *Starmerella bombicola* (formerly *Candida stellata* DBVPG 3827). [g] *Hanseniaspora uvarum* (formerly *Kloeckera apiculata*).

For its part, the information regarding the use of other yeasts such as *W. anomalus* as a source of mannoproteins is scarce [26]. This yeast has a high potential for the production of polysaccharides and other metabolites of interest, given its ability to metabolize a large variety of nitrogen sources, including nitrate [36], which would allow production costs to be reduced.

2.2. *Accelerated Release of Polysaccharides*

One disadvantage of AOL is that it is a very slow process, which requires up to 9 months to obtain the desired effects in the treated wine. Among the strategies to improve AOL, existing literature reports (a) the selection of yeast species and strains with rapid autolysis [21,39], (b) acceleration of the cell autolysis through mixed cultures involving sensitive and killer yeasts [40], mixed cultures among different yeast species, which enable the regulation of cell death [41], the addition of β-glucanase [42], and the application of ultrasound [39,43].

Several studies have reported a higher rate of polysaccharide release with ultrasound (US) treatments [39,43,44]. Lees from *S'codes ludwigii*, *S. pombe*, *M. pulcherrima*, and *S. cerevisiae*, among others, were evaluated over a 7-week aging period in a hydroalcoholic medium, applying US for 10 min a day, with *S'codes ludwigii* lees showing the highest rate of polysaccharide release after the third week (around 460 mg/L) [39]. In the same study [39], a decrease in anthocyanins content was observed in the treated red wine, without affecting the pyranoanthocyanins content [21,29], particularly with *S'codes ludwigii* lees, which also allowed to reduce proanthocyanidins content, and consequently, astringency and bitterness.

More recently, Del Fresno et al. [43] evaluated the effect of AOL (lees from *S. cerevisiae*) in the presence of oak chips in red wines. The evaluation period was 135 days at 14°C, under dark conditions, agitating the samples once a week to simulate "bâtonnage". The samples were treated with US twice a week for 5 min for the first 5 weeks. From this moment, the process was accelerated increasing to two US treatments a week (15 min per treatment). In a parallel experiment, a hydroalcoholic medium was used to adequately quantify the polysaccharides released from the lees. In general, the polysaccharide release increased after using US for 135 days, releasing around 11.8 mg/L, more than double that of the untreated samples (approx. 5.3 mg/L). Additionally, an increase in the protein content of the US-treated samples was observed after 120 days of AOL.

However, the same authors [43] reported the increase in dissolved oxygen in the treated red wines as a disadvantage, whose effect was evident in the lower anthocyanins content in the US-treated wines in the absence of lees. This finding reveals the protective effect of polysaccharides against oxidation.

3. Non-*Saccharomyces* Yeasts as a Source of Nitrogen for *Saccharomyces cerevisiae*

Yeast assimilable nitrogen (YAN) is nitrogen that yeasts can assimilate and metabolize, preferably from sources such as ammonium (NH_4^+), amino acids, and peptides of up to five amino acids [45]. Its content in the grape-must varies between 50 and 500 mg/L [46].

Among YAN's functions are reproduction and cell growth, the protein for sugar transport synthesis, and enzyme synthesis, as well as the functions accomplished by the amino acids as the precursors of aromatic compounds, mainly higher alcohols, produced by deamination and decarboxylation [47].The most predominant amino acids as NH_4^+ transporters in grape-must are α-alanine, serine, arginine, proline, glutamic acid, and glutamine [48]. During fermentation, their use varies, with higher uptake of glutamic acid, glutamine, and arginine [4], whereas proline, a proteinogenic amino acid not metabolized by *S. cerevisiae*, is among the least used [4]. Additionally, the demand for arginine is increased by lactic acid bacteria during malolactic fermentation [48].

S. cerevisiae has shown a preference for NH_4^+ and amino acids, thanks to its nitrogen catabolite repression (NCR) mechanism [49], through which the genes involved in the transport and metabolism of NH_4^+ and glutamine are activated and, once they are depleted, the genes involved in the transport and metabolism of other sources such as arginine, glutamate, and alanine, among others, become activated.

Concentrations of YAN below 150 mg/L in the grape-must carry the risk of sluggish or stuck fermentation [4], as well as a low synthesis of some aromatic compounds such as of esters, volatile fatty acids, and higher alcohols [50]. This N deficiency is remedied in the winery through the addition of extra N sources such as diammonium phosphate (DAP). However, the excessive use of DAP can lower phenylpropanoid production (affecting the complexity of the wine) [51,52], increase the wine acidification, produce high levels of residual phosphate, stimulate the production of esters such as ethyl acetate, and increase the levels of hydrogen sulfide (H_2S), especially when there is a deficit of other essential nutrients such as vitamins, minerals, and lipids; further, excessive levels of DAP can increase turbidity, promote microbial instability, and facilitate the production of unpleasant aromas and harmful compounds such as ethyl carbamate and biogenic amines [53–56].

3.1. Non-Saccharomyces Species as a Source of Nitrogen

As an alternative N source, yeast cell structures can be used in the form of hulls, hydrolyzed, or extracts, which can be produced exogenously [6], and then added to the fermentative medium. Additionally, this source of N can be obtained from the rest of the cells after the death and lysis of non-*Saccharomyces* yeasts used at the beginning of the sequential fermentations with *S. cerevisiae*.

Aureobasidium pullulans (an yeast-like fungus) is a potential source of essential amino acids [57], and it can growth in low-cost carbon sources like agricultural and food waste, due to its amylase [58], cellulase [59], lipase [60], xylanase [61], laccase [62], mannanase [63], and protease [64] activities. However, the use of *A. pullulans* for the production of protein on a large scale has still not been explored [65].

Similarly, the use of species such as *S'codes ludwigii*, *S. pombe*, *Candida stellata*, *M. pulcherrima*, *W. anomalus*, *H. vineae*, *Z. rouxii*, *Zygosaccharomyces bailii*, *L. thermotolerans*, and *Z. florentina*, among others, as potential N sources is limited, considering their high capacity for production and release of mannoproteins [21,25,26,33–35], which is up to 29% of protein (Table 1). From the little background information available, *W. anomalus* has been used as a source of single cell protein at industrial level, specifically to produce protein for fish farming [66]. This demonstrates the potential of *W. anomalus* as source of N, taking into account the ability of this yeast to use a wide range of N sources, including nitrate [36].

One aspect to consider is the depletion of nutrients during fermentation with different yeasts. Difficulties have been reported in the introduction of *S. cerevisiae* after *Hanseniaspora/Kloeckera* in sequential fermentations [67], an effect that is associated with the depletion of thiamine and calcium pantothenate [68,69], reducing the availability of these nutrients for *S. cerevisiae* (second inoculum).

Similarly, sequential fermentations can cause a depletion of YAN by first phase yeasts, especially those with high nutrient demands and a low ability to release nitrogenous compounds. This situation can be overcome with the use of yeasts such as *M. pulcherrima*, with high proteolytic activity and amino acid release as a source of N for *S. cerevisiae* (second inoculum) [70,71].

In a recent study, Prior et al. [4] carried out sequential fermentations of *L. thermotolerans/S. cerevisiae* and *T. delbrueckii/S. cerevisiae* to evaluate whether the rest of these non-*Saccharomyces* yeasts (first phase) can be used as a source of N for *S. cerevisiae*. After 48 h of fermentation with non-*Saccharomyces*, the medium was filtered and then inoculated with *S. cerevisiae*. A reduction in sugar consumption by *S. cerevisiae* was observed in the filtered medium, in other words, fermentative yield decreased.

The results obtained in the aforementioned studies show that in sequential fermentations, non-*Saccharomyces* yeast rests (first phase) after death and lysis can release various cell components into the fermentative medium to be used as N sources for *S. cerevisiae* [4]. This strategy can be used more efficiently with non-*Saccharomyces* species with high ability to release mannoproteins (Table 1) and with high β-glucanase and protease activities.

3.2. Nitrogen Requirements for Sparkling Wines Production

The selection of yeasts with high rates of nitrogenous compounds release can ensure an adequate supply of N. Two of the non-*Saccharomyces* species which have shown the greatest ability to release amino acids during the second fermentation are *S'codes ludwigii* and *S. pombe* [30], increasing the amino acids content with respect to the base wine. The advantage of these yeasts over *S. cerevisiae* is related to their amino acid release mechanisms, with their different consumption rates and their cell composition [21,47].

One of the most important properties in sparkling wine is the ability to form foam or foamability. Mannoproteins improve foam formation and stability thanks to their hydrophobicity, high glycosylation, and high molecular mass, which enable them to surround and stabilize the gas bubbles in the foam [72]. The proteins also participate in this process which, together with the peptides and amino acids, are released mainly during aging in the bottle as a result of the enzymatic degradation of the cell walls and other cell structures [7,73], contributing to the complexity of sparkling wines.

The higher production of mannoproteins in sequential fermentations by *S. bacillaris/S. cerevisiae* [74], *L. thermotolerans/S. cerevisiae*, and *T. delbrueckii/S. cerevisiae* can also be utilized, which also contributes to reducing volatile acidity and increasing 2-phenylethanol [75,76], and improving foamability and foam stability (sequential *T. delbrueckii/S. cerevisiae* and *M. pulcherrima/S. cerevisiae* fermentations) [77,78].

Finally, during AOL (in the bottle), it has been observed that the content of free amino acids and peptides depends on the yeast species and strains [73], which have an influence on the flocculation ability to facilitate lees movement during riddling [79]. A potential field for future studies is the optimization of aging and disgorging operation in sparkling wines, for example, by inserting magnetic nanoparticles to accelerate sedimentation and lees removal [80].

4. Non-*Saccharomyces* Yeasts as a Source of Exogenous Enzymes

Another potential application of non-*Saccharomyces* yeasts is enzyme production. In general, enzymes of microbial origin are considered to have greater activity and stability than those of plant and animal origin [81]. Considering that the genus *Saccharomyces* is not characterized as being a good producer of exogenous enzymes [82], there is increasing interest in finding sources of enzymes of enological interest among non-*Saccharomyces* species, some of which are summarized in Table 2.

Table 2. Summary of the enzymes produced by non-*Saccharomyces* yeasts for wine production.

Enzyme	Yeast	Application
β-glucosidase	*Lachancea thermotolerans*	Release of terpenes and thiols from their precursors: improvement of the aromatic profile [83–85].
	Torulaspora delbrueckii	Release of thiols from their cysteinylated precursors: improvement of the aromatic profile [84].
	Wickerhamomyces anomalus	- High stability of β-glucosidase excreted by *W. anomalus* MDD24 [86]. - Release of terpenes from their glycosylated precursors: improvement of the aromatic profile [85,86].
	Metschnikowia pulcherrima	- Release of terpenes from their glycosylated precursors: improvement of the aromatic profile [71,83]. - Release of thiols from their cysteinylated precursors: improvement of the aromatic profile [84]. - Foaming and release of aromatic compounds in sparkling wines [87].
	Candida stellata	Release of terpenes (β-myrcene, limonene, linalool, α-terpineol, and farnesol) from their glycosylated precursors: improvement of the aromatic profile [88].
	Hanseniaspora uvarum	- Release of terpenes and C13-norisoprenoids from their precursors: improvement of the aromatic profile [89,90]. - Activity up to 6.6-fold higher than some *S. cerevisiae* strains [90].
	Saccharomycodes ludwigii	Activity up to 46% higher than *S. cerevisiae* at 30 °C [91].
	Aureobasidium pullulans	Release of terpenes from their glycosylated precursors: improvement of the aromatic profile [92,93].
Protease	*Wickerhamomyces anomalus*	Aspartic protease WaAPr1 excreted by *W. anomalus* 227 [94].
	Metschnikowia pulcherrima	- Aspartic protease MpAPr1 excreted by *M. pulcherrima* IWBT Y1123 [95]. - Degradation of proteins: improvement of clarification and stabilization [95]. - Degradation of proteins and peptides: source of N for fermentative yeast [70,71,83]. - Improvement of grape-must extraction and clarification, wine filtration [71], and stabilization [83]. - Improvement of foam stability in sparkling wines [71,87,96]. - Increase in the amino acids content: production of aromatic compounds [83,95].
	Candida stellata	- Degradation of proteins: improvement of clarification and stabilization [97]. - Degradation of proteins and peptides: source of N for fermentative yeast [97].

Table 2. *Cont.*

Enzyme	Yeast	Application
	Hanseniaspora uvarum	Degradation of proteins: improvement of clarification and stabilization [89,97].
	Lachancea thermotolerans, Torulaspora delbrueckii, Zygosaccharomyces bailii, Pichia kluyveri	- Increase in the amino acids content: source of N for fermentative yeast [83]. - Increase in the amino acids content: production of aromatic compounds [83]. - Degradation of proteins: improvement of clarification and stabilization [83].
Glucanase	*Wickerhamomyces anomalus*	- β-1,3-glucanase excreted by *W. anomalus* AS1: viscosity reduction in grape-musts [98,99]. - β-1,3-glucanase excreted by *W. anomalus* AS1: improvement of bioactive profile by release of *trans*-resveratrol from its glycosylated precursor polydatin [98]. - Antimicrobial control against *Dekkera/Brettanomyces*: attack at the cell wall level [100].
	Schizosaccharomyces pombe	High mannoproteins-releasing during AOL [21,38,99].
	Saccharomycodes ludwigii	High mannoproteins-releasing during AOL [21,25].
	Lachancea thermotolerans, Metschnikowia pulcherrima, Debaryomyces hansenii	Release of mannoproteins [25,83].
Hydroxycinnamate decarboxylase (HCDC)	*Metschnikowia pulcherrima, Hanseniaspora guilliermondii, Hanseniaspora opuntiae, Hanseniaspora vineae, Hanseniaspora clermontiae, Pichia guillermondii*	Involved in the synthesis of vinylphenolic pyranoanthocyanins: improvement of color stability in red wines [83,101–103].
Urease	*Schizosaccharomyces pombe*	- High urease activity [104]. - Urea hydrolysis: reduced synthesis of ethyl carbamate (carcinogenic [105]).
Carboxypeptidase	*Aureobasidium pullulans*	- Degradation of ochratoxin A (OTA) [106]. - Drawback: possible action against fermentative yeasts [107].
Pectinase	*Wickerhamomyces anomalus*	Degradation of pectins: improvement of clarification and turbidity reduction [85].
	Metschnikowia pulcherrima	- Degradation of pectins: improvement of clarification and turbidity reduction [87,97]. - Improvement of extraction of anthocyanins and polyphenols from the skins of the grape berries [108].
	Candida stellata	Degradation of pectins: improvement of clarification and turbidity reduction [85,97].
	Hanseniaspora uvarum	Degradation of pectins: improvement of clarification and turbidity reduction [89,97].
	Aureobasidium pullulans	- Degradation of pectins: improvement of clarification and turbidity reduction [92,93]. - Production of cold-active and acid-tolerant pectinases, suitable for low-temperature winemaking [92].

Table 2. *Cont.*

Enzyme	Yeast	Application
Cellulase	*Lachancea thermotolerans*, *Metschnikowia pulcherrima*, *Candida stellata*, *Hanseniaspora uvarum*, *Aureobasidium pullulans*, *Debaryomyces hansenii*	- Degradation of cellulose released from the grape cell walls: improvement of extraction, filtration and clarification [83,92,97]. - Improvement of extraction of pigments and aromatic compounds from the skins of the grape berries [83,92,97].
Xylanase	*Lachancea thermotolerans*, *Candida stellata*, *Hanseniaspora uvarum*, *Aureobasidium pullulans*	Degradation of hemicellulose: improvement of wine aroma by increasing of monoterpenyl diglycoside precursors in the grape-must [83,92,97].
β-lyase	*Torulaspora delbrueckii*, *Kluyveromyces marxianus*, *Meyerozyma guilliermondii* (formerly *Pichia guilliermondii*)	Release of thiols from their cysteinylated precursors: improvement of the aromatic profile [109,110].
Lipase	*Lachancea thermotolerans*	Increase on free fatty acids concentration [83].
	Aureobasidium pullulans	Improvement of wine aroma: synthesis of ethyl esters and ethyl acetates from lipid cleavage [111].

There is little information regarding the application of purified β-glucosidase, β-lyase, xylanase, cellulase, among others enzymes, produced by non-*Saccharomyces* yeasts for winemaking processes, leaving open the possibility of future research which focuses on improving the release of terpenes, thiols, norisoprenoids, and/or their precursors, with positive impacts on the aromatic profile, especially in white wines, considering that approximately 90% of these compounds are conjugated in the grape skin [99]. This is in addition to the selection of suitable strains, because high levels of β-glucosidase can increase the synthesis of undesirable volatile phenols [101], as well as hydrolyze the anthocyanins in red wines, exposing them to oxidation and/or transformation into colorless forms [27,112].

In contrast, one problem in white wines is the protein haze (wine turbidity), which is usually corrected by removing the proteins from the grape-must with bentonite, with the disadvantage of removing other compounds of enological interest, mainly aromatic compounds. The protease activity of non-*Saccharomyces* yeasts can be used (Table 2) and thus reduce the protein content in the grape-must and therefore prevent the wine haze. The proteases hydrolyze the proteins from the grape-must into smaller molecules like peptides and amino acids, with the consequent clarification and subsequent stability of the wine obtained, and with the additional advantage of providing YAN for the fermentative yeasts [87,89].

5. Non-*Saccharomyces* as Biocontrol Agents Against Contaminating Yeasts

5.1. Antimicrobial Peptides

Some peptides produced by yeasts have shown antimicrobial effects against several grape-must/wine contaminating yeasts. In general, these peptides show lengths of up to 100 amino acids, sorted into variable sequences [113]. Their mechanism of action would be related to changes in the integrity of the cell wall of the target yeasts [114]. Peptides with molecular mass below 10 kDa have shown greater antimicrobial effects [115], such as those produced by *Candida intermedia*, especially the LAMAP1790 strain, with an effect on several strains of *Brettanomyces bruxellensis* and without affecting the growth of *S. cerevisiae* [115].

S. cerevisiae CCMI885 is another yeast that produces antimicrobial peptides with molecular mass lower than 10 kDa and with an effect on *B. bruxellensis*, *Hanseniaspora uvarum*, *Hanseniaspora guilliermondii*, *C. stellata*, *L. thermotolerans*, *Kluyveromyces marxianus*, and *T. delbrueckii* [116]. However, as these peptides have not shown total inhibition over *B. bruxellensis*, their application may require the use of other usual winemaking treatments, such as the addition of SO_2.

According to Peña et al. [117], to achieve the application of these peptides at an industrial level, it is necessary to understand their behavior in mediums with different pH values and sugar levels, as well as high alcohol levels and in the presence of other winemaking yeasts and/or bacteria. Additionally, the implementation of procedures that enable high production and, therefore, satisfy a potential industrial level demand is needed.

5.2. Killer Toxins

Some yeasts produce molecules called "killer toxins", which are glycosylated proteins with effect against sensitive yeast strains [118]. For instance, the action of killer toxins CpKT1 and CpKT2 produced by *Candida pyralidae* (YWBT Y1140 strain) on the cell wall of *B. bruxellensis* have been reported [119]. These toxins have a molecular mass of over 50 kDa, are stable at acidic pH values (3.5–4.5), temperatures between 15 and 25 °C, at high alcoholic content, and at different sugar concentrations. In other words, they are stable under normal winemaking conditions.

In the same way, a killer toxin produced by *T. delbrueckii* NPCC 1033 (TdKT) has shown potential to control yeasts such as *Brettanomyces bruxellensis*, *Pichia guilliermondii*, *Pichia mandshurica*, and *Pichia membranifaciens* [120], being the mechanism of action, the attack at the cell wall level, related to their glucanase and chitinase enzymatic activities. The toxin has a molecular mass of over 30 kDa, its killer activity is stable at pH values of 4.2–4.8, and is inactivated at temperature above 40 °C, confirming their potential use as a biocontrol tool at oenological conditions.

Similarly, Kwkt toxins produced by *Kluyveromyces wickerhamii* [121] and PMKT2 produced by *P. membranifaciens* have shown effects on *B. bruxellensis* [122], although in the case of PMKT2, effects on *S. cerevisiae* have also been observed.

W. anomalus also produces killer toxins with effects on *Dekkera/Brettanomyces*, especially the Pikt toxin [121,123] produced by the D2 and DBVPG 3003 strains, whose fungicidal effect on wine can be sustained for 10 days [121]. The mechanism of action would be the attack at the cell wall level, specifically of β-1,6 glucans. [124], in a similar manner to the *W. anomalus* Cf20 toxin (KTCf20) that binds to the β-1,3 and β-1,6 glucans. β-1,3-glucanase activity has been previously reported in toxins secreted by *W. anomalus* [100]. Additionally, several toxins secreted by this yeast has shown stability and high activity at pH values of 3.0-5.0 and temperatures up to 30 °C, i.e., these toxins are compatible with the winemaking conditions [100,125].

Other studies have reported the antimicrobial effect of *W. anomalus* on species such as *Pichia guilliermondii* or *P. membranifaciens* [125,126] during the first stages of fermentation, and even on *S. cerevisiae* [125,127]. This indicates the need for more studies to evaluate the compatibility of the *W. anomalus* strains that produce these toxins with other yeasts to avoid technological problems such as sluggish and stuck fermentations, as well as take advantage of the potential of the different non-*Saccharomyces* species involved in the fermentation process, to obtain wines with greater complexity and stability.

Although it is not a non-*Saccharomyces* yeast, *Saccharomyces eubayanus* has also shown the ability to produce the killer toxin SeKT, with effect on spoilage yeasts such as *B. bruxellensis*, *P. membranifaciens*, *Meyerozyma guilliermondii* and *P. manshurica* [128]. The mechanism of action comprises cell wall disruption through β-glucanase and chitinase activities. The toxin have a molecular mass of around 70 kDa, and has shown stability at high glucose and ethanol concentrations (300 g/L and 16% *v/v*, respectively), at SO_2 concentrations of up to 100 mg/L, and at temperatures and pH values less than 26 °C and 5.0, respectively.

The results obtained in the aforementioned studies show the potential use of the killer toxins as a biocontrol tool at oenological conditions, especially against the spoilage yeast *B. bruxellensis*, for example, during wine aging and storing.

5.3. Other Molecules as Biocontrol Agents

One feature of *M. pulcherrima* is its potential effect against different contaminating yeast species. In the context of wine production, this activity would be mainly related to the production of the pigment pulcherrimin, from the chelation of Fe in the fermentative medium [129], reducing the availability of this mineral, with harmful effects on *Brettanomyces/Dekkera*, *Pichia*, *Hanseniaspora*, *S'codes ludwigii*, and *Candida* [130].

However, the most significant advantage of this antimicrobial mechanism is the absence of harmful effects on *S. cerevisiae* [130]. In other words, it may be compatible with the main yeast used for wine production, for example, in mixed fermentations, with the consequent reduction of the SO_2 dose, usually used as an antimicrobial agent [2].

6. Future Perspectives

One of the greatest challenges related to the industrial use of non-*Saccharomyces* yeasts for AOL is achieving acceptance from producers, especially in regions with a deep-rooted winemaking tradition and taking into account the unfavorable background of these yeasts, as well as the economic impact of prolonged storage of the wines in AOL and the possible risks of contamination. The optimization of aging conditions is one aspect that requires special attention, especially those conditions which enable the acceleration of the release of polysaccharides from the lees.

Another aspect that requires more research is the use of exogenously produced lees added to the wine [6], especially for the aging of red wines, which involves the identification of species, and especially strains with high rates of polysaccharide release which, in turn, present low anthocyanin adsorption [28] and low expression of anthocyanase activity (anthocyanin-β-glucosidase) that causes hydrolysis of anthocyanins [27]. Additionally, this requires strategies, which can be simultaneously implemented with the selection of yeasts with high capacity to produce pyranoanthocyanins, more stable against degradation [27], thereby minimizing color loss. One of the yeasts that has shown these characteristics is *S. pombe*.

There is also a need for additional studies on fermentation with natural grape-musts to verify the properties of mannoproteins released by non-*Saccharomyces* yeasts in synthetic mediums, given that in these last mediums, there is no interaction between the mannoproteins and the components which are naturally present in the grape-must, and to take advantage of the potential of these yeasts for the production of new products with industrial applicability [74].

Similarly, the effect of the addition of inactive yeasts, hulls, or lees as sources of N on parameters related to protein haze and turbidity formation is still not clear. One possible alternative is the addition of protease-hydrolyzed hulls/lees, which are also produced by non-*Saccharomyces* yeasts. *M. pulcherrima* has shown high protease activity and, therefore, the release of amino acids as a source of N for *S. cerevisiae*, especially in mixed cultures. This activity has also been shown by *W. anomalus*, with the advantage of using a wide range of N sources, including nitrate [36], and by *A. pullulans*, with the ability to use low-cost carbon sources such as agricultural and food waste, and whose protein production on a large scale has still not been studied [65].

It has been reported that the species and strain of the yeast influence the free amino acids and peptide content during AOL [57,58], thus reducing the content of amino acids in aging periods of over 9 months [73]. Therefore, it is necessary to conduct studies, which simulate normal aging conditions (even up to 10 years in sparkling wines) by evaluating the impact of the proteins, mannose, glucose, and galactose present in the lees on the quality of treated wine, which, until now, has only been studied in model mediums [131]. The effect of proteins from non-*Saccharomyces* yeasts on the quality of sparkling wines, their effect on lees movement during riddling, as well as the effect on the sensorial profile must also be studied.

Finally, most of the positive contributions of non-*Saccharomyces* yeasts with regard to *S. cerevisiae* are related to a higher presence of active enzymes, which depends, in part, on the carbon and nitrogen sources present in the fermentative medium. Small changes in the composition of the medium can affect the nature, quantity, and diversity of the secreted enzymes [132]. Therefore, proper maintenance of N levels in the fermentative medium is of vital importance, which can be achieved by identifying yeast species and strains with high release rates of nitrogenous compounds, mainly amino acids.

References

1. Morata, A. Enological repercussions of non-*Saccharomyces* species in wine biotechnology. *Fermentation* **2019**, *5*, 72. [CrossRef]
2. Vejarano, R. *Saccharomycodes ludwigii*, control and potential uses in winemaking processes. *Fermentation* **2018**, *4*, 71. [CrossRef]
3. Ivit, N.N.; Kemp, B. The impact of non-*Saccharomyces* yeast on traditional method sparkling wine. *Fermentation* **2018**, *4*, 73. [CrossRef]
4. Prior, K.J.; Bauer, F.F.; Divol, B. The utilisation of nitrogenous compounds by commercial non-*Saccharomyces* yeasts associated with wine. *Food Microbiol.* **2019**, *79*, 75–84. [CrossRef]
5. Del Barrio-Galán, R.; Ortega-Heras, M.; Sánchez-Iglesias, M.; Pérez-Magariño, S. Interactions of phenolic and volatile compounds with yeast lees, commercial yeast derivatives and non toasted chips in model solutions and young red wines. *Eur. Food Res. Technol.* **2012**, *234*, 231–244. [CrossRef]
6. Suárez-Lepe, J.A.; Morata, A. Nuevo Método de Crianza Sobre Lías. Oficina Española de Patentes y Marcas ES2311372B2, 7 October 2009.
7. Alexandre, H.; Guilloux-Benatier, M. Yeast autolysis in sparkling wine—A review. *Aust. J. Grape Wine Res.* **2006**, *12*, 119–127. [CrossRef]
8. Vidal, S.; Francis, L.; Williams, P.; Kwiatkowski, M.; Gawel, R.; Cheynier, V.; Waters, E. The mouth-feel properties of polysaccharides and anthocyanins in a wine like medium. *Food Chem.* **2004**, *85*, 519–525. [CrossRef]
9. Guadalupe, Z.; Palacios, A.; Ayestarán, B. Maceration enzymes and mannoproteins: A possible strategy to increase colloidal stability and color extraction in red wines. *J. Agric. Food Chem.* **2007**, *55*, 4854–4862. [CrossRef]
10. Chalier, P.; Angot, B.; Delteil, D.; Doco, T.; Gunata, Z. Interactions between aroma compounds and whole mannoprotein isolated from *Saccharomyces cerevisiae* strains. *Food Chem.* **2007**, *100*, 22–30. [CrossRef]
11. Gonzalez-Ramos, D.; Cebollero, E.; Gonzalez, R. A recombinant *Saccharomyces cerevisiae* strain overproducing mannoproteins stabilizes wine against protein haze. *Appl. Environ. Microbiol.* **2008**, *74*, 5533–5540. [CrossRef]
12. Lubbers, S.; Leger, B.; Charpentier, C.; Feuillat, M. Effet colloide-protecteur d'extraits de parois de levures sur la stabilité tartrique d'une solution hydro-alcoolique modèle. *OENO One* **1993**, *27*, 13. [CrossRef]
13. Rodrigues, A.; Ricardo-Da-Silva, J.M.; Lucas, C.; Laureano, O. Effect of commercial mannoproteins on wine colour and tannins stability. *Food Chem.* **2012**, *131*, 907–914. [CrossRef]
14. Morata, A.; Loira, I.; Suárez-Lepe, J.A. Influence of yeasts in wine colour. In *Grape and Wine Biotechnology*; Morata, A., Loira, I., Eds.; IntechOpen: London, UK, 2016; pp. 285–305.
15. Vejarano, R.; Gil-Calderón, A.; Díaz-Silva, V.; León-Vargas, J. Improvement of the bioactive profile in wines and its incidence on human health: Technological strategies. In *Advances in Grape and Wine Biotechnology*; Morata, A., Loira, I., Eds.; IntechOpen: London, UK, 2019.
16. Loira, I.; Vejarano, R.; Morata, A.; Ricardo-Da-Silva, J.M.; Laureano, O.; González, M.C.; Suárez-Lepe, J.A. Effect of *Saccharomyces* strains on the quality of red wines aged on lees. *Food Chem.* **2013**, *139*, 1044–1051. [CrossRef]
17. Rosi, I.; Gheri, A.; Domizio, P.; Fia, G. Production de macromolecules parietals de *Saccharomyces cerevisiae* au cours de la fermentation et leur influence sur la fermentation malolactique. *Rev. des Œnologues* **2000**, *94*, 18–20.
18. Moruno, E.G.; Sanlorenzo, C.; Boccaccino, B.; Di Stefano, R. Treatment with yeast to reduce the concentration of ochratoxin A in red wine. *Am. J. Enol. Vitic.* **2005**, *56*, 73–76.
19. Moreno-Arribas, V.; Pueyo, E.; Nieto, F.J.; Martín-Álvarez, P.J.; Polo, M.C. Influence of the polysaccharides and the nitrogen compounds on foaming properties of sparkling wines. *Food Chem.* **2000**, *70*, 309–317. [CrossRef]

20. Gallardo-Chacón, J.J.; Vichi, S.; López-Tamames, E.; Buxaderas, S. Changes in the sorption of diverse volatiles by *Saccharomyces cerevisiae* lees during sparkling wine aging. *J. Agric. Food Chem.* **2010**, *58*, 12426–12430. [CrossRef]
21. Palomero, F.; Morata, A.; Benito, S.; Calderón, F.; Suárez-Lepe, J.A. New genera of yeasts for over-lees aging of red wine. *Food Chem.* **2009**, *112*, 432–441. [CrossRef]
22. Kopecká, M.; Fleet, G.H.; Phaff, H.J. Ultrastructure of the cell wall of *Schizosaccharomyces pombe* following treatment with various glucanases. *J. Struct. Biol.* **1995**, *114*, 140–152. [CrossRef]
23. Hidalgo, J. *Tratado de Enología, Tomo I*, 2nd ed.; Mundiprensa: Madrid, Spain, 2011; ISBN 84-8476-119-3.
24. Domizio, P.; Liu, Y.; Bisson, L.F.; Barile, D. Cell wall polysaccharides released during the alcoholic fermentation by *Schizosaccharomyces pombe* and *S. japonicus*: Quantification and characterization. *Food Microbiol.* **2017**, *61*, 136–149. [CrossRef]
25. Giovani, G.; Rosi, I.; Bertuccioli, M. Quantification and characterization of cell wall polysaccharides released by non-*Saccharomyces* yeast strains during alcoholic fermentation. *Int. J. Food Microbiol.* **2012**, *160*, 113–118. [CrossRef] [PubMed]
26. Domizio, P.; Romani, C.; Lencioni, L.; Comitini, F.; Gobbi, M.; Mannazzu, I.; Ciani, M. Outlining a future for non-*Saccharomyces* yeasts: Selection of putative spoilage wine strains to be used in association with *Saccharomyces cerevisiae* for grape juice fermentation. *Int. J. Food Microbiol.* **2011**, *147*, 170–180. [CrossRef] [PubMed]
27. Wightman, J.D.; Wrolstad, R.E. β-glucosidase activity in juice-processing enzymes based on anthocyanin analysis. *J. Food Sci.* **1996**, *61*, 544–548. [CrossRef]
28. Morata, A.; Loira, I.; Escott, C.; del Fresno, J.M.; Bañuelos, M.A.; Suárez-Lepe, J.A. Applications of *Metschnikowia pulcherrima* in wine biotechnology. *Fermentation* **2019**, *5*, 63. [CrossRef]
29. Morata, A.; Gómez-Cordovés, M.C.; Calderón, F.; Suárez, J.A. Effects of pH, temperature and SO_2 on the formation of pyranoanthocyanins during red wine fermentation with two species of *Saccharomyces*. *Int. J. Food Microbiol.* **2006**, *106*, 123–129. [CrossRef]
30. Ivit, N.N.; Loira, I.; Morata, A.; Benito, S.; Palomero, F.; Suárez-Lepe, J.A. Making natural sparkling wines with non-*Saccharomyces* yeasts. *Eur. Food Res. Technol.* **2018**, *244*, 925–935. [CrossRef]
31. Granchi, L.; Romano, P.; Mangani, S.; Guerrini, S.; Vincenzini, M. Production of biogenic amines by wine microorganisms. *Bull. OIV* **2005**, *78*, 595–609.
32. Marcobal, A.; Martín-Álvarez, P.J.; Polo, M.C.; Muñoz, R.; Moreno-Arribas, M.V. Formation of biogenic amines throughout the industrial manufacture of red wine. *J. Food Prot.* **2006**, *69*, 397–404. [CrossRef]
33. Martin, V. *Hanseniaspora Vineae: Caracterización y su Uso en la Vinificación*; Universidad de la Republica: Montevideo, Uruguay, 2016.
34. Domizio, P.; Liu, Y.; Bisson, L.F.; Barile, D. Use of non-*Saccharomyces* wine yeasts as novel sources of mannoproteins in wine. *Food Microbiol.* **2014**, *43*, 5–15. [CrossRef]
35. Lencioni, L.; Romani, C.; Gobbi, M.; Comitini, F.; Ciani, M.; Domizio, P. Controlled mixed fermentation at winery scale using *Zygotorulaspora florentina* and *Saccharomyces cerevisiae*. *Int. J. Food Microbiol.* **2016**, *234*, 36–44. [CrossRef]
36. Zha, Y.; Hossain, A.H.; Tobola, F.; Sedee, N.; Havekes, M.; Punt, P.J. *Pichia anomala* 29X: A resistant strain for lignocellulosic biomass hydrolysate fermentation. *FEMS Yeast Res.* **2013**, *13*, 609–617. [CrossRef] [PubMed]
37. Fugelsang, K.C.; Edwards, C.G. *Wine Microbiology: Practical Applications and Procedures*; Springer Science & Business Media: Berlin, Germany, 2007; ISBN 038733341X.
38. Loira, I.; Morata, A.; Palomero, F.; González, C.; Suárez-Lepe, J.A. *Schizosaccharomyces pombe*: A promising biotechnology for modulating wine composition. *Fermentation* **2018**, *4*, 70. [CrossRef]
39. Kulkarni, P.; Loira, I.; Morata, A.; Tesfaye, W.; González, M.C.; Suárez-Lepe, J.A. Use of non-*Saccharomyces* yeast strains coupled with ultrasound treatment as a novel technique to accelerate ageing on lees of red wines and its repercussion in sensorial parameters. *LWT* **2015**, *64*, 1262. [CrossRef]
40. Todd, B.E.N.; Fleet, G.H.; Henschke, P.A. Promotion of autolysis through the interaction of killer and sensitive yeasts: Potential application in sparkling wine production. *Am. J. Enol. Vitic.* **2000**, *51*, 65–72.
41. Englezos, V.; Rantsiou, K.; Giacosa, S.; Río Segade, S.; Rolle, L.; Cocolin, L. Cell-to-cell contact mechanism modulates *Starmerella bacillaris* death in mixed culture fermentations with *Saccharomyces cerevisiae*. *Int. J. Food Microbiol.* **2019**, *289*, 106–114. [CrossRef]

42. Palomero, F.; Morata, A.; Benito, S.; González, M.C.; Suárez-Lepe, J.A. Conventional and enzyme-assisted autolysis during ageing over lees in red wines: Influence on the release of polysaccharides from yeast cell walls and on wine monomeric anthocyanin content. *Food Chem.* **2007**, *105*, 838–846. [CrossRef]
43. Del Fresno, J.M.; Loira, I.; Morata, A.; González, C.; Suárez-Lepe, J.A.; Cuerda, R. Application of ultrasound to improve lees ageing processes in red wines. *Food Chem.* **2018**, *261*, 157–163. [CrossRef]
44. Del Fresno, J.M.; Morata, A.; Escott, C.; Loira, I.; Cuerda, R.; Suárez-Lepe, J.A. Sonication of yeast biomasses to improve the ageing on lees technique in red wines. *Molecules* **2019**, *24*, 635. [CrossRef]
45. Jackson, R.S. Fermentation. In *Wine Science: Principles and Applications*; Jackson, R.S., Ed.; Academic Press/Elsevier: Hoboken, NJ, USA, 2014; ISBN 9780123814692.
46. Bely, M.; Rinaldi, A.; Dubourdieu, D. Influence of assimilable nitrogen on volatile acidity production by *Saccharomyces cerevisiae* during high sugar fermentation. *J. Biosci. Bioeng.* **2003**, *96*, 507–512. [CrossRef]
47. Gobert, A.; Tourdot-Maréchal, R.; Morge, C.; Sparrow, C.; Liu, Y.; Quintanilla-Casas, B.; Vichi, S.; Alexandre, H. Non-*Saccharomyces* yeasts nitrogen source preferences: Impact on sequential fermentation and wine volatile compounds profile. *Front. Microbiol.* **2017**, *8*, 2175. [CrossRef]
48. Ribéreau-Gayon, P.; Glories, Y.; Maujean, A.; Dubourdieu, D. Nitrogen compounds. In *Handbook of Enology: The Chemistry of Wine Stabilization and Treatments*; John Wiley and Sons Ltd.: Hoboken, NJ, USA, 2006; ISBN 978-0-470-01037-2.
49. Cooper, T.G. Transmitting the signal of excess nitrogen in *Saccharomyces cerevisiae* from the Tor proteins to the GATA factors: Connecting the dots. *FEMS Microbiol. Rev.* **2002**, *26*, 223–238. [CrossRef] [PubMed]
50. De Koker, S. *Nitrogen Utilisation of Selected Non-Saccharomyces Yeasts and the Impact on Volatile Compound Production*; Stellenbosch University: Stellenbosch, South Africa, 2015.
51. Martin, V.; Giorello, F.; Fariña, L.; Minteguiaga, M.; Salzman, V.; Boido, E.; Aguilar, P.S.; Gaggero, C.; Dellacassa, E.; Mas, A.; et al. De novo synthesis of benzenoid compounds by the yeast *Hanseniaspora vineae* increases the flavor diversity of wines. *J. Agric. Food Chem.* **2016**, *64*, 4574–4583. [CrossRef]
52. Martin, V.; Boido, E.; Giorello, F.; Mas, A.; Dellacassa, E.; Carrau, F. Effect of yeast assimilable nitrogen on the synthesis of phenolic aroma compounds by *Hanseniaspora vineae* strains. *Yeast* **2016**, *33*, 323–328. [CrossRef] [PubMed]
53. Burin, V.M.; Gomes, T.M.; Caliari, V.; Rosier, J.P.; Bordignon Luiz, M.T. Establishment of influence the nitrogen content in musts and volatile profile of white wines associated to chemometric tools. *Microchem. J.* **2015**, *122*, 20–28. [CrossRef]
54. Mendes-Ferreira, A.; Barbosa, C.; Lage, P.; Mendes-Faia, A. The impact of nitrogen on yeast fermentation and wine quality. *Cienc. Tec. Vitivinic.* **2011**, *26*, 17–32.
55. Sturgeon, J.Q.; Bohlscheid, J.C.; Edwards, C.G. The effect of nitrogen source on yeast metabolism and H_2S formation. *J. Wine Res.* **2013**, *24*, 182–194. [CrossRef]
56. Ribereau-Gayon, P.; Dubourdieu, D.; Doneche, B.; Lonvaud, A. *Handbook of Enology: The Microbiology of Wine and Vinifications*, 2nd ed.; John Wiley & Sons, Ltd.: Chichester, UK, 2006; Volume 1, ISBN 978-0-470-01034-1.
57. Chi, Z.; Yan, K.; Gao, L.; Li, J.; Wang, X.; Wang, L. Diversity of marine yeasts with high protein content and evaluation of their nutritive compositions. *J. Mar. Biol. Assoc. United Kingdom* **2008**, *88*, 1347–1352. [CrossRef]
58. Manitchotpisit, P.; Skory, C.D.; Leathers, T.D.; Lotrakul, P.; Eveleigh, D.E.; Prasongsuk, S.; Punnapayak, H. α-Amylase activity during pullulan production and α-amylase gene analyses of *Aureobasidium pullulans*. *J. Ind. Microbiol. Biotechnol.* **2011**, *38*, 1211–1218. [CrossRef]
59. Leite, R.S.R.; Bocchini, D.A.; Martins, E.S.; Silva, D.; Gomes, E.; Da Silva, R. Production of cellulolytic and hemicellulolytic enzymes from *Aureobasidium pullulans* on solid state fermentation. *Appl. Biochem. Biotechnol.* **2007**, *137–140*, 281–288.
60. Leathers, T.D.; Rich, J.O.; Anderson, A.M.; Manitchotpisit, P. Lipase production by diverse phylogenetic clades of *Aureobasidium pullulans*. *Biotechnol. Lett.* **2013**, *35*, 1701–1706. [CrossRef]
61. Manitchotpisit, P.; Leathers, T.D.; Peterson, S.W.; Kurtzman, C.; Li, X.L.; Eveleigh, D.E.; Lotrakul, P.; Prasongsuk, S.; Dunlap, C.A.; Vermillion, K.E.; et al. Multilocus phylogenetic analyses, pullulan production and xylanase activity of tropical isolates of *Aureobasidium pullulans*. *Mycol. Res.* **2009**, *113*, 1107–1120. [CrossRef]
62. Rich, J.O.; Leathers, T.D.; Anderson, A.M.; Bischoff, K.M.; Manitchotpisit, P. Laccases from *Aureobasidium pullulans*. *Enzyme Microb. Technol.* **2013**, *53*, 33–37. [CrossRef] [PubMed]

63. Chi, Z.; Wang, F.; Chi, Z.; Yue, L.; Liu, G.; Zhang, T. Bioproducts from *Aureobasidium pullulans*, a biotechnologically important yeast. *Appl. Microbiol. Biotechnol.* **2009**, *82*, 793–804. [CrossRef] [PubMed]
64. Ni, X.; Chi, Z.; Ma, C.; Madzak, C. Cloning, characterization, and expression of the gene encoding alkaline protease in the marine yeast *Aureobasidium pullulans* 10. *Mar. Biotechnol.* **2008**, *10*, 319–327. [CrossRef] [PubMed]
65. Bozoudi, D.; Tsaltas, D. The multiple and versatile roles of *Aureobasidium pullulans* in the vitivinicultural sector. *Fermentation* **2018**, *4*, 85. [CrossRef]
66. Huyben, D.; Nyman, A.; Vidaković, A.; Passoth, V.; Moccia, R.; Kiessling, A.; Dicksved, J.; Lundh, T. Effects of dietary inclusion of the yeasts *Saccharomyces cerevisiae* and *Wickerhamomyces anomalus* on gut microbiota of rainbow trout. *Aquaculture* **2017**, *473*, 528–537. [CrossRef]
67. Medina, K.; Boido, E.; Dellacassa, E.; Carrau, F. Growth of non-*Saccharomyces* yeasts affects nutrient availability for *Saccharomyces cerevisiae* during wine fermentation. *Int. J. Food Microbiol.* **2012**, *157*, 245–250. [CrossRef]
68. Bataillon, M.; Rico, A.; Sablayrolles, J.M.; Salmon, J.M.; Barre, P. Early thiamin assimilation by yeasts under enological conditions: Impact on alcoholic fermentation kinetics. *J. Ferment. Bioeng.* **1996**, *82*, 145–150. [CrossRef]
69. Wang, X.D.; Bohlscheid, J.C.; Edwards, C.G. Fermentative activity and production of volatile compounds by *Saccharomyces* grown in synthetic grape juice media deficient in assimilable nitrogen and/or pantothenic acid. *J. Appl. Microbiol.* **2003**, *94*, 349–359. [CrossRef]
70. Romano, P.; Capece, A.; Jespersen, L. Taxonomic and ecological diversity of food and beverage yeasts. In *Yeasts in Food and Beverages*; Springer: Berlin/Heidelberg, Germany, 2006; pp. 13–53.
71. Barbosa, C.; Lage, P.; Esteves, M.; Chambel, L.; Mendes-Faia, A.; Mendes-Ferreira, A. Molecular and phenotypic characterization of *Metschnikowia pulcherrima* strains from Douro Wine Region. *Fermentation* **2018**, *4*, 8. [CrossRef]
72. Vincenzi, S.; Crapisi, A.; Curioni, A. Foamability of Prosecco wine: Cooperative effects of high molecular weight glycocompounds and wine PR-proteins. *Food Hydrocoll.* **2014**, *34*, 202–207. [CrossRef]
73. Martínez-Rodríguez, A.J.; Carrascosa, A.V.; Martín-Álvarez, P.J.; Moreno-Arribas, V.; Polo, M.C. Influence of the yeast strain on the changes of the amino acids, peptides and proteins during sparkling wine production by the traditional method. *J. Ind. Microbiol. Biotechnol.* **2002**, *29*, 314–322. [CrossRef]
74. Lemos Junior, W.J.F.; Nadai, C.; Rolle, L.; da Silva Gulao, E.; Miguez da Rocha Leão, M.H.; Giacomini, A.; Corich, V.; Vincenzi, S. Influence of the mannoproteins of different strains of *Starmenella bacillaris* used in single and sequential fermentations on foamability, tartaric and protein stabilities of wines. *OENO One* **2020**, *54*, 231–243. [CrossRef]
75. Del Fresno, J.M.; Morata, A.; Loira, I.; Bañuelos, M.A.; Escott, C.; Benito, S.; González Chamorro, C.; Suárez-Lepe, J.A. Use of non-*Saccharomyces* in single-culture, mixed and sequential fermentation to improve red wine quality. *Eur. Food Res. Technol.* **2017**, *243*, 2175–2185. [CrossRef]
76. Comitini, F.; Gobbi, M.; Domizio, P.; Romani, C.; Lencioni, L.; Mannazzu, I.; Ciani, M. Selected non-*Saccharomyces* wine yeasts in controlled multistarter fermentations with *Saccharomyces cerevisiae*. *Food Microbiol.* **2011**, *28*, 873–882. [CrossRef]
77. González-Royo, E.; Pascual, O.; Kontoudakis, N.; Esteruelas, M.; Esteve-Zarzoso, B.; Mas, A.; Canals, J.M.; Zamora, F. Oenological consequences of sequential inoculation with non-*Saccharomyces* yeasts (*Torulaspora delbrueckii* or *Metschnikowia pulcherrima*) and *Saccharomyces cerevisiae* in base wine for sparkling wine production. *Eur. Food Res. Technol.* **2015**, *240*, 999–1012. [CrossRef]
78. Medina-Trujillo, L.; González-Royo, E.; Sieczkowski, N.; Heras, J.; Canals, J.M.; Zamora, F. Effect of sequential inoculation (*Torulaspora delbrueckii/Saccharomyces cerevisiae*) in the first fermentation on the foaming properties of sparkling wine. *Eur. Food Res. Technol.* **2017**, *243*, 681–688. [CrossRef]
79. Perpetuini, G.; Di Gianvito, P.; Arfelli, G.; Schirone, M.; Corsetti, A.; Tofalo, R.; Suzzi, G. Biodiversity of autolytic ability in flocculent *Saccharomyces cerevisiae* strains suitable for traditional sparkling wine fermentation. *Yeast* **2016**, *33*, 303–312. [CrossRef]
80. Loira, I.; Morata, A.; Escott, C.; Del Fresno, J.M.; Tesfaye, W.; Palomero, F.; Suárez-Lepe, J.A. Applications of nanotechnology in the winemaking process. *Eur. Food Res. Technol.* **2020**, *246*, 1533–1541. [CrossRef]
81. Anbu, P.; Gopinath, S.C.B.; Chaulagain, B.P.; Lakshmipriya, T. Microbial enzymes and their applications in industries and medicine 2016. *Biomed Res. Int.* **2017**, *2017*, 2195808. [CrossRef]
82. García, M.; Esteve-Zarzoso, B.; Cabellos, J.M.; Arroyo, T. Advances in the study of *Candida stellata*. *Fermentation* **2018**, *4*, 74. [CrossRef]

83. Escribano, R.; González-Arenzana, L.; Garijo, P.; Berlanas, C.; López-Alfaro, I.; López, R.; Gutiérrez, A.R.; Santamaría, P. Screening of enzymatic activities within different enological non-*Saccharomyces* yeasts. *J. Food Sci. Technol.* **2017**, *54*, 1555–1564. [CrossRef] [PubMed]
84. Zott, K.; Thibon, C.; Bely, M.; Lonvaud-Funel, A.; Dubourdieu, D.; Masneuf-Pomarede, I. The grape must non-*Saccharomyces* microbial community: Impact on volatile thiol release. *Int. J. Food Microbiol.* **2011**, *151*, 210–215. [CrossRef] [PubMed]
85. Cordero-Bueso, G.; Esteve-Zarzoso, B.; Cabellos, J.M.; Gil-Díaz, M.; Arroyo, T. Biotechnological potential of non-*Saccharomyces* yeasts isolated during spontaneous fermentations of Malvar (*Vitis vinifera* cv. L.). *Eur. Food Res. Technol.* **2013**, *236*, 193–207. [CrossRef]
86. Swangkeaw, J.; Vichitphan, S.; Butzke, C.E.; Vichitphan, K. Characterization of β-glucosidases from *Hanseniaspora* sp. and *Pichia anomala* with potentially aroma-enhancing capabilities in juice and wine. *World J. Microbiol. Biotechnol.* **2011**, *27*, 423–430. [CrossRef]
87. Fernández, M.; Úbeda, J.F.; Briones, A.I. Typing of non-*Saccharomyces* yeasts with enzymatic activities of interest in wine-making. *Int. J. Food Microbiol.* **2000**, *59*, 29–36. [CrossRef]
88. Hock, R.; Benda, I.; Schreier, P. Formation of terpenes by yeasts during alcoholic fermentation. *Z. Lebensm. Unters. Forsch.* **1984**, *179*, 450–452. [CrossRef]
89. Andorrà, I.; Berradre, M.; Rozès, N.; Mas, A.; Guillamón, J.M.; Esteve-Zarzoso, B. Effect of pure and mixed cultures of the main wine yeast species on grape must fermentations. *Eur. Food Res. Technol.* **2010**, *231*, 215–224. [CrossRef]
90. Hu, K.; Jin, G.J.; Xu, Y.H.; Tao, Y.S. Wine aroma response to different participation of selected *Hanseniaspora uvarum* in mixed fermentation with *Saccharomyces cerevisiae*. *Food Res. Int.* **2018**, *108*, 119–127. [CrossRef] [PubMed]
91. Bovo, B.; Carlot, M.; Lombardi, A.; Lomolino, G.; Lante, A.; Giacomini, A.; Corich, V. Exploring the use of *Saccharomyces cerevisiae* commercial strain and *Saccharomycodes ludwigii* natural isolate for grape marc fermentation to improve sensory properties of spirits. *Food Microbiol.* **2014**, *41*, 33–41. [CrossRef] [PubMed]
92. Merín, M.G.; Morata de Ambrosini, V.I. Highly cold-active pectinases under wine-like conditions from non-*Saccharomyces* yeasts for enzymatic production during winemaking. *Lett. Appl. Microbiol.* **2015**, *60*, 467–474. [CrossRef] [PubMed]
93. Onetto, C.A.; Borneman, A.R.; Schmidt, S.A. Investigating the effects of *Aureobasidium pullulans* on grape juice composition and fermentation. *Food Microbiol.* **2020**, *90*, 103451. [CrossRef]
94. Schlander, M.; Distler, U.; Tenzer, S.; Thines, E.; Claus, H. Purification and properties of yeast proteases secreted by *Wickerhamomyces anomalus* 227 and *Metschnikovia pulcherrima* 446 during growth in a white grape juice. *Fermentation* **2017**, *3*, 2. [CrossRef]
95. Theron, L.W.; Bely, M.; Divol, B. Monitoring the impact of an aspartic protease (MpAPr1) on grape proteins and wine properties. *Appl. Microbiol. Biotechnol.* **2018**, *102*, 5173–5183. [CrossRef] [PubMed]
96. Reid, V.J.; Theron, L.W.; Toit, M.D.; Divol, B. Identification and partial characterization of extracellular aspartic protease genes from *Metschnikowia pulcherrima* IWBT Y1123 and *Candida apicola* IWBT Y1384. *Appl. Environ. Microbiol.* **2012**, *78*, 6838–6849. [CrossRef]
97. Strauss, M.L.A.; Jolly, N.P.; Lambrechts, M.G.; Van Rensburg, P. Screening for the production of extracellular hydrolytic enzymes by non-*Saccharomyces* wine yeasts. *J. Appl. Microbiol.* **2001**, *91*, 182–190. [CrossRef]
98. Schwentke, J.; Sabel, A.; Petri, A.; König, H.; Claus, H. The yeast *Wickerhamomyces anomalus* AS1 secretes a multifunctional exo-β-1,3-glucanase with implications for winemaking. *Yeast* **2014**, *31*, 349–359. [CrossRef]
99. Claus, H.; Mojsov, K. Enzymes for wine fermentation: Current and perspective applications. *Fermentation* **2018**, *4*, 52. [CrossRef]
100. Izgü, F.; Altinbay, D.; Acun, T. Killer toxin of *Pichia anomala* NCYC 432; purification, characterization and its exo-β-1,3-glucanase activity. *Enzyme Microb. Technol.* **2006**, *39*, 669–676. [CrossRef]
101. Morata, A.; Vejarano, R.; Ridolfi, G.; Benito, S.; Palomero, F.; Uthurry, C.; Tesfaye, W.; González, C.; Suárez-Lepe, J.A. Reduction of 4-ethylphenol production in red wines using HCDC+ yeasts and cinnamyl esterases. *Enzyme Microb. Technol.* **2013**, *52*, 99–104. [CrossRef]
102. Medina, K.; Boido, E.; Dellacassa, E.; Carrau, F. Effects of non-*Saccharomyces* yeasts on color, anthocyanin, and anthocyanin-derived pigments of Tannat grapes during fermentation. *Am. J. Enol. Vitic.* **2018**, *69*, 148–156. [CrossRef]

103. Benito, S.; Morata, A.; Palomero, F.; González, M.C.; Suárez-Lepe, J.A. Formation of vinylphenolic pyranoanthocyanins by *Saccharomyces cerevisiae* and *Pichia guillermondii* in red wines produced following different fermentation strategies. *Food Chem.* **2011**, *124*, 15–23. [CrossRef]
104. Lubbers, M.W.; Rodriguez, S.B.; Honey, N.K.; Thornton, R.J. Purification and characterization of urease from *Schizosaccharomyces pombe*. *Can. J. Microbiol.* **1996**, *42*, 132–140. [CrossRef] [PubMed]
105. Uthurry, C.A.; Suárez Lepe, J.A.; Lombardero, J.; García Del Hierro, J.R. Ethyl carbamate production by selected yeasts and lactic acid bacteria in red wine. *Food Chem.* **2006**, *94*, 262–270. [CrossRef]
106. De Felice, D.V.; Solfrizzo, M.; De Curtis, F.; Lima, G.; Visconti, A.; Castoria, R. Strains of *Aureobasidium pullulans* can lower ochratoxin a contamination in wine grapes. *Phytopathology* **2008**, *98*, 1261–1270. [CrossRef]
107. Zhang, H.; Apaliya, M.T.; Mahunu, G.K.; Chen, L.; Li, W. Control of ochratoxin A-producing fungi in grape berry by microbial antagonists: A review. *Trends Food Sci. Technol.* **2016**, *51*, 88–97. [CrossRef]
108. Belda, I.; Conchillo, L.B.; Ruiz, J.; Navascués, E.; Marquina, D.; Santos, A. Selection and use of pectinolytic yeasts for improving clarification and phenolic extraction in winemaking. *Int. J. Food Microbiol.* **2016**, *223*, 1–8. [CrossRef]
109. Belda, I.; Ruiz, J.; Navascués, E.; Marquina, D.; Santos, A. Improvement of aromatic thiol release through the selection of yeasts with increased β-lyase activity. *Int. J. Food Microbiol.* **2016**, *225*, 1–8. [CrossRef]
110. Belda, I.; Ruiz, J.; Alastruey-Izquierdo, A.; Navascués, E.; Marquina, D.; Santos, A. Unraveling the enzymatic basis of wine "Flavorome": A phylo-functional study of wine related yeast species. *Front. Microbiol.* **2016**, *7*, 12. [CrossRef]
111. Lin, M.M.H.; Boss, P.K.; Walker, M.E.; Sumby, K.M.; Grbin, P.R.; Jiranek, V. Evaluation of indigenous non-*Saccharomyces* yeasts isolated from a South Australian vineyard for their potential as wine starter cultures. *Int. J. Food Microbiol.* **2020**, *312*, 108373. [CrossRef]
112. Manzanares, P.; Rojas, V.; Genovés, S.; Vallés, S. A preliminary search for anthocyanin-β-D-glucosidase activity in non-*Saccharomyces* wine yeasts. *Int. J. Food Sci. Technol.* **2000**, *35*, 95–103. [CrossRef]
113. Zhang, L.J.; Gallo, R.L. Antimicrobial peptides. *Curr. Biol.* **2016**, *26*, R14–R19. [CrossRef] [PubMed]
114. Branco, P.; Viana, T.; Albergaria, H.; Arneborg, N. Antimicrobial peptides (AMPs) produced by *Saccharomyces cerevisiae* induce alterations in the intracellular pH, membrane permeability and culturability of *Hanseniaspora guilliermondii* cells. *Int. J. Food Microbiol.* **2015**, *205*, 112–118. [CrossRef] [PubMed]
115. Peña, R.; Ganga, M.A. Novel antimicrobial peptides produced by *Candida intermedia* LAMAP1790 active against the wine-spoilage yeast *Brettanomyces bruxellensis*. *Antonie van Leeuwenhoek* **2019**, *112*, 297–304.
116. Albergaria, H.; Francisco, D.; Gori, K.; Arneborg, N.; Gírio, F. *Saccharomyces cerevisiae* CCMI 885 secretes peptides that inhibit the growth of some non-*Saccharomyces* wine-related strains. *Appl. Microbiol. Biotechnol.* **2010**, *86*, 965–972. [CrossRef]
117. Peña, R.; Chávez, R.; Rodríguez, A.; Ganga, M.A. A control alternative for the hidden enemy in the wine cellar. *Fermentation* **2019**, *5*, 25. [CrossRef]
118. Schmitt, M.J.; Breinig, F. The viral killer system in yeast: From molecular biology to application. *FEMS Microbiol. Rev.* **2002**, *26*, 257–276. [CrossRef]
119. Mehlomakulu, N.N.; Setati, M.E.; Divol, B. Characterization of novel killer toxins secreted by wine-related non-*Saccharomyces* yeasts and their action on *Brettanomyces* spp. *Int. J. Food Microbiol.* **2014**, *188*, 83–91. [CrossRef]
120. Villalba, M.L.; Susana Sáez, J.; del Monaco, S.; Lopes, C.A.; Sangorrín, M.P. TdKT, a new killer toxin produced by *Torulaspora delbrueckii* effective against wine spoilage yeasts. *Int. J. Food Microbiol.* **2016**, *217*, 94–100. [CrossRef]
121. Comitini, F.; Ingeniis De, J.; Pepe, L.; Mannazzu, I.; Ciani, M. *Pichia anomala* and *Kluyveromyces wickerhamii* killer toxins as new tools against *Dekkera/Brettanomyces* spoilage yeasts. *FEMS Microbiol. Lett.* **2004**, *238*, 235–240. [CrossRef]
122. Santos, A.; San Mauro, M.; Bravo, E.; Marquina, D. PMKT2, a new killer toxin from *Pichia membranifaciens*, and its promising biotechnological properties for control of the spoilage yeast *Brettanomyces bruxellensis*. *Microbiology* **2009**, *155*, 624–634. [CrossRef] [PubMed]
123. Oro, L.; Ciani, M.; Bizzaro, D.; Comitini, F. Evaluation of damage induced by Kwkt and Pikt zymocins against *Brettanomyces/Dekkera* spoilage yeast, as compared to sulphur dioxide. *J. Appl. Microbiol.* **2016**, *121*, 207–214. [CrossRef] [PubMed]

124. De Ingeniis, J.; Raffaelli, N.; Ciani, M.; Mannazzu, I. *Pichia anomala* DBVPG 3003 secretes a ubiquitin-like protein that has antimicrobial activity. *Appl. Environ. Microbiol.* **2009**, *75*, 1129–1134. [CrossRef] [PubMed]
125. Fernández de Ullivarri, M.; Mendoza, L.M.; Raya, R.R. Characterization of the killer toxin KTCf20 from *Wickerhamomyces anomalus*, a potential biocontrol agent against wine spoilage yeasts. *Biol. Control* **2018**, *121*, 223–228. [CrossRef]
126. Lopes, C.A.; Sáez, J.S.; Sangorrín, M.P. Differential response of *Pichia guilliermondii* spoilage isolates to biological and physicochemical factors prevailing in Patagonian wine fermentations. *Can. J. Microbiol.* **2009**, *55*, 801–809. [CrossRef]
127. Csutak, O.; Vassu, T.; Corbu, V.; Cîrpici, I.; Ionescu, R. Killer activity of *Pichia anomala* CMGB 88. *Biointerface Res. Appl. Chem.* **2017**, *7*, 2085–2089.
128. Villalba, M.L.; Mazzucco, M.B.; Lopes, C.A.; Ganga, M.A.; Sangorrín, M.P. Purification and characterization of *Saccharomyces eubayanus* killer toxin: Biocontrol effectiveness against wine spoilage yeasts. *Int. J. Food Microbiol.* **2020**, *331*, 108714. [CrossRef]
129. Türkel, S.; Ener, B. Isolation and characterization of new *Metschnikowia pulcherrima* strains as producers of the antimicrobial pigment pulcherrimin. *Z. Naturforsch. C* **2009**, *64*, 405–410.
130. Oro, L.; Ciani, M.; Comitini, F. Antimicrobial activity of *Metschnikowia pulcherrima* on wine yeasts. *J. Appl. Microbiol.* **2014**, *116*, 1209–1217. [CrossRef]
131. Charpentier, C.; Escot, S.; Gonzalez, E.; Dulau, L.; Feuillat, M. The influence of yeast glycosylated proteins on tannins aggregation in model solution. *J. Int. Sci. Vigne Vin* **2004**, *38*, 209–218. [CrossRef]
132. Buerth, C.; Heilmann, C.J.; Klis, F.M.; de Koster, C.G.; Ernst, J.F.; Tielker, D. Growth-dependent secretome of *Candida utilis*. *Microbiology* **2011**, *157*, 2493–2503. [CrossRef] [PubMed]

3

SSR-Marker Analysis—A Method for *S. cerevisiae* Strain Characterization and its Application for Wineries

Friederike Rex [1,*], Adeline Hirschler [1] and Maren Scharfenberger-Schmeer [1,2]

[1] Institute for Viticulture and Oenology, Dienstleistungszentrum Ländlicher Raum (DLR) Rheinpfalz, Breitenweg 71, D-67435 Neustadt, Germany; adeline.hirschler@dlr.rlp.de (A.H.); maren.scharfenberger-schmeer@dlr.rlp.de (M.S.-S.)

[2] Hochschule Kaiserslautern, Weincampus Neustadt, Breitenweg 71, D-67435 Neustadt, Germany

* Correspondence: friederike.rex@dlr.rlp.de

Abstract: Considering that many *Saccharomyces cerevisiae* strains exist and that they have different fermentation capacities, the challenge is to select the yeast strain that generates the most interesting wine character and wine flavor for the winemaker. A method based on simple sequence repeats (SSRs) markers, occurring in the yeast genome, was developed to differentiate the collected *S. cerevisiae* strains. For the amplification of the polymorphic SSR markers performed by polymerase chain reaction (PCR), two primer sets showing different size products for different *S. cerevisiae* strains were designed. The PCR-method with gel electrophoresis was validated using capillary sequencing and then used as a service for winegrowers combined with a sensory analysis via napping. This approach can be used for the preservation of the yeast diversity associated with given terroirs and as an option for an increased safety of fermentations. The application of *S. cerevisiae* strains collected in spontaneous fermentations and used for fermentation sustains the initial character of the wine and ensures a secure fermentation at the same time.

Keywords: simple sequence repeats markers; *Saccharomyces cerevisiae*; napping

---

1. Introduction

Yeasts, eukaryotic unicellular microorganisms, have been used for thousands of years for the production of alcoholic beverages or food [1]. In the scientific classification, yeasts belong to the kingdom Fungi and are part of the Ascomycota phylum. In wine production, these organisms are essential for the fermentation process and for the character of each specific wine [2,3]. The most abundant yeasts in a vineyard are generally the yeasts of the genre *Candida*, *Hanseniaspora*, *Pichia* and *Metschnikowia* [4,5]. The specie *S. cerevisiae* is present in a smaller proportion. The freshly pressed, sugar-rich grape must offers optimal conditions to these microorganisms for their growth. During fermentation, the diversity of yeasts decreases, due to the production of ethanol, which is toxic to cells and leads to cell death [6,7]. The alcohol-tolerance limit varies depending on the genus and species of the yeast. The species *S. cerevisiae* tolerates the highest ethanol concentration [8]. Non-Saccharomyces yeasts die above an ethanol concentration of 4% (*v*/*v*), and *S. cerevisiae* becomes the dominant species [9]. To ensure a successful fermentation, the presence of *S. cerevisiae* is essential for winegrowers. For a reliable and predictable fermentation process, the must fermentation can be carried out by adding a sufficient amount of a pure culture of a selected commercially available *S. cerevisiae* yeast strain [10]. Consequently, this will prevent stuck fermentation and off-flavors that may be generated by other microorganisms naturally present on grapes [11]. In fact, by adding a commercially available wine yeast, other microorganisms present in a negligible quantity in the must will be suppressed. Contrary to such an induced fermentation, a spontaneous fermentation lets the must ferment exclusively with the yeasts and other microorganisms naturally occurring in the vineyards. These endogenic organisms live on the skin of the grapes or are part of the cellar flora. The implementation of a spontaneous fermentation allows the winegrowers to develop a unique terroir flavor and an individual wine character. In fact, the yeast flora differs from one vineyard to another, indicating that terroir-specific yeasts can contribute to the wine's individuality [12]. The marketing strategy of such wines is to reach a consumer society that is striving for more natural and local products [13]. However, in spontaneous fermentation, risks of fermentation blockage or the presence of off-flavors due to harmful microorganisms exist [14,15]. An alternative strategy for winegrowers consists in propagating yeast that originates from their own vineyard in order to inoculate the must, conjugating unique sensory attributes with safe fermentations [16].

S. cerevisiae is widely used as a model organism in cell biology and genetics. In 1996, it was the first eukaryote for which one strain's genome was sequenced [17]. Its nuclear genome, composed of 16 linear chromosomes, contains about 12 million base pairs and 6275 genes. In the genome, Simple Sequence Repeats (SSRs), also called microsatellites, are short DNA motifs (two to six bases) repeated continuously five to 50 times. Weller and Jeffreys first characterized SSR in 1984 [18] as a "polymorphic GGAT repeat" in the human myoglobin gene. The genetic transmission of these noncoding sequences follows Mendel's laws of heredity. Their patterns are very abundant throughout the genome of most eukaryotes and prokaryotes [19].

The sequence lengths of SSRs are intra- and interspecific across species. Because SSRs suffer higher rates of mutation than the rest of the genome [20], polymorphic SSRs are a powerful biological tool for the genetic differentiation of *S. cerevisiae* strains as they can be used as molecular markers. Indeed, depending on the size of the SSR, it is possible to differentiate strains from each other. SSR markers were also used for genotyping the yeast *Brettanomyces bruxellensis* at the strain level [21] and for other organisms such as the potato, whose genotype is highly heterozygous for multiple SSR alleles [22]. SSR markers are highly informative and are useful for fingerprinting and linkage studies.

This work focuses on the development of a fast method for *S. cerevisiae* strain differentiation. Strain differentiation is already possible by sequencing the genome, but this is still a cost-intensive method. Other methods such as karyotyping or the restriction fragment length polymorphism (RFLP) analysis of mDNA or the SC1/SC2 marker are used for strain differentiation [23–25] but have limits because they are hardly applicable for a routine analysis or insufficiently discriminating at the strain level [26]. On the contrary, SSR analyses were already used for the characterization of *S. cerevisiae* strains [26,27] or for *S. cerevisiae* strain diversity studies [28]. Here, a polymerase chain reaction (PCR)-based SSR-marker method combined with microvinifications and a napping was applied as an alternative method to be used as a quick and low-budget service for winegrowers [26,29,30]. The developed method is based on the analysis of polymorphic SSRs in yeast genome. For the development, 15 primers were tested [28]. These primers were designed in order to amplify specific polymorphic SSR-regions of *S. cerevisiae*. Multiplex sets were realized by combining several primers. The length of the fragment was observed by agarose gel electrophoresis (Figure 1). The aim is a winery's own yeast, which can be provided as a starter culture and applied in a winery in the Netherlands (Flevoland) following a detailed protocol (Figure 1).

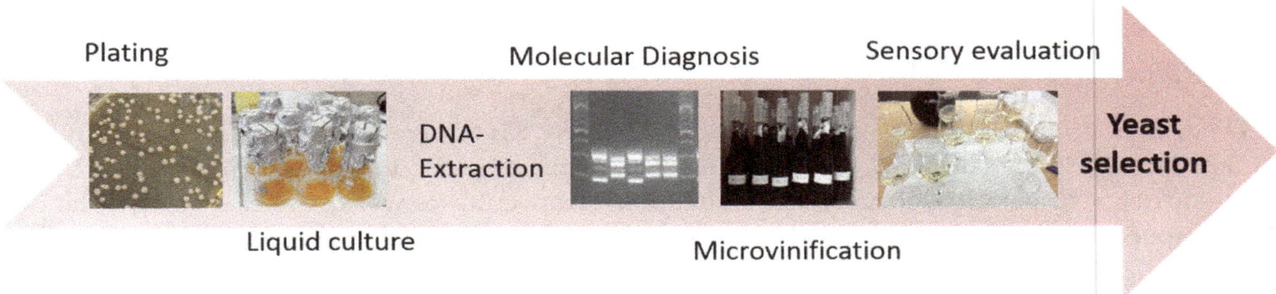

Figure 1. The progress of the yeast selection starting with the plating and ending with a yeast for a winery in order to ensure its spontaneous fermentation.

2. Materials and Methods

2.1. Material

2.1.1. Commercially Available *Saccharomyces Cerevisiae* Strains

Nine commercially available *S. cerevisiae* yeasts were used. These dry yeasts, listed in Table 1, were rehydrated and used as indicated in the manufacturer's instructions.

Table 1. Commercially available yeast strains.

Commercial Name	Abbreviation	Origin
Lalvin BM4*4	Bm4	Lallemand, Fredericia, Denmark
Oenoferm Color	Oc	Erbslöh, Geisenheim, Germany
Oenoferm Bouquet	Ob	Erbslöh, Geisenheim, Germany
Oenoferm Klosterneuburg	Ok	Erbslöh, Geisenheim, Germany
Oenoferm Freddo	Of	Erbslöh, Geisenheim, Germany
Melody	Mel	Chr Hansen, Hoersholm, Denmark
Siha Activhefe 8	Sa8	Eaton Technologies, Langenlonsheim, Germany
1895c	1895c	Swiss Wineyeast GmbH, Meilen, Switzerland
Lalvin EC 1118	Lec	Lallemand, Fredericia, Denmark

2.1.2. Wine Samples

The samples were collected in 2019 in five different spontaneous fermentations tanks of a winery. The winery was harvested in 2017 for the first time. The vineyard was isolated and located in Lelystad (Province of Flevoland) in the Netherlands. Only spontaneous fermentation were conducted in this winery.

2.1.3. Yeast Isolation and Preservation

The samples were plated on YPD-Agar (Yeast Extract-Peptone-Dextrose, Carl Roth, Karlsruhe, Germany) culture medium. For an optimal isolation, these samples were diluted in 0.9% NaCl, with dilution factors ranging from 10^3 to 10^6. After three days, 10 colonies of *S. cerevisiae* from each sample were transferred into liquid YPD-Medium. For yeast conservation, 700 µL of glycerol (AppliChem, Darmstadt, Germany) and 300 µL of the three days-old liquid culture were put in a cryogenic tube. The tubes were homogenized and immersed in liquid nitrogen in order to shock-freeze the yeasts. 10 µL of the culture were also observed under the microscope before preservation in order to verify the absence of contamination. Ten randomly chosen *S. cerevisiae* yeast colonies of each sample, and therefore a total of 50 yeasts, were isolated and preserved.

2.2. Methods:

2.2.1. DNA Extraction

DNA-extraction was done with the Qiagen DNeasy Kit (Qiagen, Hilden, Germany) following the manufacturer's instructions. Cell lysis was performed by the lyticase from *Arthrobacter luteus* (Sigma-Aldrich, Darmstadt, Germany). DNA extracted from cultures was quantified using a NanoDrop ND-1000 spectrophotometer (Thermo-Fisher Scientific Inc., Darmstadt, Germany).

2.2.2. SSR-PCR

Fifteen published primer pairs were tested [28] (Table 2). The primers were first checked in order to determine the size range of the PCR product. Primers generating product-length differences between the nine commercially available strains listed in Table 1 were selected for the multiplex combination.

Table 2. Primers' origin and sequence with the motif and names of the SSR (simple sequence repeats).

Primer	Motif	ORF Coordinates	Sequences
ScAAT2	TAA	YBL084c	FW:CAGTCTTATTGCCTTGAACGA RV:GTCTCCATCCTCCAAACAGCC
ScAAt3	TAA	YDR160w	FW:TGGGAGGAGGGAAATGGACAG RV:TTCAGTTACCCGCACAATCTA
C5	GT	VI-210250/210414	FW:TGACACAATAGCAATGGCCTTCA RV:GCAAGCGACTAGAACAACAATCACA
C3	CAA	YGL139w	FW:CTTTTTATTTACGAGCGGGCCAT RV:AAATCTCATGCCTGTGAGGGGTAT
C8	TAA	YGL014w	FW:CAGGTCGTTCTAACGTTGGTAAAATG RV:GCTGTTGCTGTTGGTAGCATTACTGT
C11	GT	X-518870/519072	FW:TTCCATCATAACCGTCTGGGATT RV:TGCCTTTTTCTTAGATGGGCTTTC
YKR072c	GAC	YKR072c	FW:AGATACAGAAGATAAGAACGAAAA RV:TTATTGATGCTTATCTATTATACC
SCYOR267c	TGT	YOR267c	FW:TACTAACGTCAACACTGCTGCCAA RV:GGATCTACTTGCAGTATACGGG
YKL172w	GAA	YKL172w	FW:CAGGACGCTACCGAAGCTCAAAAG RV:ACTTTTGGCCAATTTCTCAAGAT
ScAAT1	TTA	XIII-86902/87140	FW:AAGCGTAAGCAATGGTGTAGATACTT RV:CAAGCCTCTTCAAGCATGACCTTT
C4	TAA + TAG	XV-110701/110935	FW:AGGAGAAAAATGCTGTTTATTCTGACC RV:TTTTCCTCCGGGACGTGAAATA
C9	TAA	YOR156c	FW:AAGGGTTCGTAAACATATAACTGGCA RV:TATAAGGGAAAAGAGCACGATGGC
ScAAT5	TAA	XVI-897051/8970210	FW:AGCATAATTGGAGGCAGTAAAGCA RV:TCTCCGTCTTTTTTGTACTGCGTG
C6	CA	XVI-485898/485996	FW:GTGGCATCATATCTGTCAATTTTATCAC RV:CAATCAAGCAAAAGATCGGCCT
YPL009c	CTT	YPL009c	FW:AACCCATTGACCTCGTTACTATCGT RV:TTCGATGGCTCTGATAACTCCATTC

For the PCR, the Qiagen Multiplex PCR Kit (Qiagen, Hilden, Germany) was used following the instructions suggested for the amplification of microsatellite loci (Table 2). The PCR settings were the following: Initiation (95 °C, 3 min); Denaturation (95 °C, 30 s); Annealing (60 °C, 30 s); Extension (72 °C, 1 min); and Final Extension (72 °C, 7 min). Thirty-five cycles were performed.

2.2.3. Fragment Length Determination by Agarose Gel Electrophoresis

The result of the PCR was visualized by gel electrophoresis using a 2% agarose gel. 100 mL TAE-Buffer (AppliChem GmbH, Darmstadt, Germany) and 2 g agarose (AppliChem GmbH, Darmstadt,

Germany) were taken, as well as 10 µL of the fluorescent dye GelRed (GeneON, Ludwigshafen, Germany). The loaded samples were composed of 5 µL of PCR product added to 1 µL loading dye (Thermo Fisher Scientific, Waltham, MA, USA). An electric field (90 mA, 90 V, 8 W) was applied for 2 h.

2.2.4. Capillary Sequencing and Method Validation

For a better precision in order to validate the method of the agarose gel, the Capillary Sequencer ABI 3130xl Genetic Analyzer (Applied Biosystems, located within Geilweilerhof, Siebeldingen, Germany) was used. The PCR was carried out as described, but labeled primers were used for the detection in the sequencer. The electrophoresis profiles were captured by ABI 3130 xL Data Collection software. The reading of the size of the different amplified SSRs from the sequencer was performed using the GeneMapper v4.0 (Thermo Fisher Scientific, Waltham, MA, USA) software.

2.2.5. Fermentation and Sensory Tests

For each yeast strain, 750 mL Weißburgunder must was inoculated with 1.0×10^8 yeast cells. The determination of the yeast concentration and viability was performed by the Oculyze Fermentation Wine (FW, Oculyze GmbH, Wildau Germany). 50 µL of yeast culture were added to 50 µL 0.01% alkaline methylene violet. After 5 min, 50 µL of the reaction were taken for the measurement. After inoculating the must, the fermentation was checked weekly with a density meter DM 35 Basic (Anton Paar, Ostfildern, Germany) for a few weeks, depending on the rate of each individual yeast fermentation. For the sensory perception, a panel of 17 people was taken for a napping sensory test [29,30]. Graphic coordinates were attributed to each wine with the Fizz biosystems software, and the software FactoMineR [31] was used for the data analysis. The script was developed by Vestner 2019. The produced wines were also tasted by the owner of the winery who decided which of the identified yeasts were interesting enough for future wine-making.

3. Results

3.1. Primer Sets' Composition and Method Validation

The size ranges of the commercially available yeast's fragment length, obtained by agarose gel electrophoresis or sequencing, are shown in Table 3 for each primer set. Based on these results, two multiplex sets were created by combining primers of different size ranges. The first one combines the primers ScAAT2, ScAAt3, C5, SCYOR267c, C8 and C11. The second combines the primers YKL172w, C4, C9, ScAAT5, C6 and YPL009c. The molecular markers were combined while making sure to avoid overlapping fragment sizes and matching PCR conditions. The primers C3, ScAAT1 and YKR072c were not used for the primer set composition because they did not fulfill these preconditions.

Table 3. The table shows the fragment lengths of the marker in base pairs (bp) resulting from the DNA of commercially available yeasts and the arrangement of the primer sets.

Primer	Size Range by Agarose gel (bp)	Size Range by Capillary Sequencing (bp)	Primer Set
ScAAT2	310–390	376–385	
ScAAt3	260	244–274	
C5	130–170	/	
SCYOR267c	305–390	305–325	
C8	150	133–146	
C11	200–250	/	
YKL172w	135	120–125	
C4	240–320	248–260	
C9	100	90–100	
ScAAT5	160–180	156–162	
C6	100–120	100–102	
YPL009c	250–290	276–311	

The results of the gel electrophoresis for the commercially available yeasts after multiplex PCR are shown in Figures 2 and 3. Six different fragment length patterns were obtained for Set 1 (A′ to

F′) and five for Set 2 (1′ to 5′). Six pattern combinations for nine commercially available yeasts were obtained (A′1′, B′1′, C′2′, D′3′, E′4′ and F′5′).

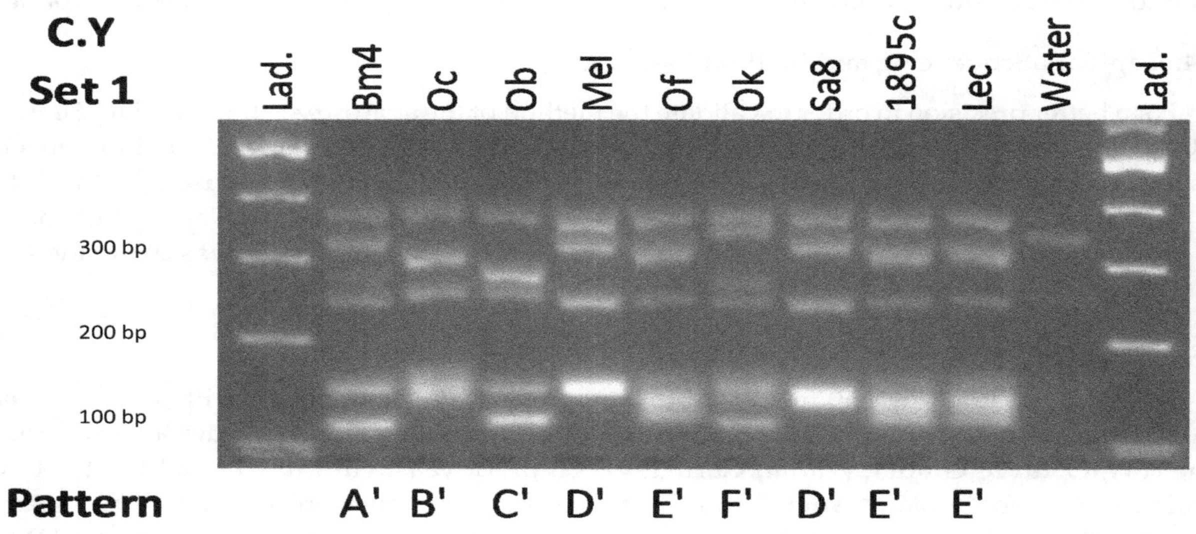

Figure 2. Results of the gel electrophoresis for the commercially available yeasts named above after multiplex PCR using Multiplex Set 1 with the nomenclature of the patterns A–F.

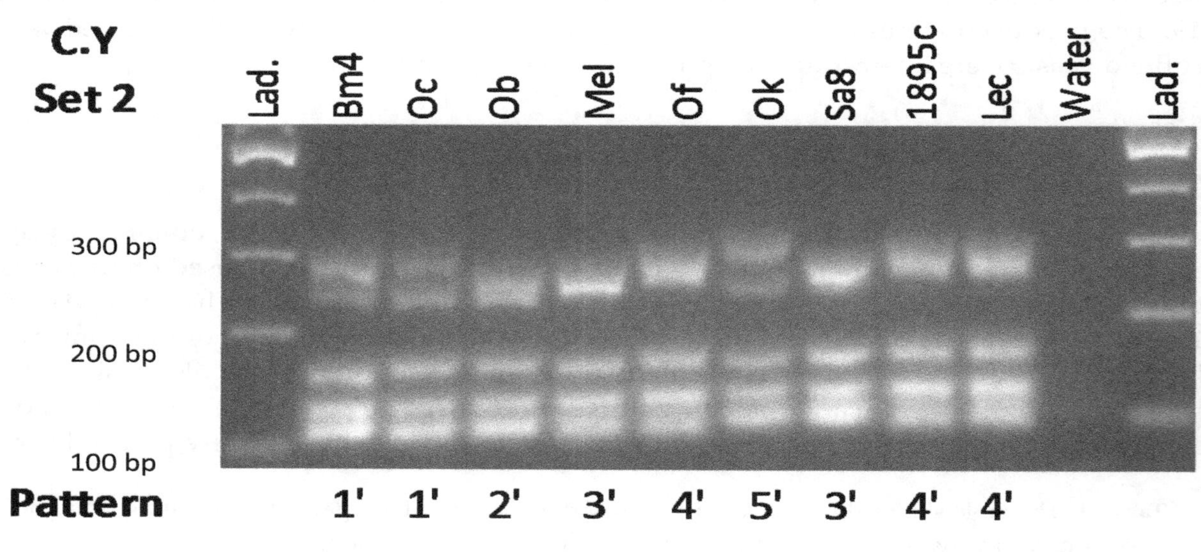

Figure 3. Results of the gel electrophoresis for the commercially available yeasts named above after multiplex PCR using Multiplex Set 2 with the patterns numbered from 1–5.

3.2. Yeast Differentiation

The results of the different patterns obtained by gel electrophoresis for the winery's own yeast after two multiplex SSR PCRs using sets 1 and 2 are summarized in Table 4. Ten different patterns (A to J) were present for the multiplex set 1, and eight different patterns (1 to 8) were present for the multiplex set 2. When combining the two sets, 21 different patterns could thus be obtained for the 50 yeasts. Examples of the patterns A and C–H are shown in Figure 4. The combination 1A shows the most recurrent combination, since it contains 18 of the 50 yeasts. Sixteen yeasts were single in their pattern category.

Table 4. Number of patterns obtained by gel electrophoresis for the winery's own yeast after SSR PCR using multiplex set 1 (1 to 8) and multiplex set 2 (A to J).

Pattern	1	2	3	4	5	6	7	8
A	18	0	0	0	1	0	0	0
B	0	1	0	0	0	0	0	1
C	4	0	1	1	0	7	0	3
D	1	0	1	0	0	0	0	0
E	0	0	0	0	1	0	0	0
F	0	1	0	0	0	1	0	0
G	2	0	0	0	0	0	0	0
H	1	0	0	0	1	1	0	0
I	0	0	0	0	0	0	1	1
J	1	0	0	0	0	0	0	0

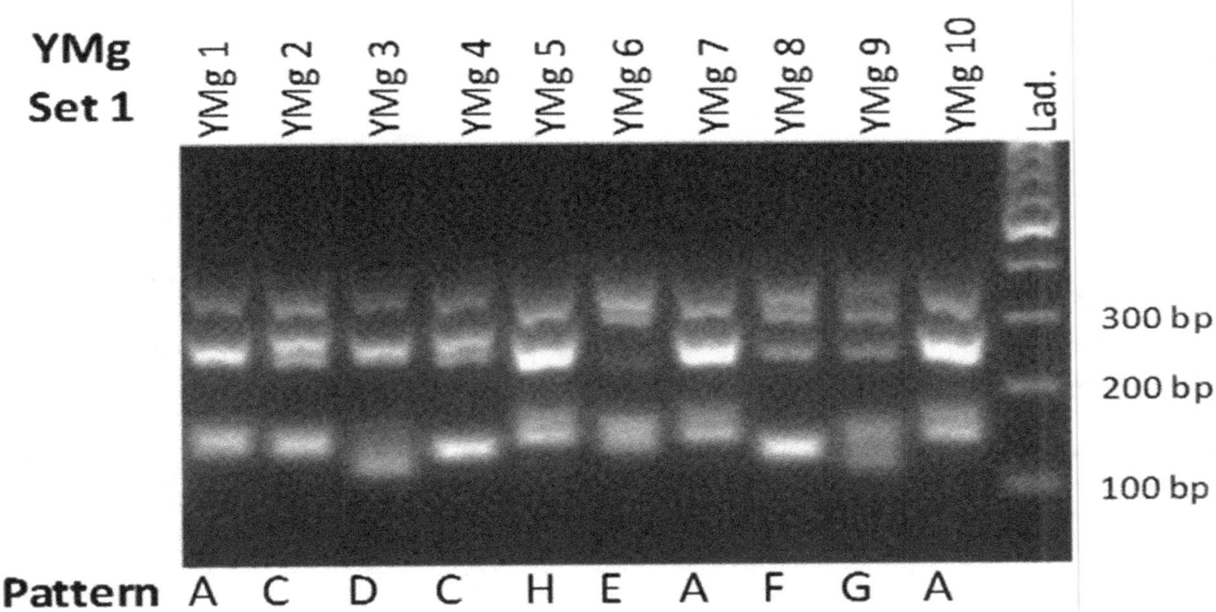

Figure 4. Example of an agarose gel showing the patterns A and C–H. The fragment lengths in base pairs (bp) for the yeast strains Ymg 1–10, amplified with the primer set 1, are shown.

3.3. Fermentative Tests

Representative yeast strains were chosen for microvinification. Seventeen yeasts identified by SSR PCR were chosen: three yeasts of group 1A, one of 5A, one of 2B, one of 3C, one of 4C, two of 6C, one of 8C, one of 3D, one of 5E, one of 2F, one of 1G, one of 5H, one of 6H and one of 7I. These yeasts were chosen since they showed a different pattern in the SSR PCR and thus seemed to be different strains. Three yeasts from group 1A were taken in order to see if similarities could be observed as they were identical yeasts in terms of the pattern of the SSR PCR. After inoculation of sterile must with each yeast, the process of fermentation was monitored by a density meter. The fermentation process of one yeast of each group is presented in Figure 5, and the fermentation time is listed in Table 5. The initial must had a density of 65° Oe. The end of fermentation is characterized by a negative Oechsle degree. The fermentation time ranges from 33 to 73 days after inoculation, forming six groups with the same fermentation time.

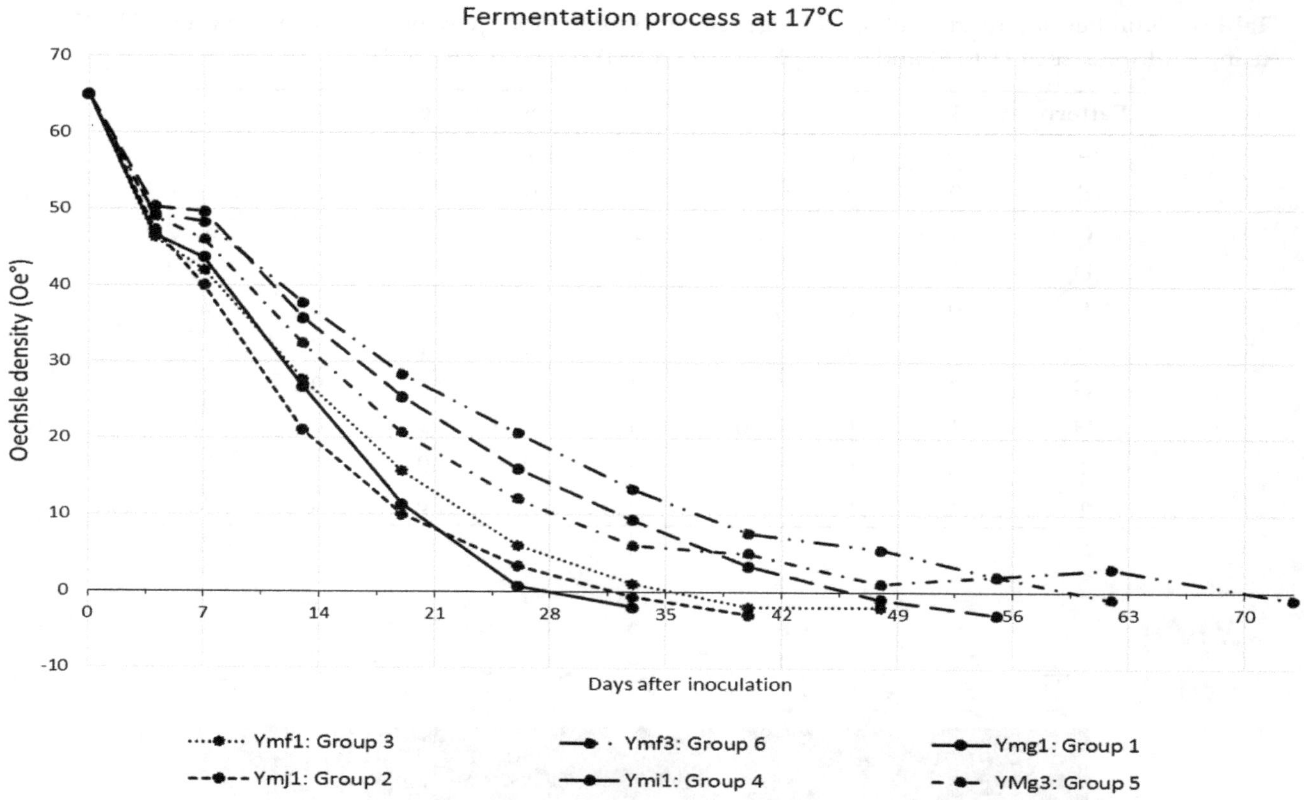

Figure 5. Fermentation process for the microvinifications. One density measurement per group is presented here from 33 to 73 days.

Table 5. Fermentation time in days for the yeast strains. The strains were arranged in six groups according to their fermentation time.

Group	Fermentation Time (days)	Yeasts	Total Number of Yeast
Group 1	33	Ymg1	1
Group 2	40	Ymj1, Ymg4, Ymg7, Ymg9, Ymj4	5
Group 3	48	Ymf1, Ymg5, Ymi3	3
Group 4	55	Ymi1, Ymg2, Ymg8, Ymj2	4
Group 5	62	Ymg3, Ymh4	2
Group 6	73	Ymg6, Ymf3	2

3.4. Sensory Differences

The ten yeasts showing the shortest fermentation time (Table 5) were taken for the sensory test. The results of the napping test are shown in Figure 6. This representation shows the sensory distance between the wines, as well as the 95% confidence ellipses. Figure 7 shows the descriptor correlations. Sensory differences as well as similarities can be observed. The wines Ymi-3 and Ymg-4 are close in taste, both exhibiting the descriptors fruity and honey, whereas the wines Ymj-1, Ymg-9, Ymg-7 and Ymg-1 all share the descriptor sour. The taste of wine Ymg-5 is quite different from other wines and corresponds to the citrus descriptor. Ymi-1 and Ymj-4 are linked as well and have a smoky taste. The wine Ymf-1 has its own sensory taste by being sweeter. The final choice was made by the vineyard owner who selected the yeast Ymf-1 because of its great harmonized taste of sweetness, fruitiness and sourness.

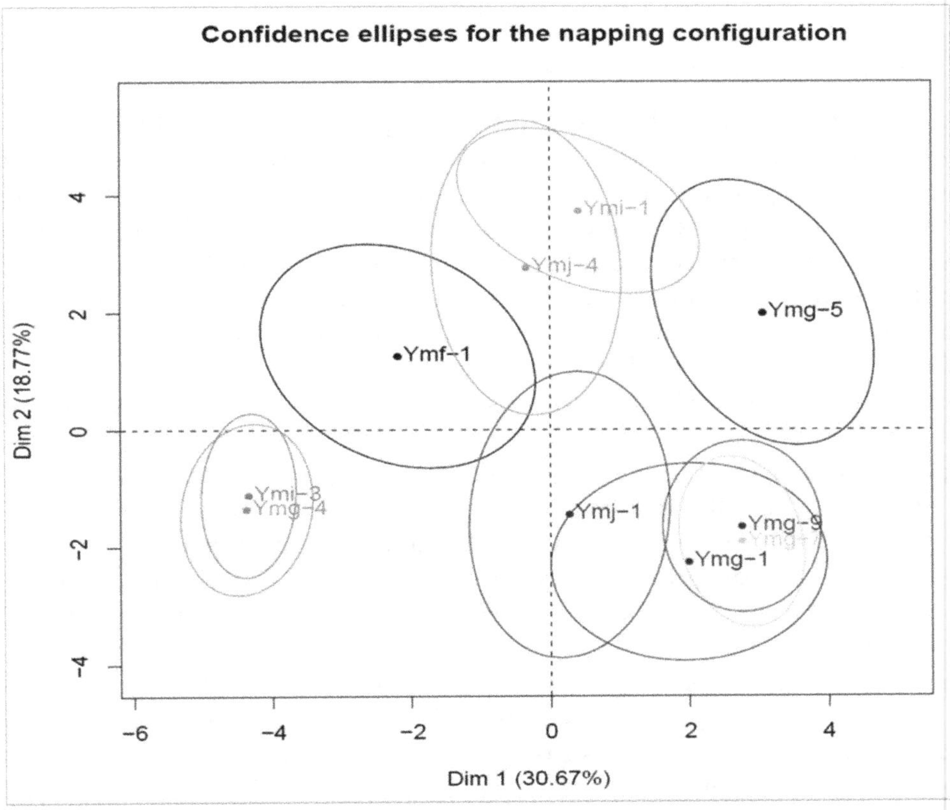

Figure 6. The results of the napping procedure. The wines are named after the yeast strain used for the fermentation, and the wines with the brighter ellipses were grouped together by the panelists.

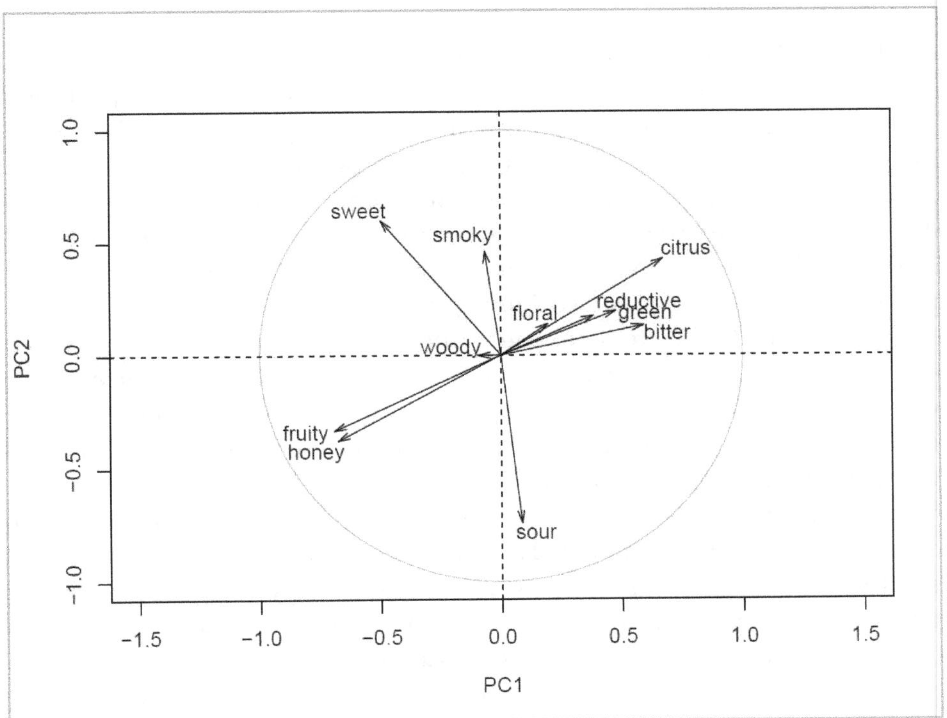

Figure 7. The figure shows the descriptors the panelists used for the wines and the correlations resulting from the arrangement of the wines.

4. Discussion

4.1. Method Validation and Primer Set Composition

The fragment lengths obtained by the capillary sequencer and by the traditional agarose gel electrophoresis were compared. It can be stated that the results correlate, so that the SSR-PCR method with the agarose gel electrophoresis is validated. However, the capillary sequencer has a higher precision than the traditional agarose gel technique (1 bp versus 20 bp). Small differences could thus have been overseen using the agarose gel: a more precise instrumentation is required in this case. The use of a capillary sequencer allowed a differentiation between all nine commercially available yeasts. However, even if the traditional agarose gel method has a lower precision, it remains a good and useful tool for routine and low-budget equipment.

The fragment lengths obtained by the primers are in the same range as the values reported in the literature [10,26,27,32,33].

Concerning the PCR, the recommended primer annealing temperature was 57 °C [28]. However, after testing this PCR-condition, the results on the agarose gel were unreadable. When increasing the annealing temperature to 60 °C, the end products were optimal for all primers except for ScAAT1 and YKR072c. Since YKR072c was described as "slightly variable" in contrast to other primers [33], and since only two out of 15 primers did not work at 60 °C, this annealing temperature was kept for the subsequent PCR experiments.

Primers were combined depending on the product size obtained after PCR, according to the literature [28]. Primers showing products with different sizes were combined, while avoiding superpositions that could mask out the results. However, these primers were designed specifically for the amplification of polymorphic SSRs of *S. cerevisiae* strains [28]. Ultimately, it was important to determine different patterns of PCR products for the subsequent differentiation of *S. cerevisiae* strains. Six different combinations of patterns could be obtained for the nine commercially available yeasts.

4.2. Winery's Own Yeast Differentiation

By combining the results of both multiplex sets, 21 different patterns could be obtained, which means that 21 different yeasts were isolated and differentiated by the developed method. However, it could happen that two different yeasts showing small (< 10 bp) fragment length differences were present in the same group due to the low precision of the agarose gel.

There are two possible reasons to explain why fewer than six bands were sometimes detected for the six primers of a set: on the one hand, some SSRs could have been absent in some yeasts or could have malfunctioned. On the other hand, the fragment length of a primer pair might have been too close to another primer pair, making them undistinguishable.

As the aim of the project was to provide winegrowers with a particular yeast from their own vineyard or cellar to affirm their own terroir character, it was important to verify that the isolated yeasts were not commercially available yeasts. In fact, the samples were collected during spontaneous fermentation, but the fermentation tank may have had leftovers from yeasts used for previous fermentation with commercially available yeasts or there may have been a nearby tank with commercially available yeast strains fermenting nearby. A comparison of the biomolecular results of the winery's own yeast to the commercially available yeasts showed that only Set 2 presented some similarities between the two yeast categories. Therefore, by combining both sets, none of the Flevo yeasts showed a similar pattern to the nine commercially available yeasts. If the comparison would have been done with Set 2 only, similarities could have been found with commercially available yeasts. It was predictable that no commercially available yeasts would be identified, as the vineyard in Flevoland never employed such yeast strains, which confirms our results.

However, there are more than 200 commercially available *S. cerevisiae* yeast strains present on the market, making it a challenge to ensure that the yeast strains found by this method are not commercial ones, especially for wineries that use or have used commercially available yeasts, unlike the winery in Flevoland. It is possible that the patterns found are similar to other commercially available yeasts.

4.3. Fermentative Tests

A correlation was observed between yeast strains of the same group, which showed the same pattern after SSR-PCR and fermentation time. Groups 1A and 6C represent the categories with the largest amount of yeast strains. The three yeast strains of group 1A have a fermentation time of 33 to 48 days and belong to the fastest ten yeasts that completely fermented the must. Both yeast strains of group 6C have a fermentation time of 55 days, which shows a good correlation between the group and fermentation time. The density discrepancy between the three replicates can be explained by the fact that these experiments were realized with living organisms. Indeed, each yeast, even if it is the same strain, can evolve differently.

Fermentation tests were realized at 17 °C in order to create a stressful environment for the yeast strains and because they approximated the conditions found in a winery. The capability and the velocity rate of the fermentation in a stressful environment is indeed a significant factor [34]. Therefore, the choice of wines used for the sensory test was made on the basis of the fermentation rate. The first 10 finished wines were sensorily tested, since a fast fermentation is a prerequisite for strain selection.

4.4. Sensory Differences

The aim of the sensory test was to show organoleptic differences between the wines fermented with different yeast strains. The ten first yeast strains that fermented the must into wine were used, given that the winegrower's interest focused on yeast strains that did not show overly high fermentation rates in order to avoid sluggish fermentations.

A good correlation was found between the placement of the wines chosen by the panelists during the napping procedure and the descriptors. Different descriptors were given for each wine, proving the differences occurring with the yeast strains. In a previous study [32], the evaluation of sensory characteristics between the wines produced using different *Saccharomyces* strains also identified differences resulting from the use of different strains [35]. However, some wines fermented with yeasts like Ymi-3 and Ymg-4 were close to each other and therefore had similar fermentation skills (both fermented the must with an interval of only eight days), even if they originated from different groups (8C and 4C, respectively). Wines fermented with the yeasts Ymg-9 and Ymg-7 were also sensorily similar: both fermented the must in 40 days, but their molecular patterns differed. Although the sensory results showed differences, the confidence ellipses were quite important, which rendered some differences insignificant.

5. Conclusions

Simple Sequence Repeats Markers are powerful biomolecular tools for the differentiation of yeast strains. *S. cerevisiae* strain differences were shown, confirming the relevance of the method for the wine industry. Since yeast sequencing is still an expensive technique, this PCR- and agarose-gel-based method is a good low-budget alternative, even if the precision is lower. Based on the comparison of SSR fragment lengths obtained after PCR, an effective differentiation could be obtained for the vineyard in Flevoland, as there were 21 different yeasts for a total of 50 isolated yeast strains. After fermentation and sensory tests, it was possible to provide a yeast strain with specific properties for the winery in Flevoland. The choice of a specific yeast to be used for further fermentations was based on the fermentation velocity and the sensory perception of the produced wine. Since the number of tested commercially available yeasts was low compared to the existing number of products (approx. 200), an improvement to the method would be to include more commercially available yeast strains. This method could also be useful for diversity and phylogenetic studies of worldwide *S. cerevisiae* strains.

Author Contributions: A.H. an F.R. carried out the experiments. F.R. wrote the manuscript with support from A.H. M.S.-S. conceived the original idea. F.R and M.S.-S. did the project administration and the funding acquisition. All authors have read and agreed to the published version of the manuscript.

Acknowledgments: The authors thank Silvia Anthony, Chateau Neuf Projectonteikkeling v.o.f. for her contribution and support of the project. We also want to thank. Jochen Vestner (DLR Rheinpfalz) for analyzing the data of the napping. We acknowledge Oliver Trapp from the JKI, Institute for grapevine breeding and Geilweilerhof, Siebeldingen for the analyses on the capillary sequencer.

References

1. Chambers, A.H.; Evans, S.A.; Folta, K.M. Methyl Anthranilate and Gamma-Decalactone Inhibit Strawberry Pathogen Growth and Achene Germination. *J. Agric. Food Chem.* **2013**, *61*, 12625–12633. [CrossRef] [PubMed]
2. Nykänen, L. Formation and Occurrence of Flavor Compounds in Wine and Distilled Alcoholic Beverages. *Am. J. Enol. Vitic.* **1986**, *37*, 84–96.
3. Lambrechts, M.G.; Pretorius, I.S. Yeast and its Importance to Wine Aroma—A Review. *S. Afr. J. Enol. Vitic.* **2019**, *21*, 97–129. [CrossRef]
4. Fleet, G.H. Yeast interactions and wine flavour. *Int. J. Food Microbiol.* **2003**, *86*, 11–22. [CrossRef]
5. Clemente-Jimenez, J.M.; Mingorance-Cazorla, L.; Martínez-Rodríguez, S.; Las Heras-Vázquez, F.J.; Rodríguez-Vico, F. Molecular characterization and oenological properties of wine yeasts isolated during spontaneous fermentation of six varieties of grape must. *Food Microbiol.* **2004**, *21*, 149–155. [CrossRef]
6. van Uden, N. Chapter 2—Ethanol Toxicity and Ethanol Tolerance in Yeasts. In *Annual Reports on Fermentation Processes*; Tsao, G.T., Ed.; Elsevier: Munich, Germany, 1985; pp. 11–58.
7. D'Amore, T.; Panchal, C.J.; Russell, I.; Stewart, G.G. A study of ethanol tolerance in yeast. *Crit. Rev. Biotechnol.* **1989**, *9*, 287–304. [CrossRef] [PubMed]
8. Stanley, D.; Bandara, A.; Fraser, S.; Chambers, P.J.; Stanley, G.A. The ethanol stress response and ethanol tolerance of Saccharomyces cerevisiae. *J. Appl. Microbiol.* **2010**, *109*, 13–24. [CrossRef] [PubMed]
9. Romano, P. Function of yeast species and strains in wine flavour. *Int. J. Food Microbiol.* **2003**, *86*, 169–180. [CrossRef]
10. Techera, A.G.; Jubany, S.; Carrau, F.; Gaggero, C. Differentiation of industrial wine yeast strains using microsatellite markers. *Lett. Appl. Microbiol.* **2001**, *33*, 71–75. [CrossRef]
11. Bartowsky, E.J.; Pretorius, I.S. Microbial formation and modification of flavor and off-flavor compounds in wine. In *Biology of Microorganisms on Grapes, in Must and in Wine*; Springer: Heidelberg, Germany, 2009; pp. 209–231.
12. Alexandre, H. Wine Yeast Terroir: Separating the Wheat from the Chaff—For an Open Debate. *Microorganisms* **2020**, *8*, 787. [CrossRef]
13. Merle, A.; Piotrowski, M. Consommer des produits alimentaires locaux—Comment et pourquoi? *Décisions Mark.* **2012**, *67*, 37–48. [CrossRef]
14. Capozzi, V.; Fragasso, M.; Romaniello, R.; Berbegal, C.; Russo, P.; Spano, G. Spontaneous Food Fermentations and Potential Risks for Human Health. *Fermentation* **2017**, *3*, 49. [CrossRef]
15. Erusso, P.; Ecapozzi, V.; Espano, G.; Ecorbo, M.R.; Esinigaglia, M.; Bevilacqua, A. Metabolites of Microbial Origin with an Impact on Health: Ochratoxin A and Biogenic Amines. *Front. Microbiol.* **2016**, *7*, 482. [CrossRef]
16. Capozzi, V.; Fragasso, M.; Russo, P. Microbiological Safety and the Management of Microbial Resources in Artisanal Foods and Beverages: The Need for a Transdisciplinary Assessment to Conciliate Actual Trends and Risks Avoidance. *Microorganisms* **2020**, *8*, 306. [CrossRef]
17. Goffeau, A.; Barrell, B.G.; Bussey, H.; Davis, R.W.; Dujon, B.; Feldmann, H.; Galibert, F.; Hoheisel, J.D.; Jacq, C.; Johnston, M.; et al. Life with 6000 Genes. *Science* **1996**, *274*, 546–567. [CrossRef] [PubMed]
18. Weller, P.; Jeffreys, A.; Wilson, V.; Blanchetot, A. Organization of the human myoglobin gene. *EMBO J.* **1984**, *3*, 439–446. [CrossRef] [PubMed]
19. Field, D.; Wills, C. Abundant microsatellite polymorphism in Saccharomyces cerevisiae, and the different distributions of microsatellites in eight prokaryotes and S. cerevisiae, result from strong mutation pressures and a variety of selective forces. *Proc. Natl. Acad. Sci. USA* **1998**, *95*, 1647–1652. [CrossRef]
20. Jarne, P.; Lagoda, P.J. Microsatellites, from molecules to populations and back. *Trends Ecol. Evol.* **1996**, *11*, 424–429. [CrossRef]

21. Albertin, W.; Panfili, A.; Miot-Sertier, C.; Goulielmakis, A.; Delcamp, A.; Salin, F.; Lonvaud-Funel, A.; Curtin, C.; Masneuf-Pomarede, I. Development of microsatellite markers for the rapid and reliable genotyping of Brettanomyces bruxellensis at strain level. *Food Microbiol.* **2014**, *42*, 188–195. [CrossRef]
22. Ghislain, M.; Spooner, D.M.; Rodríguez, F.; Villamón, F.; Núñez, J.; Vásquez, C.; Waugh, R.; Bonierbale, M. Selection of highly informative and user-friendly microsatellites (SSRs) for genotyping of cultivated potato. *Theor. Appl. Genet.* **2003**, *108*, 881–890. [CrossRef]
23. Garofalo, C.; Berbegal, C.; Grieco, F.; Tufariello, M.; Spano, G.; Capozzi, V. Selection of indigenous yeast strains for the production of sparkling wines from native Apulian grape varieties. *Int. J. Food Microbiol.* **2018**, *285*, 7–17. [CrossRef]
24. Granchi, L.; Ganucci, D.; Buscioni, G.; Mangani, S.; Guerrini, S. The Biodiversity of Saccharomyces cerevisiae in Spontaneous Wine Fermentation: The Occurrence and Persistence of Winery-Strains. *Fermentation* **2019**, *5*, 86. [CrossRef]
25. Tufariello, M.; Maiorano, G.; Rampino, P.; Spano, G.; Grieco, F.; Perrotta, C.; Capozzi, V. Selection of an autochthonous yeast starter culture for industrial production of Primitivo "Gioia del Colle" PDO/DOC in Apulia (Southern Italy). *LWT* **2019**, *99*, 188–196. [CrossRef]
26. Vaudano, E.; Garcia-Moruno, E. Discrimination of Saccharomyces cerevisiae wine strains using microsatellite multiplex PCR and band pattern analysis. *Food Microbiol.* **2008**, *25*, 56–64. [CrossRef] [PubMed]
27. Legras, J.-L.; Ruh, O.; Merdinoglu, D.; Karst, F. Selection of hypervariable microsatellite loci for the characterization of Saccharomyces cerevisiae strains. *Int. J. Food Microbiol.* **2005**, *102*, 73–83. [CrossRef] [PubMed]
28. Börlin, M.; Venet, P.; Claisse, O.; Salin, F.; Legras, J.L.; Masneuf-Pomarede, I. Cellar-Associated Saccharomyces cerevisiae Population Structure Revealed High-Level Diversity and Perennial Persistence at Sauternes Wine Estates. *Appl. Environ. Microbiol.* **2016**, *82*, 2909–2918. [CrossRef]
29. Pages, J. Recueil direct de distances sensorielles: Application à l'évaluation de dix vins blancs du Val-de-Loire. *Sci. Aliment.* **2003**, *23*, 679–688. [CrossRef]
30. Pagès, J. Collection and analysis of perceived product inter-distances using multiple factor analysis: Application to the study of 10 white wines from the Loire Valley. *Food Qual. Prefer.* **2005**, *16*, 642–649. [CrossRef]
31. Lê, S.; Josse, J.; Husson, F. FactoMineR: AnRPackage for Multivariate Analysis. *J. Stat. Softw.* **2008**, *25*, 1–18. [CrossRef]
32. Hennequin, C.; Thierry, A.; Richard, G.F.; Lecointre, G.; Nguyen, H.V.; Gaillardin, C.; Dujon, B. Microsatellite Typing as a New Tool for Identification of Saccharomyces cerevisiae Strains. *J. Clin. Microbiol.* **2001**, *39*, 551–559. [CrossRef]
33. Sihlhavy, K.; Berger, S.U.; Mandl, K.A.; Hack, R.O.; Regner, F.E. Microsatellite analysis of commercial wine yeast strains. *Mitt. Klosterneubg.* **2006**, *56*, 140–146.
34. Torija, M.J. Effects of fermentation temperature on the strain population of Saccharomyces cerevisiae. *Int. J. Food Microbiol.* **2003**, *80*, 47–53. [CrossRef]
35. Iranzo, J.U.; Magaña, F.G.; Viñas, M.G. Evaluation of the formation of volatiles and sensory characteristics in the industrial production of white wines using different commercial strains of the genus Saccharomyces. *Food Control* **2000**, *11*, 143–147. [CrossRef]

4

Identification of Yeasts with Mass Spectrometry during Wine Production

Miroslava Kačániová [1,2,*], Simona Kunová [3], Jozef Sabo [1], Eva Ivanišová [4], Jana Žiarovská [5], Soňa Felsöciová [6] and Margarita Terentjeva [7]

1 Department of Fruit Sciences, Viticulture and Enology, Faculty of Horticulture and Landscape Engineering, Slovak University of Agriculture, Tr. A. Hlinku 2, 94976 Nitra, Slovakia; sabododik@gmail.com
2 Department of Bioenergy, Food Technology and Microbiology, Institute of Food Technology and Nutrition, University of Rzeszow, Zelwerowicza St. 4, 35601 Rzeszow, Poland
3 Department of Food Hygiene and Safety, Faculty of Biotechnology and Food Sciences, Slovak University of Agriculture, Tr. A. Hlinku 2, 94976 Nitra, Slovakia; simona.kunova@uniag.sk or simona.kunova@gmail.com
4 Department of Technology and Quality of Plant Products, Faculty of Biotechnology and Food Sciences, Slovak University of Agriculture, Tr. A. Hlinku 2, 94976 Nitra, Slovakia; eva.ivanisova@uniag.sk
5 Department of Plant Genetics and Breeding, Faculty of Agrobiology and Food Resources, Slovak University of Agriculture, Tr. A. Hlinku 2, 94976 Nitra, Slovakia; jana.ziarovska@uniag.sk
6 Department of Microbiology, Faculty of Biotechnology and Food Sciences, Slovak University of Agriculture, Tr. A. Hlinku 2, 94976 Nitra, Slovakia; sona.felsociova@uniag.sk
7 Institute of Food and Environmental Hygiene, Faculty of Veterinary Medicine, Latvia University of Life Sciences and Technologies, K. Helmaņa iela 8, LV-3004 Jelgava, Latvia; margarita.terentjeva@llu.lv
* Correspondence: miroslava.kacaniova@gmail.com

Abstract: The aim of the present study was to identify yeasts in grape, new wine "federweisser" and unfiltered wine samples. A total amount of 30 grapes, 30 new wine samples and 30 wine samples (15 white and 15 red) were collected from August until September, 2018, from a local Slovak winemaker, including Green Veltliner (3), Mūller Thurgau (3), Palava (3), Rhein Riesling (3), Sauvignon Blanc (3), Alibernet (3), André (3), Blue Frankish (3), Cabernet Sauvignon (3), and Dornfelder (3) grapes; federweisser and unfiltered wine samples were also used in our study. Wort agar (WA), yeast extract peptone dextrose agar (YPDA), malt extract agar (MEA) and Sabouraud dextrose agar (SDA) were used for microbiological testing of yeasts. MALDI-TOF Mass Spectrometry (Microflex LT/SH) (Bruker Daltonics, Germany) was used for the identification of yeasts. A total of 1668 isolates were identified with mass spectrometry. The most isolated species from the grapes was *Hanseniaspora uvarum*, and from federweisser and the wine—*Saccharomyces cerevisiae*.

Keywords: yeasts; grape; federweisser; wine; microbiota identification; MALDI-TOF MS Biotyper

1. Introduction

Yeasts naturally occur in wines and vineyards and are especially common on the grapes. Population of yeast species on the grape is not constant and increases during the ripening process. *Kloeckera apiculata* is a lemon-like cell shape yeast, which colonizes the grape surface [1]. *Kloeckera apiculata* comprises more than 50% of the total healthy grape microbiota. Other yeasts like *Kloeckera* were isolated from the surface of the grapes, which included mainly genera *Metschnikowia, Candida, Cryptococcus, Pichia, Rhodotorula, Zygosaccharomyces* or *Kluyveromyces* [2]. The presence of yeasts of the genus *Aureobasidium* attracted attention as a transitional genus between yeast and microscopic fungi. All the yeasts associated with natural microbiota of grapes are wild yeast strains or non-saccharomyces. Despite the presence of those yeasts on the surface of grapes, the wine production consists of subsequent fermentation stages, which are typical for only particular yeast genera [3]. The *Saccharomyces* genus is the most important for the wine making process; however, this yeast is found on the grapes only in very small amounts. Previous studies that counted *Saccharomyces* on grapes found as little as 50 CFU/g. Mostly wild yeasts cultures could be found on the grapes and in freshly pressed must with colonization rates of 10^3 to 10^5 CFU/mL. During alcoholic fermentation, *Saccharomyces cerevisiae* is dominant, while yeasts in the *Pichia* and *Candida* genera are widespread in finished wine. The osmotolerant yeasts *Zygosaccharomyces* were reported in wines with higher content of residual sugar; yeasts of the *Brettanomyces* genus were common for wines in barrels [4,5].

The most important yeasts associated with wine production were Hanseniaspora uvarum (anamorph Kloeckera apiculata), Metschnikowia pulcherrima, Rhodotorula mucilaginosa, Rhodotorula glutinis, Aureobasidium pullulans, Cryptococcus magnus, Pichia manshurica, Pichia membranifaciens (anamorph Candida valida), Pichia fermentans, Pichia kluyveri, Pichia occidentalis (anamorph Candida sorbosa), Wickerhamomyces anomalus (anamorph Candida pelliculosa; Pichia anomala is synonymous), Cyberlindnera jadinii (Pichia jadinii is synonymous), Kregervanrija fluxuum (anamorph Candida vini), Candida stellata, Candida inconspicua, Meyerozyma guilliermondii, Zygosaccharomyces bailii, Brettanomyces bruxellensis (teleomorph Dekkera bruxellensis), Saccharomycodes ludwigii, Torulaspora delbrueckii and Saccharomyces cerevisiae. Kluyveromyces marxianus and Debaryomyces hansenii associated with grapes and are known as a contaminant in wine production. The microbiota of grapes creates better conditions for the growth of yeasts rather than bacteria. Low pH (pH 3–3.3), high content of sugars (mainly glucose) in grapes, and an anaerobic environment in must are necessary for ethanol fermentation of sugars, converting them into alcohol (ethanol) and CO_2 [5–8].

The aim of this study was to identify yeasts in grapes, federweisser and wine samples.

2. Materials and Methods

2.1. Collection of Grape, Federweisser and Wine Samples

An amount of 90 samples, including grape berries (n = 30), federweisser (n = 30) and wine (n = 30) of *Vitis vinifera* were collected aseptically in the viticultural area of Vrbové (approximately 48°37′12″ N and 017°43′25″ E) in 2018. The grape berry samples were transported on ice and stored at −20 °C until processing. The white grape varieties Green Veltliner, Mūller Thurgau, Palava, Rhein Riesling and Sauvignon Blanc as well as red grape varieties Alibernet, André, Blue Frankish, Cabernet Sauvignon and Dornfelder were collected. Three sampling points in distal spatial points of different rows were used for sampling of grape berries. Grape samples were collected in August, and processed independently.

Samples of "federweisser" were collected at the end of August 2018 and in the middle of September 2018 from the same winery as the grapes. Samples were collected into 200 mL sterile plastic bottles and stored at 8 ± 1 °C in a refrigerator. Before testing, the samples (n = 30) were diluted with sterile physiological saline (0.85%). A total of 100 μL of each dilution (10^{-1} to 10^{-5}) was used for microbiological testing.

An amount of 200 mL of each unfiltered wine (before microfiltration) and immediately after were stored at 6–8 °C in a refrigerator. Collected wine samples were fermented with *Saccharomyces cerevisiae* in the producing process. The samples were later incubated in the laboratory at room temperature (25 ± 2 °C) for one week until the laboratory testing was initiated.

2.2. Cultivation Media

Wort agar (WA) (HiMedia, Mumbai, India), yeast extract peptone dextrose agar (YPDA) (Conda, Madrid, Spain), malt extract agar (MEA) (Biomark, Maharashtra, India) and Sabouraud dextrose agar (SDA) (Conda, Madrid, Spain) were used for identification of yeasts. All media were supplemented with chloramphenicol (100 mg/L) to inhibit bacterial growth. Chloramphenicol (Biolife, Monza, Italy) was added into cultivation media before sterilization by autoclaving at 115–121 °C for 15 min. The acid base indicator bromocresol green (BG, Biolofe, Monza, Italy) (20 mg/L) (pH range: 3.8–5.4) was added into the MEA and WA cultivation media before sterilization. Media for yeast cultivation were inoculated with 100 μL of the sample suspension. Inoculated agars were incubated at 25 °C for 3–5 days and the yeasts were identified by colony morphology (colour, surface, edge and elevation) and reinoculated onto trypton soya agar (TSA) (Oxoid, Basingstoke, UK). Yeast species were identified with a MALDI-TOF MS Biotyper.

2.3. Identification of Isolates with Mass Spectrometry

Qualitative analysis of yeasts isolates was performed with MALDI-TOF mass spectrometry (Bruker Daltonics, Bremen, Germany). Isolates were put in 300 μL of distilled water and 900 μL of ethanol, and the suspension centrifuged for 2 min at 14,000 rpm. The pellet was centrifuged repeatedly and allowed to dry. An amount of 30 μL of 70% formic acid was added to the pellet and 30 μL of acetonitrile. Tubes were centrifuged for 2 min at 14,000 rpm and 1 μL of the supernatant was used for MALDI identification. Once dry, every spot was overlaid with 1 μL of an HCCA matrix and left to dry at room temperature before analysis. Generated spectra were analyzed on a MALDI-TOF Microflex LT (Bruker Daltonics, Bremen, Germany) instrument using Flex Control 3.4 software and Biotyper Realtime Classification 3.1 with BC-specific software. Criteria for reliable identification were a score of ≥2.0 at species level [9].

2.4. Statistical Analysis

The statistical processing of the data obtained from each evaluation was done with Statgraphics Plus version 5.1 (AV Trading, Umex, Dresden, Germany). For each replication the mean was calculated, and the data set were log transformed. Descriptive statistics and logical-cognitive methods and one-way analysis ANOVA were used in the evaluation and statistical analysis.

3. Results and Discussion

Grapes are inhabited by versatile microbial groups and have a complex microbial ecology, including filamentous fungi, yeasts and bacteria. These microorganisms pose different physiological characteristics and may affect the wine quality. Some species of parasitic fungi or environmental bacteria might be only found in grapes, while other microorganisms like yeast, lactic acid and acetic acid bacteria occur during the winemaking process [10].

The yeast count in grape ranged from 2.34 (Greener Veltliner) to 2.67 (Dornfelder) log CFU/g on MEA, from 2.19 (Mūller Thurgau) to 2.38 (Dornfelder) log CFU/g on WA, from 2.46 (Greener Veltliner) to 2.66 (Dornfelder) log CFU/g on YPDA, and from 1.55 (Greener Veltliner) to 1.88 (Dornfelder) log CFU/g on SDA. The colonization of grapes with yeasts is shown in Table 1.

Table 1. Yeasts counts in grape berries on different media.

Sample	MEA	WA	YPDA	SDA
	Microbial Counts log CFU/g			
Green Veltliner	2.37 ± 0.14	2.22 ± 0.06	2.46 ± 0.05	1.55 ± 0.14
Mūller Thurgau	2.34 ± 0.09	2.19 ± 0.04	2.49 ± 0.04	1.66 ± 0.22
Palava	2.36 ± 0.13	2.20 ± 0.07	2.52 ± 0.01	1.69 ± 0.17
Rhein Riesling	2.43 ± 0.01	2.17 ± 0.06	2.51 ± 0.02	1.67 ± 0.16
Sauvignon Blanc	2.40 ± 0.03	2.19 ± 0.04	2.49 ± 0.02	1.65 ± 0.12
Alibernet	2.64 ± 0.10	2.26 ± 0.08	2.53 ± 0.03	1.73 ± 0.16
André	2.66 ± 0.07	2.33 ± 0.10	2.57 ± 0.06	1.79 ± 0.05
Blue Frankish	2.64 ± 0.03	2.36 ± 0.06	2.59 ± 0.06	1.82 ± 0.04
Cabernet Sauvignon	2.66 ± 0.03	2.34 ± 0.09	2.64 ± 0.01	1.84 ± 0.02
Dornfelder	2.67 ± 0.05	2.38 ± 0.04	2.66 ± 0.04	1.88 ± 0.06

WA—wort agar; YPDA—yeast extract peptone dextrose agar; MEA—malt extract agar; SDA—Sabouraud dextrose agar.

ANOVA analysis was performed to inspect the significant differences among the microbial count for individual wine varieties when different cultivation media were used (Table 2).

Table 2. One-way ANOVA for analyzed wine varieties—grapes.

Cultivation Media	Source	Sum of Squares	Degrees of Freedom	Mean Square	F Statistic	p-Value
MEA	treatment	0.5779	9	0.0642	9.55	1.65×10^{-5}
	error	0.1347	20	0.0067		
	total	0.7125	29			
WA	treatment	0.1755	9	0.0195	4.26	0.0025
	error	0.0868	20	0.0043		
	total	0.2623	29			
YPDA	treatment	0.1179	9	0.0131	8.74	3.77×10^{-5}
	error	0.0307	20	0.0015		
	total	0.1487	29			
SDA	treatment	0.2863	9	0.0318	1.13	0.1287
	error	0.3513	20	0.0176		
	total	0.6376	29			

WA—wort agar; YPDA—yeast extract peptone dextrose agar; MEA—malt extract agar; SDA—Sabouraud dextrose agar.

Statistically significant differences among microbial counts for individual cultivation media were found in three of the four cultivation media used (Table 3).

Table 3. Significant differences among analyzed grape varieties for individual cultivation media.

Treatments Pair	Tukey HSD p-Value	Tukey HSD Inferfence
MEA		
A vs. F	0.0180304	* $p < 0.05$
A vs. G	0.0085086	** $p < 0.01$
A vs. H	0.0180304	* $p < 0.05$
A vs. I	0.0076352	** $p < 0.01$
A vs. J	0.0068497	** $p < 0.01$
B vs. F	0.0085086	** $p < 0.01$
B vs. G	0.0039776	** $p < 0.01$
B vs. H	0.0085086	** $p < 0.01$
B vs. I	0.0035659	** $p < 0.01$
B vs. J	0.0032002	** $p < 0.01$
C vs. F	0.0130929	* $p < 0.05$
C vs. G	0.0061450	** $p < 0.01$
C vs. H	0.0130929	* $p < 0.05$
C vs. I	0.0055127	** $p < 0.01$
C vs. J	0.0049454	** $p < 0.01$
E vs. G	0.0247440	* $p < 0.05$
E vs. I	0.0222763	* $p < 0.05$
E vs. J	0.0200452	* $p < 0.05$
WA		
D vs. H	0.0395384	* $p < 0.05$
D vs. J	0.0206542	* $p < 0.05$
YPDA		
A vs. H	0.0262122	* $p < 0.05$
A vs. I	0.0010053	** $p < 0.01$
A vs. J	0.0010053	** $p < 0.01$
B vs. I	0.0068426	** $p < 0.01$

Table 3. *Cont.*

Treatments Pair	Tukey HSD *p*-Value	Tukey HSD Inferfence
YPDA		
B vs. J	0.0011070	** $p < 0.01$
C vs. J	0.0085840	** $p < 0.01$
D vs. I	0.0210350	* $p < 0.05$
D vs. J	0.0034566	** $p < 0.01$
E vs. I	0.0043409	** $p < 0.01$
E vs. J	0.0010053	** $p < 0.01$
F vs. J	0.0168484	* $p < 0.05$

A—Green Veltliner; B—Műller Thurgau; C—Palava; D—Rhein Riesling; E—Sauvignon Blanc; F—Alibernet; G—André; H—Blue Frankish; I—Cabernet Sauvignon; J—Dornfelder; WA—wort agar; YPDA—yeast extract peptone dextrose agar; MEA—malt extract agar; SDA—Sabouraud dextrose agar.

Different studies have evaluated the surface microbiota of grape berries due to a possible impact on the hygienic state of the grapes and the direct influence on the winemaking process and wine quality [11–18].

The yeasts count in "federweisser" ranged from 3.51 in Greener Veltliner and Palava to 3.80 log CFU/mL in Dornfelder on MEA. On WA, the yeasts count from 3.30 in Palava to 3.53 log CFU/mL in Dornfelder were observed. On YPDA, the yeasts count varied from 3.24 in Rhein Riesling to 3.45 log CFU/mL in Dornfelder, and from 3.13 (Sauvignon Blanc) to 3.33 (Dornfelder) log CFU/mL on SDA. Yeasts counts in federweisser are summarized in Table 4.

Table 4. Yeast counts in "federweisser" on different media.

Sample	MEA	WA	YPDA	SDA
	log CFU/g			
Green Veltliner	3.51 ± 0.15	3.41 ± 0.06	3.37 ± 0.14	3.21 ± 0.01
Műller Thurgau	3.54 ± 0.10	3.38 ± 0.06	3.30 ± 0.04	3.19 ± 0.03
Palava	3.51 ± 0.05	3.30 ± 0.06	3.27 ± 0.04	3.16 ± 0.05
Rhein Riesling	3.58 ± 0.06	3.34 ± 0.01	3.24 ± 0.01	3.14 ± 0.02
Sauvignon Blanc	3.56 ± 0.10	3.36 ± 0.05	3.27 ± 0.05	3.13 ± 0.02
Alibernet	3.67 ± 0.08	3.40 ± 0.06	3.29 ± 0.06	3.17 ± 0.03
André	3.70 ± 0.07	3.43 ± 0.02	3.36 ± 0.05	3.18 ± 0.06
Blue Frankish	3.74 ± 0.02	3.46 ± 0.05	3.39 ± 0.05	3.24 ± 0.09
Cabernet Sauvignon	3.76 ± 0.05	3.48 ± 0.06	3.41 ± 0.01	3.30 ± 0.04
Dornfelder	3.80 ± 0.07	3.53 ± 0.03	3.45 ± 0.06	3.33 ± 0.01

WA—wort agar; YPDA—yeast extract peptone dextrose agar; MEA—malt extract agar; SDA—Sabouraud dextrose agar.

In study in Slovakia [19], the highest yeasts counts were on MEA for Pinot Noir—6.43 log CFU/mL and the lowest for Moravian Muscat—4.62 log CFU/mL. The highest yeasts count on WA were in Pinot Noir—6.39 log CFU/mL, but the lowest in Irsai Oliver—5.38 log CFU/mL. The highest count of yeasts on wild yeast medium (WYM) was in Blue Frankish 6.33 log CFU/mL and the lowest in Dornfelder 4.20 log CFU/mL [19].

As the results show, a higher number of yeasts were detected in "federweisser" than in grape. The young wine is a product of fermentation where *S. cerevisiae* was mostly found. Other species like *Hanseniaspora uvarum*, *Metschnikowia pulcherrima* or the genera *Pichia* or *Candida* may be present during the individual fermentation stages when the alcohol content do not exceed 4–6% [5,20]. The main microbiota of the grape is the yeast *Hanseniaspora uvarum* followed by *Metschnikowia pulcherrima* [4]. These species also initiate the pre-alcoholic fermentation but are being replaced by the dominant *S. cerevisiae* 3–4 days after fermentation. *Saccharomyces cerevisiae* starts to multiply within 20 days after inoculation into the must [21].

ANOVA analysis was performed to inspect the significant differences among the microbial count for individual wine varieties when different cultivation media were used (Table 5).

Table 5. One-way ANOVA results for the analyzed wine varieties—federweisser.

Cultivation Media	Source	Sum of Squares	Degrees of Freedom	Mean Square	F Statistic	*p*-Value
MEA	treatment	0.3717	9	0.0413	6.36	0.0002
	error	0.1234	20	0.0062		
	total	0.4951	29			
WA	treatment	0.1305	9	0.0145	5.23	0.0007
	error	0.0525	20	0.0026		
	total	0.1831	29			
YPDA	treatment	0.1426	9	0.0158	3.36	0.0056
	error	0.0818	20	0.0041		
	total	0.2244	29			
SDA	treatment	0.1194	9	0.0133	6.91	0.0002
	error	0.0397	20	0.0020		
	total	0.1591	29			

WA—wort agar; YPDA—yeast extract peptone dextrose agar; MEA—malt extract agar; SDA—Sabouraud dextrose agar.

Statistically significant differences among microbial counts for individual cultivation media were found in three of the four cultivation media used (Table 6).

Table 6. Significant differences among analyzed federweisser samples for individual cultivation media.

Reatments Pair	Tukey HSD *p*-Value	Tukey HSD Inferfence
MEA		
A vs. H	0.0368638	* $p < 0.05$
A vs. I	0.0238533	* $p < 0.05$
A vs. J	0.0055561	** $p < 0.01$
B vs. H	0.0456379	* $p < 0.05$
B vs. I	0.0296858	* $p < 0.05$
B vs. J	0.0069685	** $p < 0.01$
C vs. H	0.0410343	* $p < 0.05$
C vs. I	0.0266151	* $p < 0.05$
C vs. J	0.0062207	** $p < 0.01$
E vs. H	0.0456379	* $p < 0.05$
E vs. I	0.0296858	* $p < 0.05$
E vs. J	0.0069685	** $p < 0.01$
WA		
C vs. H	0.0329699	* $p < 0.05$
C vs. I	0.0100494	* $p < 0.05$
C vs. J	0.0010463	** $p < 0.01$
D vs. J	0.0084520	** $p < 0.01$
E vs. J	0.0071077	** $p < 0.01$
YPDA		
D vs. J	0.0105986	* $p < 0.05$
E vs. J	0.0359336	* $p < 0.05$
SDA		
B vs. J	0.0319805	* $p < 0.05$
C vs. I	0.0217427	* $p < 0.05$
C vs. J	0.0044497	** $p < 0.01$

Table 6. *Cont.*

Reatments Pair	Tukey HSD *p*-Value	Tukey HSD Inferfence
SDA		
D vs. I	0.0081105	** $p < 0.01$
D vs. J	0.0016306	** $p < 0.01$
E vs. I	0.0036406	** $p < 0.01$
E vs. J	0.0010053	** $p < 0.01$
F vs. I	0.0466723	* $p < 0.05$
F vs. J	0.0098975	** $p < 0.01$
G vs. J	0.0178947	* $p < 0.05$

A—Green Veltliner; B—Mũller Thurgau; C—Palava; D—Rhein Riesling; E—Sauvignon Blanc; F—Alibernet; G—André; H—Blue Frankish; I—Cabernet Sauvignon; J—Dornfelder; WA—wort agar; YPDA—yeast extract peptone dextrose agar; MEA—malt extract agar; SDA—Sabouraud dextrose agar.

The yeast counts in the unfiltered wines are summarized in Table 7. The yeast counts in wine ranged from 1.51 (Greener Veltliner) to 3.23 (Dornfelder) log CFU/mL on MEA, from 1.43 (Greener Veltliner) to 2.89 (Dornfelder) log CFU/mL on WA, from 1.18 (Greener Veltliner) to 2.65 (Dornfelder) log CFU/mL on YPDA and from 1.09 (Rhein Riesling) to 2.21 (Dornfelder) log CFU/mL on SDA.

Table 7. Yeast counts in wine on different media.

Sample	MEA	WA	YPDA	SDA
	log CFU/g			
Green Veltliner	1.51 ± 0.27	1.43 ± 0.15	1.18 ± 0.06	1.13 ± 0.03
Mũller Thurgau	1.55 ± 0.32	1.52 ± 0.01	1.21 ± 0.06	1.18 ± 0.06
Palava	1.62 ± 0.34	1.49 ± 0.03	1.27 ± 0.11	1.12 ± 0.02
Rhein Riesling	1.73 ± 0.17	1.46 ± 0.05	1.27 ± 0.14	1.09 ± 0.06
Sauvignon Blanc	1.76 ± 0.11	1.48 ± 0.06	1.32 ± 0.23	1.14 ± 0.02
Alibernet	2.57 ± 0.50	2.37 ± 0.14	2.21 ± 0.05	1.51 ± 0.64
André	2.67 ± 0.42	2.41 ± 0.10	2.31 ± 0.13	1.83 ± 0.62
Blue Frankish	2.59 ± 0.28	2.44 ± 0.16	2.38 ± 0.11	2.13 ± 0.12
Cabernet Sauvignon	2.89 ± 0.37	2.51 ± 0.14	2.47 ± 0.04	2.17 ± 0.07
Dornfelder	3.23 ± 0.02	2.89 ± 0.27	2.65 ± 0.22	2.21 ± 0.06

WA—wort agar; YPDA—yeast extract peptone dextrose agar; MEA—malt extract agar; SDA—Sabouraud dextrose agar.

ANOVA analysis was performed to inspect the significant differences among the microbial count for individual wine varieties when different cultivation media were used (Table 8).

Table 8. One-way ANOVA results for analyzed wine varieties—unfiltered wine.

Cultivation Media	Source	Sum of Squares	Degrees of Freedom	Mean Square	F Statistic	*p*-Value
MEA	treatment	11.0908	9	1.23	12.60	1.95×10^{-6}
	error	1.45	20	0.0982		
	total	13.0552	29			
WA	treatment	8.05	9	0.9745	54.8807	3.84×10^{-12}
	error	0.3551	20	0.0178		
	total	9.1256	29			
YPDA	treatment	10.74	9	1.1542	65.9142	6.70×10^{-13}
	error	0.3502	20	0.0175		
	total	10.76	29			
SDA	treatment	6.61	9	0.7118	8.13	3.51×10^{-5}
	error	1.31	20	0.0827		
	total	8.0592	29			

WA—wort agar; YPDA—yeast extract peptone dextrose agar; MEA—malt extract agar; SDA—Sabouraud dextrose agar.

Statistically significant differences among microbial count for individual cultivation media were found in three of the four cultivation media used (Table 9).

Table 9. Significant differences among unfiltered wine samples for individual cultivation media.

Treatments Pair	Tukey HSD p-Value	Tukey HSD Inferfence
MEA		
A vs. F	0.0141715	* $p < 0.05$
A vs. G	0.0064264	** $p < 0.01$
A vs. H	0.0119733	* $p < 0.05$
A vs. I	0.0010053	** $p < 0.01$
A vs. J	0.0010053	** $p < 0.01$
B vs. F	0.0192566	* $p < 0.05$
B vs. G	0.0087792	** $p < 0.01$
B vs. H	0.0162964	* $p < 0.05$
B vs. I	0.0012652	** $p < 0.01$
B vs. J	0.0010053	** $p < 0.01$
C vs. F	0.0342577	* $p < 0.05$
C vs. G	0.0158481	* $p < 0.05$
C vs. H	0.0290959	* $p < 0.05$
C vs. I	0.0023022	** $p < 0.01$
C vs. J	0.0010053	** $p < 0.01$
D vs. G	0.0381648	* $p < 0.05$
D vs. I	0.0057354	** $p < 0.01$
D vs. J	0.0010053	** $p < 0.01$
E vs. I	0.0078360	** $p < 0.01$
E vs. J	0.0010053	** $p < 0.01$
WA		
A vs. F	0.0010053	** $p < 0.01$
A vs. G	0.0010053	** $p < 0.01$
A vs. H	0.0010053	** $p < 0.01$
A vs. I	0.0010053	** $p < 0.01$
A vs. J	0.0010053	** $p < 0.01$
B vs. F	0.0010053	** $p < 0.01$
B vs. G	0.0010053	** $p < 0.01$
B vs. H	0.0010053	** $p < 0.01$
B vs. I	0.0010053	** $p < 0.01$
B vs. J	0.0010053	** $p < 0.01$
C vs. F	0.0010053	** $p < 0.01$
C vs. G	0.0010053	** $p < 0.01$
C vs. H	0.0010053	** $p < 0.01$
C vs. I	0.0010053	** $p < 0.01$
C vs. J	0.0010053	** $p < 0.01$
D vs. F	0.0010053	** $p < 0.01$
D vs. G	0.0010053	** $p < 0.01$
D vs. H	0.0010053	** $p < 0.01$
D vs. I	0.0010053	** $p < 0.01$
D vs. J	0.0010053	** $p < 0.01$
E vs. F	0.0010053	** $p < 0.01$
E vs. G	0.0010053	** $p < 0.01$
E vs. H	0.0010053	** $p < 0.01$
E vs. I	0.0010053	** $p < 0.01$
E vs. J	0.0010053	** $p < 0.01$
F vs. J	0.0037886	** $p < 0.01$
G vs. J	0.0079079	** $p < 0.01$
H vs. J	0.0143649	* $p < 0.05$

Table 9. *Cont.*

Treatments Pair	Tukey HSD p-Value	Tukey HSD Inferfence
YPDA		
A vs. F	0.0010053	** $p < 0.01$
A vs. G	0.0010053	** $p < 0.01$
A vs. H	0.0010053	** $p < 0.01$
A vs. I	0.0010053	** $p < 0.01$
A vs. J	0.0010053	** $p < 0.01$
B vs. F	0.0010053	** $p < 0.01$
B vs. G	0.0010053	** $p < 0.01$
B vs. H	0.0010053	** $p < 0.01$
B vs. I	0.0010053	** $p < 0.01$
B vs. J	0.0010053	** $p < 0.01$
C vs. F	0.0010053	** $p < 0.01$
C vs. G	0.0010053	** $p < 0.01$
C vs. H	0.0010053	** $p < 0.01$
C vs. I	0.0010053	** $p < 0.01$
C vs. J	0.0010053	** $p < 0.01$
D vs. F	0.0010053	** $p < 0.01$
D vs. G	0.0010053	** $p < 0.01$
D vs. H	0.0010053	** $p < 0.01$
D vs. I	0.0010053	** $p < 0.01$
D vs. J	0.0010053	** $p < 0.01$
E vs. F	0.0010053	** $p < 0.01$
E vs. G	0.0010053	** $p < 0.01$
E vs. H	0.0010053	** $p < 0.01$
E vs. I	0.0010053	** $p < 0.01$
E vs. J	0.0010053	** $p < 0.01$
F vs. J	0.0154088	* $p < 0.05$
SDA		
A vs. H	0.0103393	* $p < 0.05$
A vs. I	0.0071377	** $p < 0.01$
A vs. J	0.0044827	** $p < 0.01$
B vs. H	0.0168657	* $p < 0.05$
B vs. I	0.0116918	* $p < 0.05$
B vs. J	0.0073639	** $p < 0.01$
C vs. H	0.0100244	* $p < 0.05$
C vs. I	0.0069202	** $p < 0.01$
C vs. J	0.0043452	** $p < 0.01$
D vs. H	0.0075929	** $p < 0.01$
D vs. I	0.0052344	** $p < 0.01$
D vs. J	0.0032871	** $p < 0.01$
E vs. H	0.0116918	* $p < 0.05$
E vs. I	0.0080777	** $p < 0.01$
E vs. J	0.0050746	** $p < 0.01$

A—Green Veltliner; B—Műller Thurgau; C—Palava; D—Rhein Riesling; E—Sauvignon Blanc; F—Alibernet; G—André; H—Blue Frankish; I—Cabernet Sauvignon; J—Dornfelder; WA—wort agar; YPDA—yeast extract peptone dextrose agar; MEA—malt extract agar; SDA—Sabouraud dextrose agar.

Altogether, 1668 isolates were identified with mass spectrometry with a score of ≥2.0 (Table 10). The most isolated species from grape was *Hanseniaspora uvarum* (70 isolates), and from "federweisser" and wine *S. cerevisiae* (85 and 120 isolates, respectively). Yeasts species of grape, "frederweisser" and wine are shown in Figures 1–3.

Table 10. Yeasts species in grape, "federweisser" and wine.

Yeast Species	Grape	"Federweisser"	Wine
		No. of Isolates	
Aureobasidium pullulans	25	0	0
Candida inconspicua	5	0	5
Candida parapsilosis	5	0	10
Candida saitoana	5	0	5
Candida sake	5	0	5
Cyberlindnera jadinii	0	0	8
Debaryomyces hansenii	0	0	15
Dekkera bruxellensis	0	0	25
Filobasidium magnum	30	0	0
Hanseniaspora uvarum	70	25	0
Issatchenkia orientalis	38	0	0
Kazachstania exigua	33	0	0
Kluyveromyces marxianus	35	0	32
Kregervanrija fluxuum	0	0	25
Metschnikowia pulcherrima	28	50	55
Meyerozyma guilliermondii	0	0	52
Naganishia diffluens	25	0	0
Pichia fermentans	10	0	58
Pichia kluyveri	12	49	50
Pichia mandshurica	10	0	25
Pichia membranifaciens	25	0	15
Pichia norvegensis	10	0	5
Pichia occidentalis	10	35	45
Rhodotorula glutinis	40	0	20
Rhodotorula mucilaginosa	25	0	45
Saccharomyces cerevisiae	0	85	120
Starmerella magnolia	30	0	15
Torulaspora delbrueckii	12	0	39
Wickerhamomyces anomalus	15	0	75
Yarrowia lipolytica	20	0	10
Zygosaccharomyces bailii	15	0	52
Zygotorulaspora florentina	25	0	50
Total	563	244	861

Brettanomyces bruxellensis, *Candida stellata*, *Saccharomyces cerevisiae* and *Zygosaccharomyces bailii* were the yeasts identified in wine [22–25]. In our study, *Pichia mandshurica*—the main contaminant of wines—was present in 66% samples of white wines (10 out of 15) and in seven samples of red wines (46%). *Pichia membranifaciens* was isolated from five samples of white (33%) and five samples of red wines (33%). *Saccharomyces cerevisiae* was isolated from all white and red wines (100%). *Zygosaccharomyces bailii* was found in 14 samples of white (93%) and two samples of red (13%) wines. Our study shows that *Z. bailii* and *P. mandshurica* were isolated more frequently from white than from red wines, while *S. cerevisiae* was identified in white and red wines. The occurrence of *Pichia manshurica* and *S. cerevisiae* was different between the wine samples. According to Thomas [26], the presence of *Zygosaccharomyces* in wine is unacceptable in terms of wine quality. The author has stated that the minimum number of yeast present in wine spoils the product under appropriate conditions [26]. *Saccharomyces cerevisiae, Debaryomyces hansenii, Wickerhamomyces anomalus (Pichia anomala), Pichia membranifaciens, Rhodotorula glutinis, Rhodotorula mucilaginosa, Torulaspora delbrueckii, Kluyveromyces marxianus, Issatchenkia orientalis, Zygosaccharomyces bailii parapsilosis, Pichia fermentans* and *Hanseniaspora uvarum* are frequent contaminants of wines as well [27,28]. However, Renous [29] did not describe associations between wine and *Pichia manshurica, Kregervanrija fluxuum (Candida vini), Candida inconspicua* and *Zygotorulaspora florentina*. Saez [30] found that *S. cerevisiae* (13.93%),

Wickerhamomyces anomalus (8.72%), *Pichia fermentans* (6.74%) and *Metschnikowia pulcherrima* (6.39%) were the most abundant in wine. *Pichia* (*Pichia manshurica*, *P. membranifaciens*) and *Brettanomyces* are producing volatile phenols, thereby affecting the quality of the wine [30].

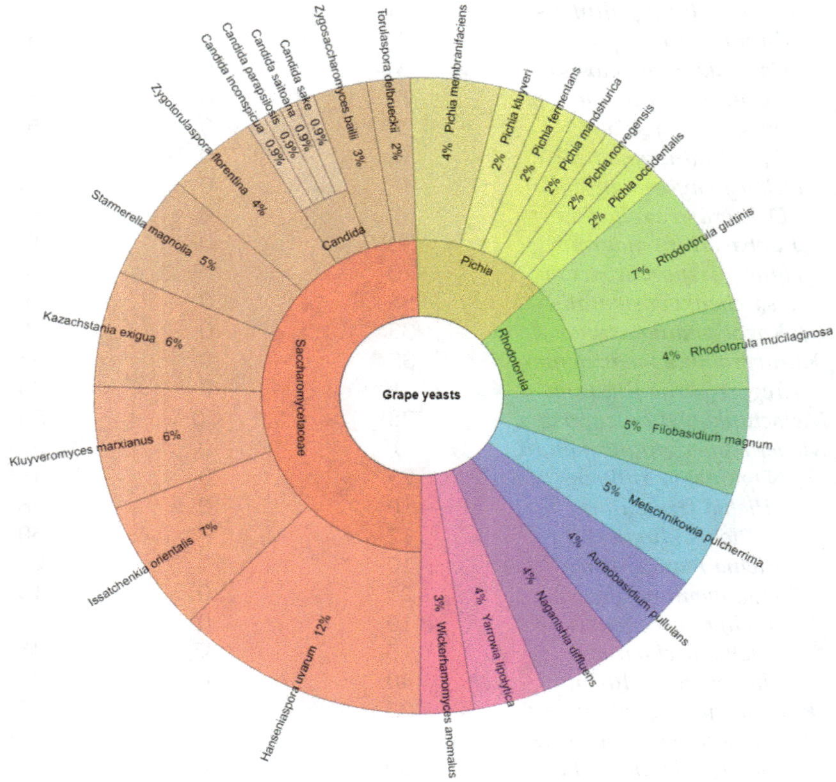

Figure 1. Yeasts isolated from the grapes.

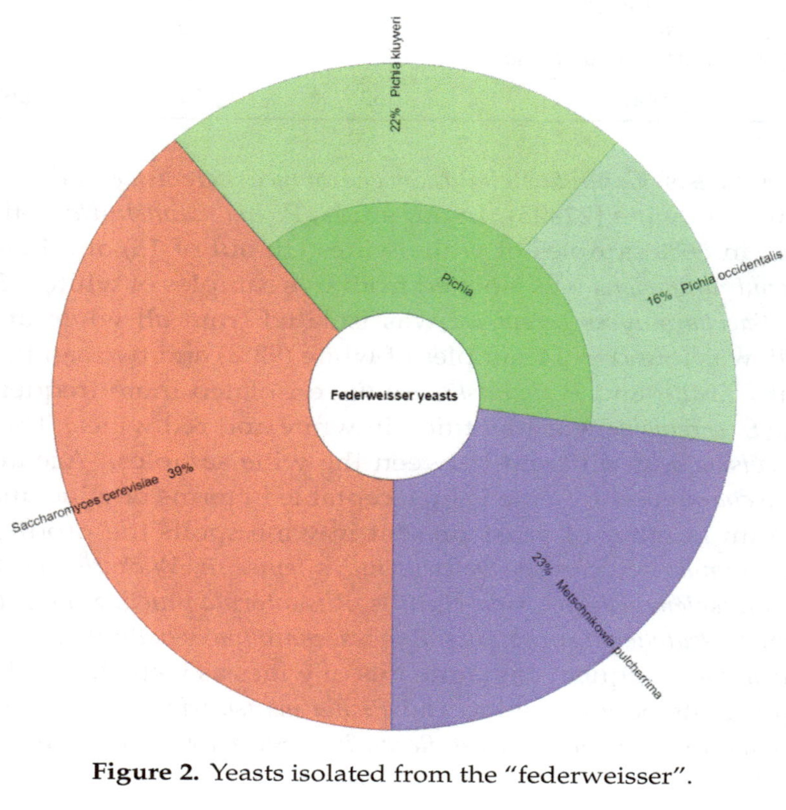

Figure 2. Yeasts isolated from the "federweisser".

Figure 3. Yeasts isolated from the wine.

Sporadically, *Candida inconspicua* (5 isolates, 0.58%), *Candida saitoana* (5 isolates, 0.58%), *Candida sake* (5 isolates, 0.58%), *Pichia norvegensis* (5 isolates, 0.58%) and other species were isolated. Jolly et al. [31] noticed the importance of *Candida*, *Cryptococcus*, *Kloeckera* and *Rhodotorula* species in the wine making process. *Candida* was considered as the dominant genus, including their teleomorphic stages—*Candida pulcherrima* (*Metschnikowia pulcherrima*), *Candida vini* (*Kregervanrija fluxuum*) and *Candida valida* (*Pichia membranifaciens*) [31].

4. Conclusions

A total of 90 samples (30 from grapes, 30 of "federweisser" and 30 of wine) was studied for characterization of the yeast species. The mass spectrometry method was used for identification of 1668 grape, "federweisser" and wine isolates. From grape, 26 species of 17 genera within 9 families, and in "federweisser" 4 species of 3 genera and families were found. In wine, 26 species of 17 genera within 6 families were identified. *Rhodoturulla* species were not included in any family and they were classified as incertae sedis (not belonging anywhere).

Author Contributions: M.K., M.T., J.Ž. were responsible for the design of the study; M.K., S.K., J.S., S.F., conducted the study and collected the samples; M.K., S.K., J.S., J.Ž. performed the laboratory analysis; M.K., S.K., E.I., M.T. were responsible for writing and editing the manuscript; all authors have carefully revised and approved the final version of the manuscript. All authors have read and agreed to the published version of the manuscript.

Acknowledgments: The Paper was supported by the project: The research leading to these results has received funding from the European Community under project no. 26220220180: Building Research Centre "AgroBioTech".

References

1. Padilla, B.; Gil, J.V.; Manzanares, P. Past and future of non-saccharomyces yeasts: From spoilage microorganisms to biotechnological tools for improving wine aroma complexity. *Front. Microbiol.* **2016**, *7*, 411. [CrossRef]
2. Kantor, A.; Kačániová, M.; Kluz, M. Natural microflora of wine grape berries. *J. Microbiol. Biotech. Food Sci.* **2015**, *4*, 32–36. [CrossRef]
3. Bozoudi, D.; Tsaltas, D. The multiple and versatile roles of *Aureobasidium pullulans* in the vitivinicultural sector. *Fermentation* **2018**, *4*, 85. [CrossRef]
4. Fugelsang, K.C.; Edwards, C.G. *Wine Microbiology: Practical Applications and Procedures*, 2nd ed.; Springer: New York, NY, USA, 2007; p. 393.
5. König, H.; Fröhlich, J.; Unden, G. *Biology of Microorganisms on Grapes, in Must and in Wine*; Springer: Berlin/Heidelberg, Germany, 2009; p. 522.
6. Fleet, G.H. Wine. In *Food Microbiology Fundamentals & Frontiers*; Doyle, M.P., Beuchat, L.R., Montville, T.J., Eds.; ASM Press: Washington, DC, USA, 2001; pp. 747–772.
7. Romano, P.; Fiore, C.; Paraggio, M.; Caruso, M.; Capece, A. Function of yeast species and strains in wine flavor. *Int. J. Food Microbiol.* **2003**, *86*, 169–180. [CrossRef]
8. Kurtzman, C.P.; Fell, J.W.; Boekhout, T. *The Yeasts: A Taxonomic Study*, 5th ed.; Elsevier Science: Amsterdam, The Netherlands, 2011; p. 2384.
9. Kačániová, M.; Mellen, M.; Vukovic, N.L.; Kluz, M.; Puchalski, C.; Haščík, P.; Kunová, S. Combined effect of vacuum packaging, fennel and svory essential oil treatment on the quality of chicken thighs. *Microorganisms* **2019**, *7*, 134. [CrossRef] [PubMed]
10. Barata, A.; Malfeito-Ferreira, M.; Loureiro, V. The microbial ecology of wine grape berries. *Int. J. Food Microbiol.* **2012**, *153*, 243–259. [CrossRef]
11. Martins, G.; Miot-Sertier, C.; Lauga, B.; Claisse, O.; Lonvaud-Funel, A.; Soulas, G.; Masneuf-Pomarède, I. Grape berry bacterial microbiota: Impact of the ripening process and the farming system. *Int. J. Food Microbiol.* **2012**, *158*, 93–100. [CrossRef]
12. Martins, G.; Vallance, J.; Mercier, A.; Albertin, W.; Stamatopoulos, P.; Rey, P.; Lonvaud, A.; Masneuf-Pomarède, I. Influence of the farming system on the epiphytic yeasts and yeast-like fungi colonizing grape berries during the ripening process. *Int. J. Food Microbiol.* **2014**, *177*, 21–28. [CrossRef]
13. Setati, M.E.; Jacobson, D.; Andong, U.C.; Bauer, F. The vineyard yeast microbiome, a mixed model microbial map. *PLoS ONE* **2012**, *7*, e52609. [CrossRef]
14. Bokulich, N.A.; Thorngate, J.H.; Richardson, P.M.; Mills, D.A. Microbial biogeography of wine grapes is conditioned by cultivar, vintage, and climate. *Proc. Natl. Acad. Sci. USA* **2014**, *111*, E139–E148. [CrossRef]
15. Portillo, M.; Del, C.; Franquès, J.; Araque, I.; Reguant, C.; Bordons, A. Bacterial diversity of Grenache and Carignan grape surface from different vineyards at Priorat wine region (Catalonia, Spain). *Int. J. Food Microbiol.* **2016**, *219*, 56–63. [CrossRef] [PubMed]
16. Salvetti, E.; Campanaro, S.; Campedelli, I.; Fracchetti, F.; Gobbi, A.; Tornielli, G.B.; Torriani, S.; Felis, G.E. Whole-metagenome-sequencing-based community profiles of *Vitis vinifera* L. cv. Corvina berries withered in two post-harvest conditions. *Front. Microbiol.* **2016**, *7*, 937. [CrossRef]
17. Grangeteau, C.; Roullier-Gall, C.; Rousseaux, S.; Gougeon, R.D.; Schmitt-Kopplin, P.; Alexandre, H.; Guilloux-Benatier, M. Wine microbiology is driven by vineyard and winery anthropogenic factors. *Microb. Biotechnol.* **2017**, *10*, 354–370. [CrossRef] [PubMed]
18. Oliveira, M.; Arenas, M.; Lage, O.; Cunha, M.; Amorim, M.I. Epiphytic fungal community in *Vitis vinifera* of the Portuguese wine regions. *Lett. Appl. Microbiol.* **2018**, *66*, 93–102. [CrossRef] [PubMed]
19. Kántor, A.; Petrová, J.; Hutková, J.; Kačániová, M. Yeast diversity in new, still fermenting wine "federweisser". *Potravinarstvo* **2016**, *10*, 120–125. [CrossRef]
20. Malík, F.; Furdíková, K.; Roman, T.; Malík, F. *Vinársky Lexicon*, 1st ed.; Fedor Malík a syn: Modra, Slovakia, 2012; p. 144.

21. Ribéreau-Gayon, P.; Dubourdieu, D.; Donèche, B.; Lonvaud, A. *Handbook of Enology, the Microbiology of Wine and Vinifications*; John Wiley & Sons Ltd.: West Sussex, UK, 2006; Volume 1, p. 512.
22. Millet, V.; Lonvaud-Funel, A. The viable but non-culturable state of wine microorganisms during storage. *Lett. Appl. Microbiol.* **2000**, *30*, 136–141. [CrossRef]
23. Divol, B.; Lonvaud-Funel, A. Evidence for viable but non-culturable yeasts in botrytis-affected wine. *J. Appl. Microbiol.* **2005**, *99*, 85–93. [CrossRef]
24. Du Toit, W.J.; Pretorius, I.S.; Lonvaud-Funel, A. The effect of suphur dioxide and oxygen on the viability and culturability of a strain of Acetobacter pasteurianus and a strain of *Brettanomyces bruxellensis* isolated from wine. *J. Appl. Microbiol.* **2005**, *98*, 862–871. [CrossRef]
25. Oliver, J.D. The viable but non-culturable state in bacteria. *J. Microbiol.* **2005**, *43*, 93–100.
26. Thomas, D.S. Yeasts as spoilage organisms in beverages. In *The Yeasts*, 2nd ed.; Rose, A.H., Harrison, J.S., Eds.; Academic Press: London, UK, 1993; Volume 5, pp. 517–561.
27. Deák, T.; Beuchat, L.R. Yeasts associated with fruit juice concentrates. *J. Food Protect.* **1993**, *56*, 777–782. [CrossRef]
28. Loureiro, V.; Malfeito-Ferreira, M. Spoilage yeasts in the wine industry (review). *Int. J. Food Microbiol.* **2003**, *86*, 23–50. [CrossRef]
29. Renouf, V.; Lonvaud-Funel, A. Development of an enrichment medium to detect *Dekkera/Brettanomyces bruxellensis*, a spoilage wine yeast, on the surface of grape berries. *Microbiol. Res.* **2007**, *162*, 154–167. [CrossRef] [PubMed]
30. Saez, J.S.; Lopes, C.H.A.; Kirs, V.E.; Sangorrín, M. Production of volatile phenols by *Pichia manshurica* and *Pichia membranifaciens* isolated from spoiled wines and cellar environment in Patagonia. *Food Microbiol.* **2011**, *28*, 503–509. [CrossRef] [PubMed]
31. Jolly, N.P.; Augustyn, O.P.H.; Pretorius, I.S. The occurrence of non-Saccharomyces yeast strains over three vintages in four vineyards and grape musts 130 from four production regions of the Western Cape, South Africa. *S. Afr. J. Enol. Vitic.* **2003**, *24*, 35–42.

5

A Microtiter Plate Assay as a Reliable Method to Assure the Identification and Classification of the Veil-Forming Yeasts during Sherry Wines Ageing

Marina Ruíz-Muñoz, Maria del Carmen Bernal-Grande, Gustavo Cordero-Bueso *, Mónica González, David Hughes-Herrera and Jesús Manuel Cantoral

Department of Biomedicine, Biotechnology and Public Health, University of Cádiz, Cádiz 11510, Spain; marina.ruizm@gmail.com (M.R.-M.); mariadelcarmen.bernal@uca.es (M.d.C.B.-G.); monica.gonzalez@uca.es (M.G.); david.hugheshererra@alum.uca.es (D.H.-H.); jesusmanuel.cantoral@uca.es (J.M.C.)

* Correspondence: gustavo.cordero@uca.es

Abstract: Yeasts involved in veil formation during biological ageing of Sherry wines are mainly *Saccharomyces cerevisiae*, and they have traditionally been divided into four races or varieties: beticus, cheresiensis, montuliensis and rouxii. Recent progress in molecular biology has led to the development of several techniques for yeast identification, based on similarity or dissimilarity of DNA, RNA or proteins. In view of the latest yeast taxonomy, there are no more races. However, molecular techniques are not enough to understand the real veil-forming yeast diversity and dynamics in Sherry wines. We propose a reliable method, using a microtiter reader, to evaluate the fermentation and assimilation of carbon and nitrogen sources, the osmotolerance and the antibiotic resistance, using 18 *S. cerevisiae* and 5 non-*Saccharomyces* yeast strains, to allow correct identification and classification of the yeast strains present in the velum of flor complex.

Keywords: flor yeast; microplate; fermentation; assimilation; *Saccharomyces cerevisiae*; non-*Saccharomyces*; velum of flor

---

1. Introduction

Several types of wines are characterized by growing of a yeast film on its surface, which is known as "veil or velum of flor". Two of these wines are Fino Sherry and Manzanilla, both produced from Palomino Fino (*Vitis vinifera* cv. *vinifera*) in the Jerez-Xérès-Sherry D.O., in Southern Spain [1,2]. The production of these types of wines consists of two successive processes. First, alcoholic fermentation, an anaerobic process in which sugars are converted in energy, produces as a result, ethanol and carbon dioxide gas. After the process is finished, wine is fortified to 15% (v/v) and it is put into oak barrels for storage. The process of biological ageing begins at this moment, using the "Solera" system [3].

Yeasts involved in veil formation during biological ageing are mainly *Saccharomyces cerevisiae*, and they have traditionally been divided into four races: beticus, cheresiensis, montuliensis and rouxii. The early methods related to taxonomy involved morphological studies and biochemical analyses of differences on fermentation and assimilation tests, where the substrates used were mainly dextrose, lactose, maltose, melibiose, raffinose, galactose and sucrose. These physiological properties were traditionally tested using Durham tubes in a liquid medium, which filled up with gas if fermentation occurred [4]. Nevertheless, both morphological and physiological characteristics may be influenced by culture conditions and spoilages and can provide ambiguous results. Using classical techniques can lead, in some cases, to incorrect classification of species or misidentification of strains. Moreover, the methodology often requires evaluation of different parameters by a complex, laborious and time-consuming process that does not include the most important characteristics of the yeasts from an industrial point of view. On the other hand, sugar fermentation and assimilation tests using Durham tubes are not too reliable because some of the yeast strains do not released CO_2 immediately (these are known as slow fermentative yeasts). These yeasts were considered for a long time as non-fermentative, but they have been recently re-classified [5]. In recent years, new methods have been developed, such as rapid kits, for yeast identification (API® strips), analyses of total cell proteins and long-chain fatty acids, using gas chromatography or the determination of chemical compounds formed by the yeasts [6,7], but their cost, reproducibility and suitability to veil-forming yeast strain classification can lead to mistakes [8,9]. Therefore, it is necessary to use other identification methods to analyse them.

Recent progress in molecular biology has led to the development of several techniques for yeast identification based on similarities or dissimilarities of DNA, RNA or proteins. These include electrophoretic karyotyping [10], RFLP (Restriction Fragment Length Polymorphism) of mitochondrial DNA [11], random amplified polymorphic DNA analysis (RAPD), PCR-based techniques such as ribosomal internal transcribed spacers (ITS-PCR) [8], surveys of simple sequences repeats (SSR-Multiplex PCR) [12] or genomic DNA sequencing of the D1/D2 region of 26S rDNA. Some of these methods have been found to be of great interest in enology, due to the high level of resolution in yeast strain characterization, and for making it possible to establish a correlation between the genetic variability and the most important industrial properties of the strains.

A study carried out by [8] showed that *S. cerevisiae flor* yeasts responsible for the biological ageing of Sherry-type wines present a 24-bp deletion, located in the ITS1 region. These results have extended the analysis of this DNA region to flor yeast isolated in the Montilla-Moriles D.O. region [9]. Both studies corroborated that this deletion is fixed in *flor yeast* of the species, *S. cerevisiae*. By using molecular analysis techniques, different properties of industrial interest have been detected, such as physiologic, genetic and metabolic properties in the different identified veil-forming yeast strains. Moreover, in view of the latest yeast taxonomy, it is important to underline that there are no more races. *S. cerevisiae* (beticus) and *S. cerevisiae* (cheresiensis) are no longer considered as races or subspecies of *S. cerevisiae*, according to the last taxonomic study. Indeed, all these previously named races or subspecies are now considered as *S. cerevisiae* synonyms, based on nuclear DNA relatedness [13]. The former race, *S. cerevisiae* (rouxii), has been classified as *Zygosaccharomyces rouxii* and *S. cerevisiae* (montuliensis) is now considered as *Torulaspora delbrueckii* [14]. These latter ones are now classified as ethanol-tolerant non-*Saccharomyces* species and they also should contain the 24-bp deletion in the 131–154 ITS1 region. Other yeast species of genera different to *Saccharomyces* have been isolated and identified by molecular and/or sequencing methods from Sherry-type wines (from the initial phase of the fermentation until the ageing phase), such as *Wickerhamomyces anomalus*, *Pichia membranaefaciens*, *Zygosaccharomyces bailii* or the undesirable species *Dekkera bruxellensis* (anamorph *Brettanomyces bruxellensis*) [3,8,15]. However, the identification strategy should be used more precisely because many database sequences of these species have identical D1/D2 sequences [16]. The comparison of the D1/D2 sequences and ITS sequences between the isolated strains and those of the type strains of the species (most can be found in the CBS database). However, many authors have simply used the first hit in the blast search as the proof of taxonomic affiliation of their isolate. This strategy can be misleading because database curators do not check submissions for the correctness of taxonomic affiliation. In this case, using classical biochemical analyses, in conjunction with molecular methods, can help us to identify and classify the yeast diversity forming the velum of flor in Sherry wines. Therefore, we propose a reliable, reproducible, inexpensive and efficient biochemical test, using polysterene 96-well plates that could reinforce the identification of the different veil-forming yeast strains.

2. Materials and Methods

2.1. Yeast Strains

Six hundred yeast strains were isolated previously from the veil of flor of "Manzanilla" and "Fino" wines, from the Sanlúcar de Barrameda and Jerez wineries (Cádiz, Spain), respectively [2,17]. After a first screening, using a Mitochondrial DNA restriction analysis and electrophoretic karyotyping by pulsed-field-gel electrophoresis (PFGE) and SSR-Multiplex PCR, eighteen different profiles of *Saccharomyces* strains were shown and then grouped according to their similarities—eight from "Manzanilla", denoted as MI to MVI, MX and MXVIII, and ten from "Fino", coded as FI to FX, respectively. Additionally, five potential non-*Saccharomyces*, isolated from "Fino" veils, which showed different karyotypes to *S. cerevisiae*, were subjected to identification by ITS-PCR and sequencing of the D1/D2 region of 26S rDNA, as stated in [18]. At least two representative members from each ITS-RFLP genotype group were randomly selected for sequencing of the LSU sRNA gene D1/D2 domain.

2.2. Mitochondrial DNA (mtDNA) Restriction Analysis

Total DNA extraction and mtDNA restriction analyses of the *Saccharomyces* yeast strains were performed by the method of [11]. Yeast DNA was digested with a *Hinf*I restriction enzyme, by incubation at 37 °C, for 3 h. Fragments were separated by electrophoresis on a 1% agarose gel in 1× TBE buffer (45 mM tris-borate, 1 mM EDTA) and with 0.5 μg/mL of ethidium bromide. The image of the gel was digitalized in a Molecular Imager apparatus (Gel-Doc XR, Hercules, CA, USA) and analyzed using Quantity One 1-D software (Bio-Rad, Hercules, CA, USA).

2.3. Electrophoretic Karyotype

Electrophoretic karyotypes of the *Saccharomyces* and *non-Saccharomyces* yeast strains were obtained by PFGE. The colonies were embedded in agarose plugs and treated following a procedure based on the protocol of [19], and chromosomes were separated using a CHEF-DR III apparatus (Bio-Rad). Gels were made of 1% pulsed-field certified agarose (Bio-Rad, Hercules, CA, USA) in 0.5× TBE electrophoresis buffer. The electrophoretic conditions were as follows: an initial pulse of 60 s and a final pulse of 120 s, during 22 h, at 14 °C and 6 V/cm, with an angle of 120°. Gels were stained with ethidium bromide solution (0.5 μg/mL ethidium bromide, 0.5× TBE buffer) after electrophoresis, and the image was captured under UV light with a camera (Bio-Rad Imaging System, Hercules, CA, USA). The image of the gel was digitalized in a Molecular Imager apparatus (Gel-Doc) and analyzed using Quantity One-1D software (Bio-Rad). Results were analyzed by comparing the electrophoretic mobility of the chromosomes with a reference strain (*Saccharomyces cerevisiae* YNN195).

2.4. Multiplex PCR-Microsatellite (SSR) Analysis

To perform multiplex PCR of microsatellite loci, three loci were simultaneously amplified, using the primers SC8132X (Fw: CTGCTCAACTTGTGATGGGTTTTGG; Rv: CCTCGTTACTATC GTCTTCATCTTGC), YOR267C (Fw: GGTGACTCTAACGGCAGAGTGG; Rv: GGATCTACTTGCA GTATACGGG) and SCPTSY7 (Fw: AAAAGCGTAAGCAATGG TGTAGAT; Rv: AAATGATGCC AATATTGAAAAGGT) (MWG Biotech AG, Ebersberg, Germany), proposed by Vaudano and García-Moruno [12]. PCR amplifications were performed in an Applied Biosystems SimplyAmp Termal Cycler in 50 μL volumes, constituted by: 2 μL of DNA solution (30–90 ng of DNA), 2.5 mM $MgCl_2$, 200 μM dNTPs, 10 μL of 5× PCR buffer, 2 U of GoTaq® Flexi DNA Polymerase (Promega) and 10 pmol of primers for loci, SCYOR267C and SC8132X, and 40 pmol of primers for locus SCPTSY7. By using the protocol—4 min at 94 °C, 28 cycles of 30 s at 94 °C, 45 s at 56 °C and 30 s at 72 °C, and 10 min at 72 °C—PCR products were separated on an agarose gel (2.5% w/v) with 5 μg/mL of ethidium bromide in 1× TBE buffer at 115 V for 45 min. DNA fragment sizes were determined by comparison with a molecular ladder marker of 50 bp (Biolab, London, UK). The gel was digitalized in a Molecular Imager apparatus (Gel-Doc XR) and analyzed using Quantity One 1-D software (Bio-Rad).

2.5. *Physiological and Biochemical Tests*

To perform a quantitative analysis of carbon fermentation and assimilation, nitrogen assimilation, osmotolerance, and resistance to cycloheximide, different solutions, filtered through 0.22-µm Stericup filters (Millipore, Burlington, MA, USA) prior to their addition, were incorporated into the wells of polysterene 96-well microplates (Nunc™ 96-well polystyrene conical bottom Microwell™, Thermofisher, Denmark). The basal medium for carbohydrate fermentation and assimilation tests (BMC) was as follows: sterile deionized water 100 mL, 1 g yeast nitrogen base (YNB, Difco) was prepared with 2 g (*w*/*v*) solutions of fructose, glucose, maltose, lactose and sucrose and 4 g (*w*/*v*) of raffinose for fermentation tests, and 0.78 g of glucose, galactose, sorbose, glucosamine, ribose, xylose, arabinose, rhamnose, sucrose, maltose, α-threalose, cellobiose, salicin, arbutin, melibiose, lactose, raffinose, melezitose, inulin, starch, erythrytol, ribitol, xylitol, L-arabinitol, D-glucitol, D-mannitol, myo-inositol, D-Glucono-1,5-lactone, 5-Keto-D-gluconate, D-gluconate, D-galacturonate, DL-lactate, succinate, citrate and L-tartaric acid, and 0.78 mL of glycerol, propane 1,2-diol, butane 2,3-diol, Tween 20, Tween 80 for assimilation tests. Nitrogen sources for assimilation were prepared by supplementing 100 mL of an autoclaved basal assimilation medium yeast carbon base (YNB, DIfco) (0.17 g YNB without amino acids, 0.5 g ammonium sulphate, 2 g glucose) with filter-sterilized solutions containing 0.5 g of ammonium sulphate, 0.18 g of ammonium citrate, 0.11 g of nitrate, 0.04 g of nitrite, 0.1 g of ethylamine and creatine, 0.087 g of creatinine, 0.11 g of glucosamine, 0.1 g of urea, and D-proline. Then, 180 µL of each substrate was dropped into the wells of the trays. Also, 180 µL of 2%, 20%, 30%, 40% and 50% of a 1:1 mixture of glucose and fructose solution were incorporated into the wells as an osmotolerance test. Wells for testing the resistance to cycloheximide were prepared by adding 180 µL of a sterilized 0.1% *w*/*v* cycloheximide solution. The medium in each well can be dehydrated by drying the plate with the lid in a Speed-Vac vacuum oven (Eppendorf), which forms a small visible residue at the base of the well. After drying, individual trays can be sealed by heat with a polypropylene-aluminium sealing sheet and can be stored at 4 °C for six months, or at −20 °C for one year until subsequent use, rehydrating them with the corresponding basal medium.

The selected yeast strains were previously grown for 24–48 h in 15 mL of YPD (1% yeast extract, 2% peptone, 2% glucose) at 28 °C overnight; when achieved, these were washed twice with sterile, distilled-water. Twenty microliters of inoculum (McFarland standard 2, diluted by a factor of 10) were introduced into each well using a multi-channel pipette. McFarland standards are used as a reference to adjust the turbidity of yeast suspensions, so the number of yeast will be within a given range, to standardize biochemical tests. A 0.5 McFarland standard was prepared by mixing 0.05 mL of 1.175% barium chloride dehydrate with 9.95 mL of 1% sulfuric acid. The standards can be compared by eye to a suspension of yeasts in sterile water. If the yeast suspension is too turbid, it can be diluted. If the suspension is not turbid enough, more yeasts can be added (Table 1).

Table 1. McFarland standard values used as a reference to adjust the turbidity of yeast suspensions.

McFarland Standard	0.5	1	2	3	4
1% Barium chloride (mL)	0.05	0.1	0.2	0.3	0.4
1% Sulfuric acid (mL)	9.95	9.9	9.8	9.7	9.6
Cell density $\pm 10^8$ cells/mL	1.5	3	6	9	12
Absorbance (610 nn)	0.1	0.25	0.45	0.60	0.67

Microplates were sealed with Breathe-Easy membranes (Sigma, Sant Louis, MI, USA) and placed into an automatic microplate reader (Multiskan go reader, Thermo Fischer Scientific, Waltham, MA, USA). The optical density of the mix was read at 610 nm every hour (one minute of shaking just before automatic reading by the microplate reader) at 25 °C over 72 h. Data were collected and processed with the software, SkanIt RE for Multiskan GO 3.2. The absorbance data were exported to MS Excel for processing. A reading of optical density (OD_{610}) = 1, measured by the microtiter reader, corresponds to 6–7 × 10^7 cells/mL. Absorbance values at 610 nm were transformed into negative values when OD < 0.4 and positive when OD > 0.4, after 72 h.

2.6. Validation and Reproducibility of the Tests in Microtiter Plates

The yeast strains used in this study were also subjected to a standard test in Durham tubes, following the protocol proposed by [20] and under the same conditions of temperature and time. These tests were repeated with the microplate tests six times with each yeast strain. The stability of the carbon and nitrogen medium of the microtiter plates in storage was also examined. The same yeast species were analyzed for their reactions in microtitre trays that had been freshly prepared and in those that had been stored for six months at 4 °C and for one year at −20 °C.

3. Results

A total of 18 strains belonging to *Saccharomyces cerevisiae* species were characterized using the pulsed-field-gel electrophoresis (PFGE) technique and mitochondrial DNA restriction analysis. Moreover, analyses of polymorphic microsatellite loci and the proposed biochemical test, using 96-well plates, were used. Chromosomal profiles, obtained from the studied strains, showed fourteen different patterns. Specifically, strains FII and FIII, FV and FVI, FVIII and FIX, MX and MXVIII showed common patterns. In regard to the SSR analysis, fourteen different patterns were also obtained, which complied with the results obtained in the PFGE (Figure 1).

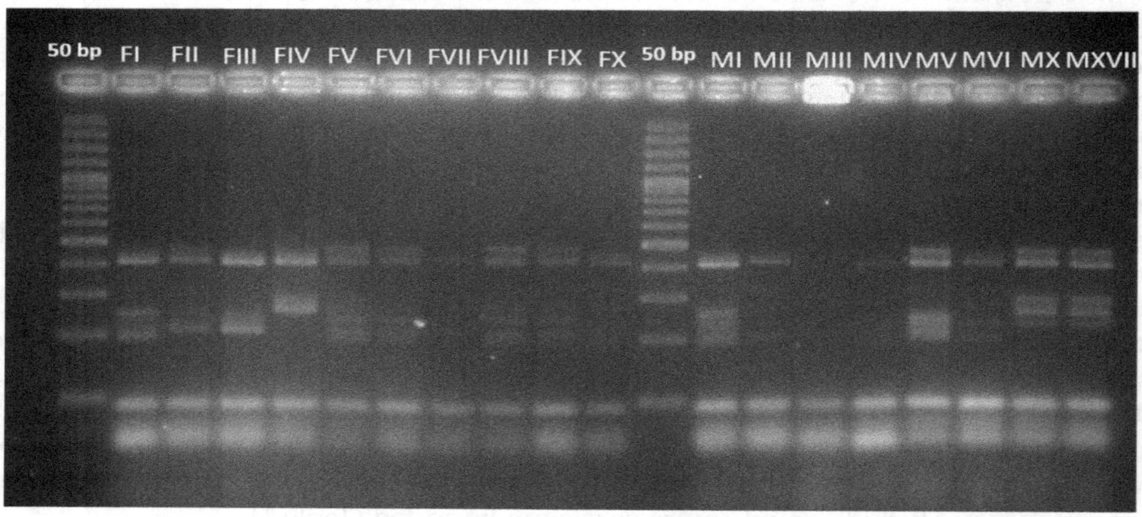

Figure 1. Patterns obtained by the simple sequences repeats (SSR)-Multiplex PCR of the *S. cerevisiae* flor yeasts, selected after the first screening. Strains FII and FIII, FV and FVI, FVIII and FIX, MX and MXVIII showed common patterns.

On the other hand, using mitochondrial DNA restriction analyses with the enzyme *Hinf*I, three different patterns were obtained. The most common pattern was H1 (corresponding to FIV, FV, FIX, MI, MII, MIII, MVIII and MX), the second was H2 (corresponding to FI, FII, MIV and MV) and the pattern H3 only corresponded to one strain (MVI).

Fifteen different strains were determined by establishing the correspondence between the three patterns obtained from the analysis of mtRFLP with *Hinf*I—coded as H1, H2 and H3—and fourteen patterns were obtained from the polymorphism study of the tested karyotypes and microsatellite loci. These fifteen different strains are FI, FII, FIII, FIV, FV, FVI, FIX, FX, MI, MII, MIII, MIV, MV, MVI and MX, according to the different combinations observed. The potential five non-*Saccharomycces* species, analyzed by the molecular method mentioned in the material and methods section (Figure 2) and after sequencing of the D1/D2 region, have been identified as *Zygosaccharomyces bailii* (NsI),

Torulaspora delbrueckii (NsII), *Zygosaccharomyces rouxii* (NsIII), *Rhodotorula mucilaginosa* (NsIV) and *Rhodotorula minuta* (NsV). However, the species, *Z. bailii* and *Z. rouxii*, and *R. mucilaginosa* and *R. minuta* showed very similar (less than 1%), or identical, D1/D2 sequences to those found in the databases.

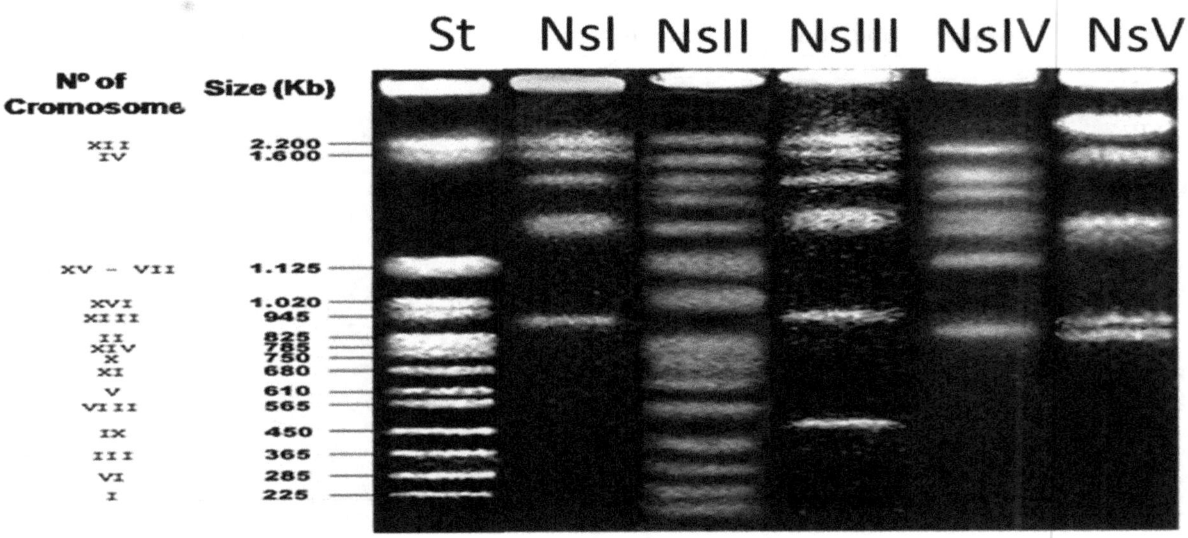

Figure 2. Karyotypes obtained by pulsed-field-gel electrophoresis (PFGE) of the *non-Saccharoymyces* veil-forming yeasts isolated in "Fino" Sherry wines. The species were identified as *Zygosaccharomyces bailii* (NsI), *Torulaspora delbrueckii* (NsII), *Zygosaccharomyces rouxii* (NsIII), *Rhodotorula mucilaginosa* (NsIV) and *Rhodotorula minuta* (NsV).

The selected strains were also analyzed to determine their fermentation and assimilation abilities, to evaluate if they shared physiological characteristics. Tables 2 and 3, show the results of those which reached a positive result, among all tested carbon and nitrogen sources.

Table 2. Fermentation of different carbohydrates by selected *S. cerevisiae* flor yeasts, analyzed with microtiter technology. Values are expressed as (+) if yeast strains were able to ferment the carbon/nitrogen source and (−) if yeast strains were not able to ferment them.

Strain Designation	Sugar Fermentation					
	Fructose	Glucose	Lactose	Maltose	Rafinose	Sucrose
FI	−	+	−	−	−	−
FII	+	+	−	−	+	+
FIII	+	+	−	−	−	−
FIV	+	+	−	−	−	+
FV	+	+	−	−	−	+
FVI	+	+	−	−	+	+
FIX	−	+	−	−	+	+
FX	−	+	+	−	−	−
MI	+	−	−	−	−	+
MII	+	+	+	−	+	+
MIII	−	−	+	−	−	−
MIV	+	+	+	−	+	−
MV	−	+	−	+	−	−
MVI	+	+	−	−	+	+
MX	+	+	−	−	+	+

Table 3. Assimilation of different carbon/nitrogen compounds by selected *S. cerevisiae* flor yeasts, analyzed with microtiter technology. Values are expressed as (+) if yeast strains were able to assimilate the carbon/nitrogen source and (−) if yeast strains were not able to assimilate them.

Strain Designation	Carbon/Nitrogen Compounds						
	Ammonium Citrate	Glucose	Inulin	Melibiose	Nitrite	Starch	Urea
FI	+	−	−	−	−	−	−
FII	+	−	−	−	−	·	−
FIII	+	−	−	−	−	−	−
FIV	+	−	−	−	−	−	−
FV	+	−	−	−	−	−	−
FVI	−	−	−	−	−	−	−
FIX	−	+	+	−	−	−	−
FX	−	−	−	−	−	−	−
MI	+	−	−	−	−	−	−
MII	+	−	+	−	−	−	+
MIII	+	−	+	−	−	−	+
MIV	+	−	−	−	−	−	−
MV	−	+	−	+	+	−	−
MVI	+	−	−	−	−	+	−
MX	−	−	+	−	−	+	−

These tests, repeated six times, were also carried out in Durham tubes, obtaining the same results in all cases, with the exception of a couple of cases (fermentation of raffinose by yeast strain FII and fermentation of sucrose by yeast strain MVI), in which Durham tubes showed questionable results. However, these results obtained by the standard method, had positive results in the microtiter-read assay (OD > 0.4). Although molecular analyses have shown similarities between some strains, the diversity of the fermentation and assimilation abilities were significantly higher (Tables 2 and 3). In the case of the sequenced non-*Saccharomyces* yeast strains, *Z. bailii* did not grow on maltose, which is usually assimilated by *Z. rouxii* and *R. mucilaginosa* isolates grew on maltose, sucrose, lactose, which are not assimilated by *R. minuta*. These results are in accordance with those published in [14,21]. Thus, by applying the biochemical tests, we have ensured that these yeast strains have been satisfactorily identified (Table 4).

Table 4. Assimilation of different carbon/nitrogen compounds by non-*Saccharomyces* species, isolated from "Fino" Sherry wine, analyzed with microtiter technology. Values are expressed as (+) if yeast strains were able to assimilate the carbon/nitrogen source and (−) if yeast strains were not able to assimilate them.

Strain Designation	Carbon/Nitrogen Compounds							
	Raffinose	Galactose	Inulin	Melibiose	Maltose	Lactose	L-Sorbose	Ethanol
NsI	−	+	−	−	−	−	−	+
NsII	+	+	+	−	+	−	−	+
NsIII	−	+	−	−	+	−	−	+
NsIV	+	+	−	−	+	−	−	+
NsV	−	−	−	−	−	+	+	+

Regarding the results obtained for osmotolerance, only two isolates grew in the media supplemented with 50% of glucose:fructose, but neither belonged to the *Saccharomyces* species (Table 5). The most osmotolerant species was *Torulaspora delbrueckii*. The tests for cycloheximide resistance gave the same results when tested by either the standard or microtitre method (Table 5).

Table 5. Osmotolerance and cycloheximide resistance of the veil-forming yeasts used in this study and analyzed by microtiter technology. Values are expressed as (+) if yeast strains were able to growth in a particular concentration of glucose/fructose solution and (−) if yeast strains were not able to growth.

Strain Designation	Osmotolerance: Growth on Media Supplemented with 1:1 (Glucose:Fructose) and Cycloheximide Resistance						
	2%	20%	30%	40%	50%	Cycloheximide 0.01%	Cycloheximide 0.1%
FI	+	+	+	−	−	−	−
FII	+	+	−	−	−	−	−
FIII	+	+	+	−	−	−	−
FIV	+	+	+	−	−	−	−
FV	+	+	+	−	−	−	−
FVI	+	+	+	−	−	−	−
FVII	+	+	−	−	−	−	−
FVIII	+	+	−	−	−	−	−
FIX	+	+	+	−	−	−	−
FX	+	+	+	−	−	−	−
MI	+	+	+	−	−	−	−
MII	+	+	+	−	−	−	−
MIII	+	+	+	−	−	−	−
MIV	+	+	+	−	−	−	−
MV	+	+	+	−	−	−	−
MVI	+	+	+	−	−	+	−
MX	+	+	+	−	−	+	−
MXVIII	+	+	+	−	−	−	−
NsI	+	+	−	−	−	−	−
NsII	+	+	+	+	+	−	−
NsIII	+	+	+	+	+	−	−
NsIV	+	+	−	−	−	+	−
NsV	+	+	−	−	−	+	−

The reproducibility of the tests in microtiter plates was assessed, using the selected *Saccharomyces* and non-*Saccharomyces* identified in this study. They were examined six times in microtiter plates. Responses to the identification tests for each species were the same in all replications. Moreover, the stored 96-well plates gave the same reaction responses, indicating that the multiplates could be stored for at least one year at −20 °C without affecting identification results.

4. Discussion

In recent years, multiplex microsatellite analysis has been implemented as one of the most decisive techniques in determining polymorphisms within the same species, but it has not often been applied to flor yeast analyses, since it could show few different patterns. However, in our study microsatellite analysis have shown fourteen different patterns, obtaining the same results as electrophoretic karyotyping. In any case, by combining multiplex microsatellite and RFLP of mtDNA analyses, when required, it is possible to obtain the same number of patterns as using electrophoretic karyotyping; these results are in agreement with those found in [9].

Pulsed-field gel electrophoresis (PFGE) has a good reproducibility and high discriminatory power, but it is an expensive and time-consuming technique. On the other hand, multiplex PCR of microsatellite loci [12] also has a high discriminatory power, sample preparation is easier and it is faster than PFGE.

Results obtained in the characterization of flor strains are highly variable, since some authors [8] have identified a higher degree of polymorphism between flor strains in mtDNA restriction than karyotyping. However, our study revealed the opposite: mtDNA restriction showed only three different patterns and PFGE showed fourteen different patterns. These results are similar to [9]. Mitochondrial DNA restriction analysis showed less polymorphisms due to the conditions of the wineries, since these DNA could be subjected to positive selection pressure based on their ethanol tolerance (15% *v*/*v*) [22] and other environmental factors, such as nutrient composition, growth temperature, and presence of toxic drugs, heavy metals, oxidizing agents, and osmotic/ionic stresses [23]. This might be the reason why the mtDNA restriction analysis showed only three different patterns of all the strains that we analyzed.

Biochemical tests offer supplementary information for characterizing flor strains. Traditionally, veil-forming yeast strains have been classified into four races belonging to the *S. cerevisiae* species, currently considered synonyms of the *S. cerevisiae* species [14]; another race, previously referred to as *S. cerevisiae* var. montuliensis, that has recently been reclassified as *Torulaspora delbrueckii* [14,24], and a last race named *S. cerevisiae rouxii*, is now *Zygosaccharomyces rouxii* [14]. This classification was based on morphological and metabolic studies of fermentation and assimilation of different carbon and nitrogen sources that differentiate these four main breeds [25]. The studies of assimilation and fermentation of sugars that we have carried out in this work have shown a high variability, so we were not able to differentiate races within *S. cerevisiae* strains. We can conclude that biochemical tests are necessary for identifying and classifying the different yeast strains that form the veil of flor of Sherry wine, because we believe that each yeast strain brings to the Sherry wines a different organoleptical characteristic. Thus, the application of this simplified method, using polysterene 96-well plates, is an inexpensive, low-time consuming and efficient technique, that can be used in combination with the molecular methods described above, to ensure veil-forming yeast diversity in Sherry wines.

Acknowledgments: This work was financed by the project UCA18DGUEII02, and by the European Union's Seventh Framework Programme via the Marie Curie Action, "Co-funding of Regional, National and International Programs" to stimulate research activities without mobility restrictions, co-financed by the 'Junta de Andalucía' and the European Commission under grant agreement No. 291780.

Author Contributions: Jesús Manuel Cantoral and Gustavo Cordero-Bueso contributed to the conception and design of experiments; Marina Ruíz-Muñoz, Gustavo Cordero-Bueso, Mónica González, David Hughes-Herrera and Maria del Carmen Bernal-Grande performed the experiments; Marina Ruíz-Muñoz, Gustavo Cordero-Bueso and Maria del Carmen Bernal-Grande analyzed the data; Marina Ruíz-Muñoz and Gustavo Cordero-Bueso contributed to writing the paper.

References

1. Rodríguez, M.E.; Infante, J.J.; Mesa, J.J.; Rebordinos, L.; Cantoral, J.M. Enological behavior of biofilms formed by genetically-characterized strains of sherry flor yeast. *Open Biotechnol. J.* **2013**, *7*, 23–29. [CrossRef]
2. Mesa, J.J.; Infante, J.J.; Rebordinos, L.; Cantoral, J.M. Characterisation of yeasts involved in the biological ageing of sherry wines. *LWT-Food Sci. Technol.* **1999**, *32*, 114–120. [CrossRef]
3. Esteve-Zarzoso, B.; Peris-Torán, M.J.; García-Maiquez, E.; Uruburu, F.; Querol, A. Yeast population dynamics during the fermentation and biological aging of sherry wines. *Appl. Environ. Microbiol.* **2001**, *67*, 2056–2061. [CrossRef] [PubMed]
4. Martínez, P.; Codon, A.C.; Pérez, L.; Benítez, T. Physiological and molecular characterization of flor yeasts: Polymorphism of flor yeast population. *Yeast* **1995**, *11*, 1399–1411. [CrossRef] [PubMed]
5. Boekhout, T.; Kurtzman, C.P. Principles and methods used in yeast classification, and an overview of currently accepted yeast genera. In *Nonconventional Yeasts in Biotechnology*; Springer: Berlin/Heidelberg, Germany, 1996; pp. 1–81. ISBN 978-3-642-79856-6.
6. Vancanneyt, B.P.; Hennebert, G.; Kersters, K. Differentiation of yeast species based on electrophoretic whole-cell protein patters. *Syst. Appl. Microbiol.* **1991**, *14*, 23–32. [CrossRef]
7. Silva, M.M.D.; Malfeito-Ferreira, M.; Loureiro, V.; Aubyn, A.S. Long-chain fatty acid composition as a criterion for yeast distinction in the brewing industry. *J. Inst. Brew.* **1994**, *100*, 17–22. [CrossRef]
8. Esteve-Zarzoso, B.; Fernández-Espinar, M.T.; Querol, A. Authentication and identification of *Saccharomyces cerevisiae* "flor" yeast races involved in sherry ageing. *Antonie van Leewenhoek* **2004**, *85*, 151–158. [CrossRef] [PubMed]
9. Marin-Menguiano, M.; Romero-Sánchez, S.; Barrales, R.R.; Ibeas, J.I. Population analysis of biofilm yeasts during fino sherry wine aging in the Montilla Moriles D.O. region. *Int. J. Food Microbiol.* **2017**, *244*, 67–73. [CrossRef] [PubMed]
10. Rodríguez, M.E.; Infante, J.J.; Molina, M.; Rebordinos, L.; Cantoral, J.M. Using RFLP-mtDNA for the rapid monitoring of the dominant inoculated yeast strains in industrial wine fermentation. *Int. J. Food Microbiol.* **2011**, *145*, 331–335. [CrossRef] [PubMed]
11. Querol, A.; Barrio, E.; Huerta, T.; Ramón, D. Molecular monitoring of wine fermentatios conducted by active dry yeast strains. *Appl. Environ. Microbiol.* **1992**, *58*, 2948–2953. [PubMed]

12. Vaudano, E.; García-Moruno, E. Discrimination of *Saccharomyces cerevisiae* wine strains using microsatellite multiplex PCR and band pattern analysis. *Food Microbiol.* **2008**, *25*, 56–64. [CrossRef] [PubMed]
13. Legras, J.C.; Moreno-García, J.; Zara, S.; Zara, G.; García-Martínez, T.; Mauricio, J.C.; Mannazzu, I.; Coi, A.L.; Zeidan, M.B.; Dequin, S.; et al. Flor yeast: New perspectives beyond wine aging. *Front. Microbiol.* **2016**, *7*, 503. [CrossRef] [PubMed]
14. Kurtzman, C.; Fell, J.W.; Boekhout, T. *The Yeasts: A Taxonomic Study*; Elvesier: London, UK, 2011; p. 2363.
15. Fernández-Espinar, M.T.; Esteve-Zarzoso, B.; Querol, A.; Barrio, E. RFLP analysis of the ribosomal internal transcribed spacers and the 5.8 S rRNA gene region of the genus *Saccharomyces*: A fast method for species identification and the differentiation of flor yeasts. *Antonie van Leeuwenhoek* **2000**, *78*, 87–97. [CrossRef]
16. Lopandic, K.; Pfliegler, W.P.; Tiefenbrunner, W.; Gangl, H.; Sipiczki, M.; Sterlinger, K. Genotypic and phenotypic evolution of yeast interspecies hybrids during high-sugar fermentation. *Appl. Microbiol. Biotechnol.* **2016**, *100*, 6331–6343. [CrossRef] [PubMed]
17. Espinazo-Romeu, M.; Cantoral, J.M.; Matallana, E.; Aranda, A. Btn2p is involved in ethanol and biofilm formation in flor yeast. *FEMS Yeast Res.* **2008**, *8*, 1127–1136. [CrossRef] [PubMed]
18. Cordero-Bueso, G.; Rodríguez, M.E.; Garrido, C.; Cantoral, J.M. Rapid and not culture-dependent assay based on multiplex PCR-SSR analysis for monitoring inoculated yeast strains in industrial wine fermentations. *Arch. Microbiol.* **2017**, *199*, 135–143. [CrossRef] [PubMed]
19. Rodríguez, M.E.; Infante, J.J.; Molina, M.; Domínguez, M.; Rebordinos, L.; Cantoral, J.M. Genomic characterization and selection of wine yeast to conduct industrial fermentations of a white wine produced in a SW Spain winery. *J. Appl. Microbiol.* **2010**, *108*, 1292–1302. [CrossRef] [PubMed]
20. Van Der Walt, J.P.; Yarrow, D. Methods for the isolation, maintenance, classification and identification of yeasts. In *The Yeasts: A Taxonomic Study*, 3rd ed.; Kreger-van Rij, N.J.W., Ed.; Elvesier Science Publishers: Amsterdan, The Netherlands, 1984; pp. 45–104.
21. Sipiczki, M. Overwintering of vineyard yeasts: Survival of interacting yeast communities in grapes mummified on vines. *Front. Microbiol.* **2016**, *7*, 212. [CrossRef] [PubMed]
22. Ibeas, J.I.; Jiménez, J. Mitochondrial DNA loss caused by etanol in *Saccharomyces* flor yeasts. *Appl. Environ. Microbiol.* **1997**, *63*, 7–12. [PubMed]
23. Kvitek, D.J.; Will, J.L.; Gasch, A.P. Variations in Stress Sensitivity and Genomic Expression in Diverse *S. cerevisiae* Isolates. *PLoS Genet.* **2008**, *4*, e1000223. [CrossRef] [PubMed]
24. Alexandre, H. Flor yeasts of *Saccharomyces cerevisiae*—Their ecology, genetics and metabolism. *Int. J. Food Microbiol.* **2013**, *167*, 269–275. [CrossRef] [PubMed]
25. Barnett, J.A.; Payne, R.W.; Yarrow, D. *Yeasts: Characterisation and Identification*, 3rd ed.; Cambridge University Press: Cambridge, UK, 2000; p. 1139.

6

Batch Fermentation Options for High Titer Bioethanol Production from a SPORL Pretreated Douglas-Fir Forest Residue without Detoxification

Mingyan Yang [1,2,3], Hairui Ji [2,4] and J.Y. Zhu [2,*]

1 School Environment Science Engineering, Chang'an University, Xi'an 710064, China; yangmingyan67@163.com
2 USDA Forest Service, Forest Products Laboratory, Madison, WI 53726, USA; jzhu@fs.fed.us
3 Key Laboratory of Subsurface Hydrology and Ecological Effects in Arid Region, Ministry of Education, Chang'an University, Xi'an 710064, China
4 College of Life Science and Technology, Beijing University of Chemical Technology, Beijing 100029, China
* Correspondence: jzhu@fs.fed.us

Academic Editor: Ronnie G. Willaert

Abstract: This study evaluated batch fermentation modes, namely, separate hydrolysis and fermentation (SHF), quasi-simultaneous saccharification and fermentation (Q-SSF), and simultaneous saccharification and fermentation (SSF), and fermentation conditions, i.e., enzyme and yeast loadings, nutrient supplementation and sterilization, on high titer bioethanol production from SPORL-pretreated Douglas-fir forest residue without detoxification. The results indicated that Q-SSF and SSF were obviously superior to SHF operation in terms of ethanol yield. Enzyme loading had a strong positive correlation with ethanol yield in the range studied. Nutrient supplementation and sterility were not necessary for ethanol production from SPORL-pretreated Douglas-fir. Yeast loading had no substantial influence on ethanol yield for typical SSF conditions. After 96 h fermentation at 38 °C on shake flask at 150 rpm, terminal ethanol titer of 43.2 g/L, or 75.1% theoretical based on untreated feedstock glucan, mannan, and xylan content was achieved, when SSF was conducted at whole slurry solids loading of 15% with enzyme and yeast loading of 20 FPU/g glucan and 1.8 g/kg (wet), respectively, without nutrition supplementation and sterilization. It is believed that with mechanical mixing, enzyme loading can be reduced without reducing ethanol yield with extended fermentation duration.

Keywords: forest residue; pretreatment; liquefaction; enzymatic hydrolysis/saccharification; fermentation; high titer bioethanol; detoxification

---

1. Introduction

Fermentation of sugars from lignocelluloses has been proposed as a viable pathway for the production of renewable biofuels to supplement petro-fuels for sustainable economic development [1]. Feedstock is a major cost factor in producing cellulosic biofuels. Using low cost and underutilized feedstock such as harvest forest residues can be a winning proposition [2]. Forest residue can be sustainably produced in large quantities in various regions of the world [3,4], however, requires extensive pretreatment to remove its strong recalcitrance to enzymatic sugar production [5]. Severe pretreatments were often applied to remove this recalcitrance but resulted in the production of fermentation inhibitors [6,7]. Pretreatment optimization and strategies can substantially reduce inhibitor formation to facilitate high solids saccharification and fermentation without detoxification [8–12]. Proper management of inhibitor and sugar profiles in fermentation can improve ethanol productivity as xylose fermentation using *Saccharomyces cerevisiae* is especially sensitive to inhibitor profile [13,14]. Some studies suggest that initial sugar concentration can affect ethanol productivity [10,15]. Therefore, it is worthwhile to study the potential batch fermentation options to maximize ethanol production.

The objective of the present study is therefore to evaluate several batch fermentation approaches under various conditions for ethanol production from a softwood forest residue. The forest residue was pretreated by Sulfite Pretreatment to Overcome the Recalcitrance of Lignocelluloses (SPORL) [16]. SPORL was chosen for its robust performance in bioconversion of softwood biomass to sugars and biofuel [11,17]. Few processes have demonstrated good performance in effective removal of the recalcitrance of softwood biomass [5,18]. A SPORL pretreated Douglas-fir forest residue whole slurry from a previous study [10] at high SO_2 loading was used because of its low inhibitors, relatively short pretreatment residence time of 1 h, and potentially low metallurgy requirement due to the low pretreatment temperature of 140 °C. Separated enzymatic hydrolysis and fermentation (SHF), simultaneous enzymatic saccharification and fermentation (SSF), as well as quasi-simultaneous enzymatic saccharification and fermentation (Q-SSF) were evaluated. In Q-SSF, the pretreated whole slurry solids (WSS) was first liquefied through a pre-hydrolysis period at elevated temperature optimized for the cellulase used, i.e., 50 °C for the present study. Fermentation was then performed after cooling down the liquefied WSS to the desired temperature for the yeast, i.e., 38 °C in the present study. Ranges of liquefaction time, cellulase and yeast loadings were all evaluated in this study.

2. Materials and Methods

2.1. Materials

Douglas-fir forest residue was collected from a regeneration harvest in a primarily Douglas-fir stand in Lane County, OR, USA. The forest residue was ground and screened as described previously [9,19] to reduce bark and ash content and dead load in transportation. The accept forest residues were labelled as FS-10 and shipped to USDA Forest Products Laboratory, Madison, WI, after air drying to a moisture content of approximately 15%.

A commercial complex cellulase enzyme, Cellic® CTec3 (abbreviated CTec3), was complimentary provided by Novozymes North America (Franklinton, North Carolina, USA). The cellulase activity was 217 FPU/mL as calibrated by a literature method [20]. All the chemicals used were ACS reagent grade and purchased from Sigma-Aldrich (St. Louis, MO, USA).

Saccharomyces cerevisiae YRH400, an engineered fungal strain for xylose fermentation, was obtained from USDA Agriculture Research Service [21]. The strain was grown at 30 °C for 2 days on YPD agar plates containing 10 g/L yeast extract, 20 g/L peptone, 20 g/L glucose, and 20 g/L agar. A colony from the plate was transferred by loop to 50 mL liquid YPD medium in a 125 mL flask and cultured for overnight at 30 °C with agitation at 150 rpm on a shaking bed incubator (Thermo Fisher Scientific, model 4450, Waltham, MA, USA). The biomass concentration was monitored using optical density at 600 nm (OD_{600}) by a UV-vis spectrophotometer (Model 8453, Agilent Technologies, Palo Alto, CA, USA). The yeast seed at logarithmic phase with an average OD_{600} of approximately 18 was harvested and centrifuged at 5000 rpm, wet solid pellet was collected for fermentation.

2.2. Pretreatment

SPORL pretreatment of Douglas fir in a pilot scale wood pulping digester of 390-L was described previously [10]. The same pretreated FS-10 whole slurry sample labeled as C-t60 was used in this study. The C-t60 was produced by pretreating FS-10 at 140 °C with a liquor to wood ratio (L/W) of 4:1 (L/kg) and a targeted total SO_2 concentration of 80 g/L and a combined (with magnesium oxide) SO_2 loading of 11 g/L in the pretreatment liquor. These chemical loadings were chosen to simulate the chemistry in a commercial sulfite pulp mill in the US that recovers magnesium in operation. The pretreatment lasted 60 min after 32 min of ramping to 140 °C. Liquor circulation provided good mixing in reaction.

The collected solids and neutralized liquor were disk milled together. The measured whole slurries had a total solids content of 21.6%.

2.3. Enzymatic Hydrolysis

Enzymatic hydrolysis of the whole slurry solid (WSS) was carried out in 250 mL Erlenmeyer on a shaking incubator (Thermo Fisher Scientific, Model 4450, Waltham, MA, USA) to evaluate the effect of cellulase loading on WSS saccharification efficiency. Solid calcium carbonate was used to adjust the pH of the whole slurry of total solids of 15% (*w*/*w*) to approximately 5.5. Elevated pH of 5.5 was used to reduce nonproductive cellulase binding to lignin [22,23]. Hydrolysis was conducted at 50 °C and incubator shaking frequency of 200 rpm for 72 h. CTec3 loadings were 5, 10, 20 FPU/g glucan based on glucan in water insoluble solids (WIS). Hydrolysates were sampled periodically. At the end of enzymatic hydrolysis, solid residues were separated by centrifugation at 13,000× *g* for 5 min. Hydrolysates were analyzed for glucose. Duplicate hydrolysis runs were performed. Means and standard deviations were used as error bars in plots.

2.4. Enzymatic Saccharification and Fermentation

Enzymatic saccharification and fermentation experiments of the SPORL whole slurry C-t60 were also carried out in 125 mL Erlenmeyer flasks on the same shaker incubator described in the previous section. The pH of the slurry was again adjusted to 5.5 using solid calcium carbonate. The enzymatic hydrolysis was conducted at 15% total solids loading (*w*/*w*) and with CTec3 loading of 20 FPU/g of glucan. Hydrolysis was carried out at 50 °C and 200 rpm for 72 h, 8 h, and 0 h, respectively, before subsequent separate saccharification and fermentation (SHF), quasi-simultaneous saccharification and fermentation (Q-SSF), and true simultaneous saccharification and fermentation (SSF). No sterilization was applied prior to all these fermentation runs. The cultured yeast broth with OD_{600} = 18 was directly applied to inoculate the enzyme loaded pretreated whole slurry at volumetric loadings of 6, 8, and 10%, corresponding to wet yeast pellet (obtained by centrifuging at 5000 rpm for 20 min) of 1.3, 1.8, and 2.2 g/kg, respectively. The hydrolyzed or liquefied whole slurries were cooled down to 30 °C for SHF and 38 °C for Q-SSF and SSF.

Different enzyme loadings of 5, 10 and 20 FPU/g of glucan in WIS were used to evaluate the effect of enzyme dosage on ethanol production in SSF without pre-hydrolysis for liquefaction of solids.

Control fermentation runs were conducted in SSF mode at CTec3 loading of 20 FPU/g and yeast volumetric loading of 8% without sterilization but with the supplementation of nutrients: Yeast extract = 5 g/L, $(NH4)_2SO_4$ = 2 g/L, NaH_2PO_4 5 = g/L. The effect of without nutrient supplementation and the effect of sterilization were also studied. The whole slurry C-t60 was autoclaved at 121 °C for 15 min prior to SSF fermentation for sterilization. The results were compared with the control run without sterilization.

All fermentation runs were carried out in duplicate. Samples were withdrawn at 4, 12, 24, 48, 72, 96, and 120 h, and centrifuged at 13000× *g* for 5 min. The supernatants were used for sugar and ethanol analyses. Furan concentrations were undetectable. Replicate analyses were conducted and the means and standard deviations of the duplicate fermentation runs were reported. Reported ethanol yield in percent theoretical was calculated based on the glucan, mannan, and xylan content in the untreated *Douglas-fir* residue FS-10 as expressed in the following equation:

$$y_{EtOH\ Theoretical}\ (\%) = \frac{C_{EtOH} \times V_{Broth}}{0.511[\frac{C_{glu}+C_{Man}}{0.9} + \frac{C_{Xyl}}{0.88}]} \times \frac{Y_{Pretrta\ Total\ Solids}}{m_{Pretreated\ Total\ Solids}} \times 100 \quad (1)$$

where C_{EtOH} *and* V_{Broth} are the terminal ethanol concentration (g/L) and the volume (L) of the fermentation broth; C_{Glu}, C_{Man}, C_{Xyl} are glucan, mannan, and xylan content (g/kg), respectively, in untreated woody biomass; $Y_{Pretreated\ Total\ Solids}$ is the yield of total solids (in oven dry weight) from pretreating one kg of woody biomass (g/kg); $m_{Pretrta\ Total\ Solids}$ is the total solids in oven dry weight of the sample used in the fermentation experiment (g).

2.5. Analytical Methods

Untreated and SPORL-pretreated FS-10 samples were first Wiley milled (Model No. 2; Arthur Thomas Co, Philadelphia, PA, USA) to 20 mesh and then hydrolyzed in two stages using sulfuric acid of 72% (*v*/*v*) at 30°C for 1 h and 3.6% (*v*/*v*) at 120°C for 1 h, respectively, as described previously [24]. Carbohydrates of the acid hydrolysates were analyzed by high performance anion exchange chromatography with pulsed amperometric detection (Dionex ICS-5000, Thermo Scientific, Sunnyvale, CA, USA). Klason lignin was quantified gravimetrically.

The enzymatic hydrolysates and fermentation broths were analyzed for monosaccharides, ethanol, furans, and organic acids as described previously [10] using two Dionex HPLC systems (Ultimate 3000, Thermo Scientific, Sunnyvale, CA, USA) equipped with a RI (RI-101) and a UV (VWD-3400RS) detector, respectively. For fast analysis, glucose in the enzymatic hydrolysates was measured using a commercial glucose analyzer (YSI 2700S, YSI Inc., Yellow Springs, OH, USA).

3. Results and Discussion

3.1. Compositional Analysis of SPORL Pretreated Douglas-Fir Residue

The chemical composition of the untreated FS-10 is listed in Table 1 along with component recoveries from the SPORL pretreated washed water insoluble solids (WIS) and sugar and degradation products concentrations in the pretreatment spent liquor. Total of 186.88 kg spent liquor were collected which represent a loss of water of 13.12 kg (200 kg of total water was used including moisture in the pretreatment with liquor to wood ratio of 4:1 L/kg) due to hot blow of the pretreated materials in the reactor at the end of pretreatment.

Table 1. Chemical composition of untreated and SPORL-pretreated unwashed wet solids and freely drainable spent liquor and total component recovery from the whole slurry solids.

	Untreated FS-10	Pretreated FS-10 Wet Solids	Freely Drainable Liquor [1]	Total Recovery (%)
Wet weight (kg)	66.41	126.50	94.35	88.1
Oven dry solids (kg)	50	43.97	7.25	101.0 [2]
Klason lignin (kg)	14.65	8.67	5.27	95.2
Arabinan (kg)	0.51	0.32		
Galatan (kg)	1.00	1.13		
Glucan (kg)	20.49	17.27	0.63 (6.72)	87.4
Mannan (kg)	4.84	3.57	1.67 (17.66)	108.3
Xylan (kg)	2.85	1.76	0.78 (8.29)	89.1

[1] numbers in parenthesis are component concentration in g/L; [2] calculated including the amount of 0.607 kg of $Mg(OH)_2$ applied.

3.2. Evaluation of Pretreatment Effectiveness and Effect of Enzyme Loading on Saccharification

Enzymatic hydrolysis of whole slurry solids (WSS) was conducted to evaluate the effectiveness of SPORL pretreatment for saccharification. Time-dependent substrate enzymatic digestibility (SED) shown in Figure 1 was defined as the percentage of glucan in WSS enzymatically hydrolyzed to glucose. Glucose release with 20 FPU/g glucan CTec3 loading increased sharply with time and reached over 90% within 24h with a glucose titer of 65 g/L and then increased slowly and reached 93% in 72 h with a titer of 67.7 g/L. The glucose release at the end of 72 h hydrolysis were 76.7 and 59.6% with a titer of 55.4 and 43.1 g/L at 48 h at 10 and 5 FPU/g glucan loadings, respectively. These results suggested SPORL pretreatment was effective for maximal saccharification of *Douglas-fir* residue FS-10, in agreement with a previous study that evaluate the water insoluble solids (WIS) of the same pretreated FS-10 [10]. Even with a 4-fold reduction in cellulase loading from 20 to 5 FPU/g glucan, the enzymatic saccharification efficiency remained at approximately 60%. Most of the fermentation experiments, nevertheless, were carried out using cellulase loading of 20 FPU/g glucan to facilitate liquefaction because mixing was done by shaking on a bed without mechanical mixing.

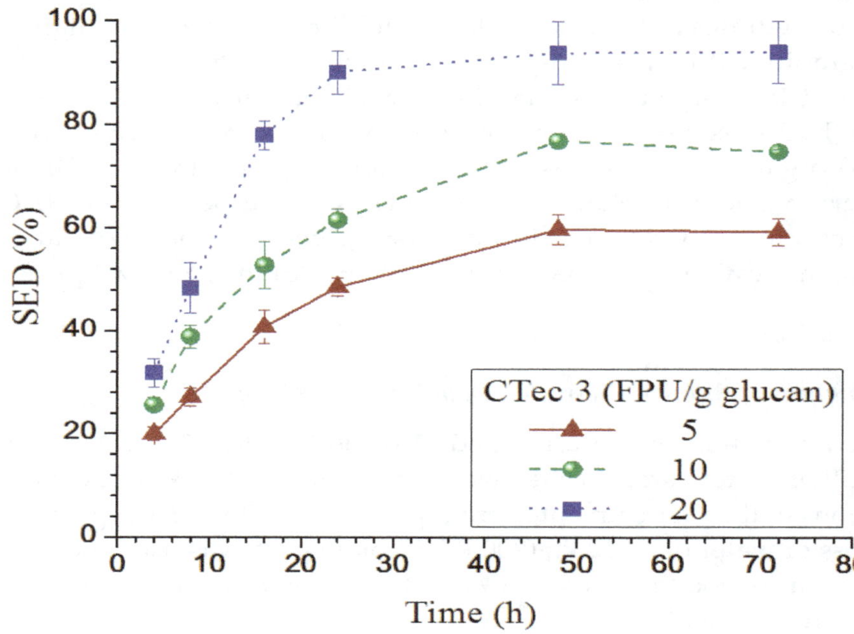

Figure 1. Time-dependent substrate enzymatic digestibility of SPORL-pretreated FS-10 WSS at 15% total solids and three CTec3 loadings.

3.3. *Comprison of Three Different Fermentation Options on Ethanol Production*

Ethanol yield in a traditional SHF was low, probably due to end-product inhibition of the hydrolysis. Glucose inhibition becomes important when glucose concentration reaches over 50 g/L [25]. The concept of performing SSF was proposed in the 1970s [26] to eliminate end product inhibition. Furthermore, there are several additional potential advantages for using SSF, the combination of hydrolysis and fermentation decreases the number of vessels needed and, therefore investment costs. The decrease in capital investment has been estimated to be greater than 20%. This is quite important since the capital costs can be expected to be comparable to the raw material costs in ethanol production from lignocelluloses [27]. In Q-SSF, lignocellulosic slurry was pre-hydrolyzed using enzymes to liquefy solids to produce a certain amount of monomeric and oligomeric sugars prior to fermentation using microorganisms. This approach avoids the disadvantage of low temperature hydrolysis required for yeast survival during liquefaction. The elevated liquefaction temperature, not only improved enzymatic activity, but also reduced the viscosity of the hydrolysate, which facilitates mixing for saccharification and fermentation.

Three fermentation strategies of SHF, Q-SSF and SSF of C-t60 WSS were carried at 15% (*w*/*w*) total solids loading with the supplementation of nutrient. The whole slurry was pre-hydrolyzed for 72 h, 8 h, and 0 h at 50 °C and 200 rpm on a shaking bed, which resulted in glucose concentrations of 67.8, 32.1 and 10 g/L, respectively. The yeast loading was 8% *v*/*v*. The results showed both Q-SSF and SSF produced a higher terminal ethanol concentration than SHF (Figure 2a). An obvious lagging phase in ethanol production can be clearly observed in SHF in the first 12 h, probably due to the high glucose concentration, which resulted in a terminal ethanol concentration of 38.6 g/L at 96 h. Compared with SHF, no visible lagging phase was observed in both Q-SFF and SFF, and ethanol production started immediately after inoculation though slower than SHF between 12 and 24 h, perhaps due to lower glucose concentration. In the first 36 h, Q-SSF had higher ethanol productivity than SSF perhaps due to the availability of more glucose and the reduced viscosity in the Q-SSF slurry due to pre-hydrolysis and liquefaction. However, Q-SSF and SSF reached the same terminal ethanol concentration of 42.3 g/L at 72 h. Glucose concentration continued to increase for the SSF and Q-SSF runs even at the end of 96 h, suggesting continued saccharification and better managing yeast activity or adding more yeast at this stage could further increase ethanol yield. Continued saccharification at the end of fermentation was also observed in a previous study [11].

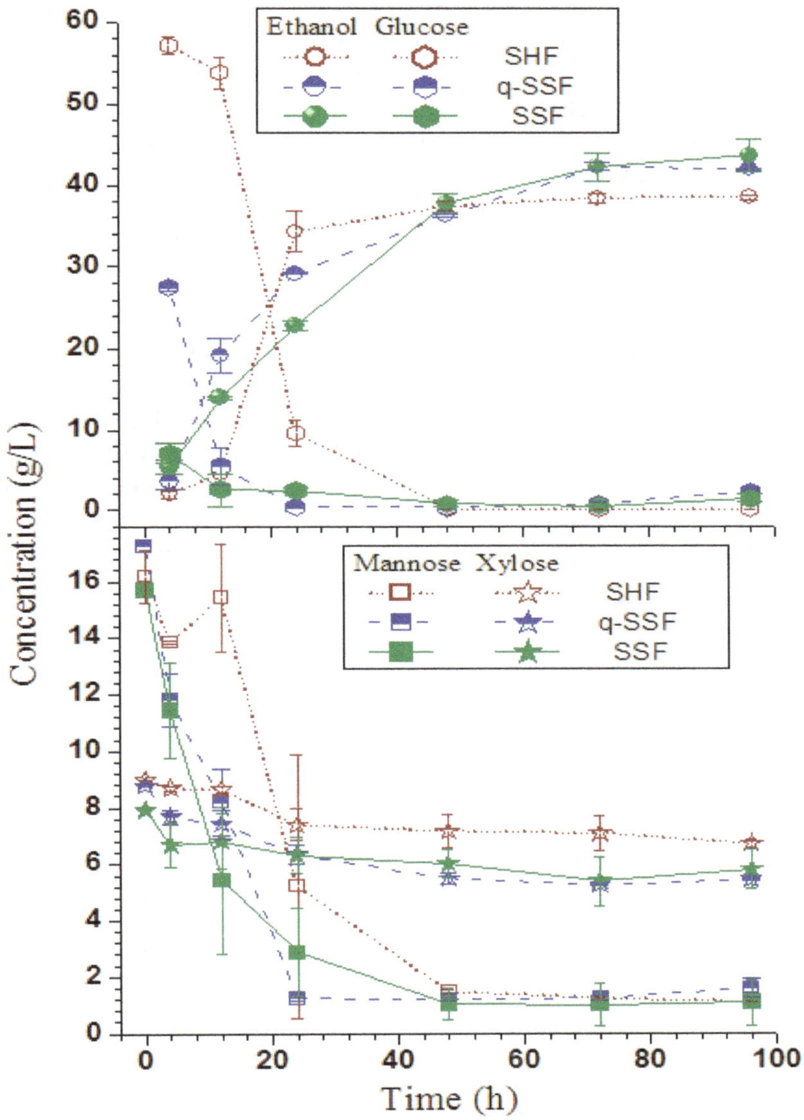

Figure 2. Comparisons of time-dependent fermentation results of SPORL pretreated *Douglas-fir* residue FS-10 among different fermentation options at CTec 3 of 20 FPU/g glucan and 8% (*v*/*v*) yeast loading. (**a**) Ethanol and glucose concentrations; (**b**) Mannose and xylose concentrations.

Time-dependent hemicellulosic sugar consumptions are shown in Figure 2b. Mannose was consumed rapidly by YRH400 in the first 48 h, especially in the SSF and Q-SSF runs. This is probably because glucose concentration was lower in these two fermentations (Figure 2a) than that in the SHF run, which resulted in better utilization of mannose and even xylose to some extent. After 48 h, glucose concentrations approaches zero and hemicellulosic sugar consumptions were pretty much completed in all three runs (Figure 2a,b). It appears that the yeast was starved and under stress, as discussed above, and incapable of consuming hemicellulosic sugars at very low concentrations. The final utilization of mannose in three fermentation runs, however, were very high, over 90%. The final xylose consumption was 38% and 32% in q-SFF and SSF, respectively, higher than the 25% in SHF.

3.4. Comprison of Different Enzyme Loadings in SSF Fermentation

The results of enzymatic hydrolysis of WSS at different enzyme loadings indicated insufficient cellulose saccharification at 5 and 10 FPU/g glucan at 15% loading of WSS (Figure 1). However, enzyme is a very important cost factor in bioethanol production [27]. SSF of WSS was carried out at enzyme loadings of 5, 10 and 20 FPU/g glucan to evaluate the effect of enzyme loading on ethanol production. The results shown in Figure 3a clearly indicated a strong positive correlation between enzyme loading and terminal ethanol yield. Ethanol concentration from 20 FPU/g glucan was much higher than 5 and 10 FPU/g glucan. It is interesting to notice that glucose concentration was highest at the highest enzyme loading of 20 FPU/g glucan during 4 to 24 h, followed by loading of 10 FPU/g glucan, and lowest at 5 FPU/g glucan. This perhaps suggests rapid cellulose saccharification at high cellulase loadings and initial (0–12 h) fermentation may be slower (Figure 2a). Glucose concentration was increased after 96 h at enzyme loadings of 10 FPU/g glucan (Figure 3a) similar to that observed in Figure 2a at 20 FPU/g glucan and in a previous study [11]. This observation indicates that better managing yeast activity can improve ethanol yield. Furthermore, ethanol concentration was not plateaued at the two low CTec 3 loadings of 5 and 10 FPU/g glucan (Figure 3a), suggesting extending fermentation time can improve ethanol yield even at low cellulase loadings. One can balance capital and enzyme costs by using low cellulase loadings with extended fermentation time.

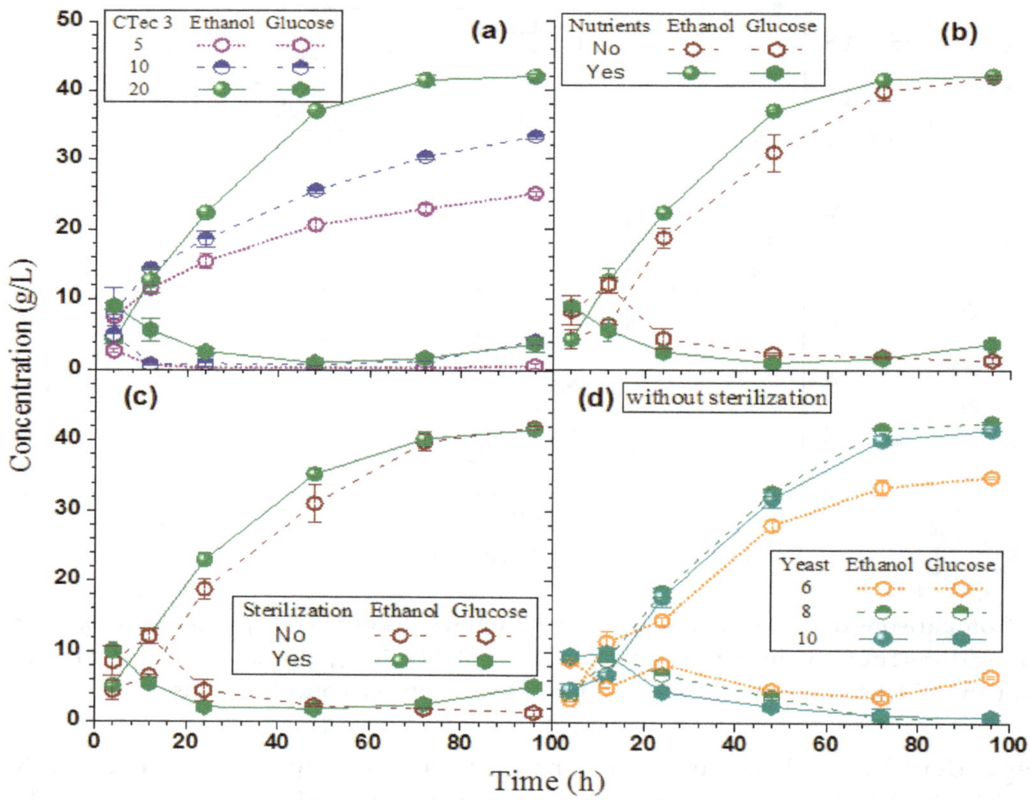

Figure 3. Effects of (**a**) cellulase loading; (**b**) nutrient supplementation; (**c**) sterilization, and (**d**) yeast loadings on time-dependent fermentation results of SPORL pretreated Douglas fir residue FS-10 WSS.

3.5. *Effect of Nutrient Supplementation on the Ethanol Production*

Fermentation runs using SSF were carried out with and without the supplementation of nutrients at CTec3 loading of 20 FPU/g glucan without sterilization. Ethanol production with nutrients supplementation was higher than that without before 96 h (Figure 3b), probably because the yeast needed more time to acclimatize in the media without nutrients. Terminal ethanol production at 96 h, however, was the same (43 g/L) with and without nutrients.

3.6. *Effect of Sterilization*

In general, the SPORL pretreated whole slurry C-t60 was quite sterile. However the pretreated FS-10 sample has been stored in a cold room for quite a few months. It is worth studying whether or not sterilization help ethanol productivity. The SSF results clearly showed a lagging phase within the first 12 h for the fermentation run without sterilization (Figure 3c). Glucose consumption was much faster when the sample was autoclaved at 121 °C for 15 min. This perhaps due to the fact that growth of *Saccharomyces cerevisiae* YRH400 was suppressed by other microorganisms that existed in the sample. After 48 h, these microorganisms may not be able to survive due to elevated ethanol concentrations. The inoculated strain YRH400 became the dominant strain. As a result, terminal ethanol concentration at 96 h was not affected due to the lack of sterilization.

3.7. *Effect of Yeast Loading on the Ethanol Production*

In industrial practice, enzyme and yeast cell concentrations should be appropriately balanced to minimize the costs for yeast and enzyme production. A higher yeast concentration in the SSF can result in a lower overall ethanol yield when the cost for yeast production is taken into account [28]. However, lowering the yeast concentration may low the volumetric productivity, and lead to a stuck fermentation. Three different yeast volumetric loadings of 6%, 8%, and 10% were inoculated in SSF with (not shown) and without (Figure 3d) sterilization. With sterilization, glucose consumption and the ethanol production were slightly slower at a lower yeast loading of 6 *v/v* % in the first 24 h, the ethanol production, however, reached the same concentration of approximately 43 g/L at the two higher yeast loadings after 96 h fermentation. These results are in agreement with a study from the literature using SO_2 steam explosion pretreated spruce [28]. When sterilization was not applied, however, glucose consumption and ethanol production were obviously slower at the lowest volumetric yeast loading of 6% than the two higher loadings of 8% and 10%. The final ethanol concentration at 96 h was 35 g/L at the lower yeast loading of 6%, substantially lower than the approximately 42 g/L at the two higher yeast loadings. This suggests sterilization can suppress other microorganism growth, which allows low yeast loading in ethanol fermentation.

4. Conclusions

The whole slurry solids (WSS) obtained from SPORL-pretreated *Douglas-fir* was evaluated for the production of bioethanol using *Saccharomyces cerevisiae* YRH400. The production of ethanol was investigated in a batch fermentation using modes, i.e., separate hydrolysis and fermentation (SHF), Quasi-simultaneous saccharification and fermentation (Q-SSF), and simultaneous saccharification and fermentation (SSF). Under SSF, the effects of different enzyme and yeast loadings, nutrient supplementation, and sterilization of the substrate on ethanol production were investigated. A final ethanol titer of 43.2 g/L was achieved when SSF was conducted at whole slurry solids loading of 15%, enzyme and yeast loading of 20 FPU/g glucan and 1.8 g/kg yeast loading (wet cell), respectively, with sterilization but without nutrition supplementation.

Acknowledgments: The authors would like to acknowledge the financial support of the Agriculture and Food Research Initiative (AFRI) Competitive grant (No. 2011-68005-30416), USDA National Institute of Food and Agriculture (NIFA) through the Northwest Advanced Renewables Alliance (NARA). The authors also appreciate Novozymes North America for providing the CTec3 enzymes; Fred Matt of USFS-FPL for conducting the chemical composition analysis of forest residue samples; Bruce Dien and Ron Hector of USDA-ARS for providing the

YRH-400 strain. Rolland Gleisner (USDA Forest Products Lab (FPL)) and William Gilles (Formerly with FPL) conducted pilot scale SPORL pretreatment. Funding was also received from The Chinese Scholarship Council for the visiting appointment of Mingyan Yang at the USDA Forest Products Lab.

Author Contributions: Mingyan Yang and J.Y. Zhu conceived and designed the experiments; Mingyan Yang conducted the experiments; Hairui Ji conducted analytical work and calculations. Mingyan Yang and J.Y. Zhu analyzed the results and wrote the paper.

References

1. U.S. DOE. Breaking the Biological Barriers to Cellulosic Ethanol: A Joint Research Agenda. *A Research Road Map Resulting from the Biomass to Biofuel Workshop Sponsored by The Department of Energy*; U.S. Department of Energy: Rockville, MD, USA, 2005.
2. National Research Council. *Renewable Fuel Standard: Potential Economic and Environmental Effects of US Biofuel Policy*; The Natinal Academies Press: Washington, WA, USA, 2011.
3. Gan, J.; Smith, C.T. Availability of logging residues and potential for electricity production and carbon displacement in the USA. *Biomass Bioenerg.* **2006**, *30*, 1011–1020. [CrossRef]
4. Perlack, R.D.; Stokes, B.J.; DOE. *2011. U.S. Billion-Ton Update: Biomass Supply for a Bioenergy and Bioproducts Industry*; Oakridge National Laboratory: Oak Ridge, TN, USA, 2011.
5. Zhu, J.Y.; Pan, X.J. Woody Biomass Pretreatment for Cellulosic Ethanol Production: Technology and Energy Consumption Evaluation. *Bioresour. Technol.* **2010**, *101*, 4992–5002. [CrossRef] [PubMed]
6. Larsson, S.; Palmqvist, E.; Hahn-Hagerdal, B.; Tengborg, C.; Stenberg, K.; Zacchi, G.; Nilvebrant, N.-O. The generation of fermentation inhibitors during dilute acid hydrolysis of softwood. *Enzyme Microb. Technol.* **1999**, *24*, 151–159. [CrossRef]
7. Zhou, H.; Leu, S.-Y.; Wu, X.; Zhu, J.Y.; Gleisner, R.; Yang, D.; Qiu, X.; Horn, E. Comparisons of high titer ethanol production and lignosulfonate properties by SPORL pretreatment of lodgepole pine at two temperatures. *RSC Adv.* **2014**, *4*, 27033–27038. [CrossRef]
8. Zhang, C.; Houtman, C.J.; Zhu, J.Y. Using low temperature to balance enzymatic saccharification and furan formation in SPORL pretreatment of Douglas-fir. *Process Biochem.* **2014**, *49*, 466–473. [CrossRef]
9. Cheng, J.; Leu, S.-Y.; Zhu, J.Y.; Gleisner, R. High titer and yield ethanol production from undetoxified whole slurry of Douglas-fir forest residue using pH-profiling in SPORL. *Biotechnol. Biofuels* **2015**, *8*, 22. [CrossRef] [PubMed]
10. Gu, F.; Gilles, W.; Gleisner, R.; Zhu, J.Y. Fermentative high titer ethanol production from a Douglas-fir forest residue without detoxification using SPORL: High SO_2 loading at a low temperature. *Ind. Biotechnol.* **2016**, *12*, 168–175. [CrossRef]
11. Zhu, J.Y.; Chandra, M.S.; Gu, F.; Gleisner, R.; Reiner, R.; Sessions, J.; Marrs, G.; Gao, J.; Anderson, D. Using sulfite chemistry for robust bioconversion of Douglas-fir forest residue to bioethanol at high titer and lignosulfonate: A pilot-scale evaluation. *Bioresour. Technol.* **2015**, *179*, 390–397. [CrossRef] [PubMed]
12. Zhou, H.; Zhu, J.Y.; Gleisner, R.; Qiu, X.; Horn, E.; Negron, J. Pilot-scale demonstration of SPORL for bioconversion of lodgepole pine to bio-ethanol and lignosulfonate. *Holzforschung* **2016**, *70*, 21–30.
13. Zhou, H.; Lan, T.; Dien, B.S.; Hector, R.E.; Zhu, J.Y. Comparisons of Five *Saccharomyces cerevisiae* strains for Ethanol Production from SPORL Pretreated Lodgepole Pine. *Biotechnol. Prog.* **2014**, *30*, 1076–1083. [CrossRef] [PubMed]
14. Almeida, J.R.M.; Runquist, D.; Sànchez Nogué, V.; Lidén, G.; Gorwa-Grauslund, M.F. Stress-related challenges in pentose fermentation to ethanol by the yeast Saccharomyces cerevisiae. *Biotechnol. J.* **2011**, *6*, 286–299. [CrossRef] [PubMed]
15. Hoyer, K.; Galbe, M.; Zacchi, G. The effect of prehydrolysis and improved mixing on high-solids batch simultaneous saccharification and fermentation of spruce to ethanol. *Process Biochem.* **2013**, *48*, 289–293. [CrossRef]
16. Zhu, J.Y.; Pan, X.J.; Wang, G.S.; Gleisner, R. Sulfite pretreatment (SPORL) for robust enzymatic saccharification of spruce and red pine. *Bioresour. Technol.* **2009**, *100*, 2411–2418. [CrossRef] [PubMed]
17. Leu, S.-Y.; Gleisner, R.; Zhu, J.Y.; Sessions, J.; Marrs, G. Robust Enzymatic Saccharification of a Douglas-fir Forest Harvest Residue by SPORL. *Biomass Bioenerg.* **2013**, *59*, 393–401. [CrossRef]
18. Yamamoto, M.; Niskanen, T.; Iakovlev, M.; Ojamo, H.; van Heiningen, A. The effect of bark on sulfur dioxide-ethanol-water fractionation and enzymatic hydrolysis of forest biomass. *Bioresour. Technol.* **2014**, *167*, 390–397. [CrossRef] [PubMed]

19. Zhang, C.; Zhu, J.Y.; Gleisner, R.; Sessions, J. Fractionation of Forest Residues of Douglas-fir for Fermentable Sugar Production by SPORL Pretreatment. *Bioenerg. Res.* **2012**, *5*, 978–988. [CrossRef]
20. Wood, T.M.; Bhat, M. Methods for Measuring Cellulase Activities. In *Methods in Enzymology*; Colowick, S.P., Kaplan, N.O., Eds.; Academic Press, Inc.: New York, NY, USA, 1988; pp. 87–112.
21. Hector, R.E.; Dien, B.S.; Cotta, M.A.; Qureshi, N. Engineering industrial Saccharomyces cerevisiae strains for xylose fermentation and comparison for switchgrass conversion. *J. Ind. Microbio. Biotechnol.* **2011**, *38*, 1193–1202. [CrossRef] [PubMed]
22. Lan, T.Q.; Lou, H.; Zhu, J.Y. Enzymatic saccharification of lignocelluloses should be conducted at elevated pH 5.2–6.2. *Bioenerg. Res.* **2013**, *6*, 476–485. [CrossRef]
23. Lou, H.; Zhu, J.Y.; Lan, T.Q.; Lai, H.; Qiu, X. pH-induced lignin surface modification to reduce nonspecific cellulase binding and enhance enzymatic saccharification of lignocelluloses. *ChemSusChem* **2013**, *6*, 919–927. [CrossRef] [PubMed]
24. Luo, X.; Gleisner, R.; Tian, S.; Negron, J.; Horn, E.; Pan, X.J.; Zhu, J.Y. Evaluation of mountain beetle infested lodgepole pine for cellulosic ethanol production by SPORL pretreatment. *Ind. Eng. Chem. Res.* **2010**, *49*, 8258–8266. [CrossRef]
25. Holtzapple, M.; Cognata, M.; Shu, Y.; Hendrickson, C. Inhibition of *Trichoderma reesei* cellulase by sugars and solvents. *Biotechnol. Bioeng.* **1990**, *36*, 275–287. [CrossRef] [PubMed]
26. Gauss, W.F.; Suzuki, S.; Takagi, M. Manufacture of Alcohol from Cellulosic Materials Using Plural Ferments. U.S. Patent 3990944 A, 9 November 1976.
27. Olofsson, K.; Bertilsson, M.; Lidén, G. A short review on SSF—An interesting process option for ethanol production from lignocellulosic feedstocks. *Biotechnol. Biofuels* **2008**, *1*, 7. [CrossRef] [PubMed]
28. Sassner, P.; Galbe, M.; Zacchi, G. Bioethanol production based on simultaneous saccharification and fermentation of steam-pretreated Salix at high dry-matter content. *Enzyme Microb. Technol.* **2006**, *39*, 756–762. [CrossRef]

7

The use of CRISPR-Cas9 Genome Editing to Determine the Importance of Glycerol Uptake in Wine Yeast during Icewine Fermentation

Jared Muysson [1], Laurianne Miller [2], Robert Allie [1] and Debra L. Inglis [1,2,3,*]

1 Centre for Biotechnology, Brock University, St. Catharines, ON L2S 3A1, Canada; jared.muysson@gmail.com (J.M.); ra18rs@brocku.ca (R.A.)

2 Department of Biological Sciences, Brock University, St. Catharines, ON L2S 3A1, Canada; 22laurianne@gmail.com

3 Cool Climate Oenology and Viticulture Institute, Brock University, St. Catharines, ON L2S 3A1, Canada

* Correspondence: dinglis@brocku.ca

Abstract: The high concentration of sugars in Icewine juice causes formidable stress for the fermenting *Saccharomyces cerevisiae,* causing cells to lose water and shrink in size. Yeast can combat this stress by increasing the internal concentration of glycerol by activating the high osmolarity glycerol response to synthesize glycerol and by actively transporting glycerol into the cell from the environment. The H^+/glycerol symporter, Stl1p, has been previously characterized as being glucose repressed and inactivated, despite osmotic stress induction. To further investigate the role of Stl1p in Icewine fermentations, we developed a rapid single plasmid CRISPR-Cas9-based genome editing method to construct a strain of the common Icewine yeast, *S. cerevisiae* K1-V1116, that lacks *STL1*. In an Icewine fermentation, the Δ*STL1* strain had reduced fermentation performance, and elevated glycerol and acetic acid production compared to the parent. These results demonstrate that glycerol uptake by Stl1p has a significant role during osmotically challenging Icewine fermentations in K1-V1116 despite potential glucose downregulation.

Keywords: Icewine; *Saccharomyces cerevisiae*; hyperosmotic stress; CRISPR-Cas9; glycerol transport; *STL1*

---

1. Introduction

Icewine is a signature sweet dessert wine that is fermented from the juice of naturally frozen grapes. Since much of the water in the grape is frozen at harvest, the pressed Icewine juice is highly concentrated in solutes such as sugars, organic acids, and nitrogenous compounds. For Icewine juice produced in the Niagara Peninsula of Ontario, Canada, the average sugar concentration for Vidal Icewine juice was reported at 39.3 Brix for 298 Icewine juice samples (approximately 450 g/L sugar) [1]. The high sugar concentration presents a significant osmotic stress for the fermenting *Saccharomyces cerevisiae*. High osmotic stress, commonly referred to as hyperosmotic stress, causes prolonged fermentation times, less than ideal alcohol yields, reduced biomass growth, and elevated levels of undesired metabolites [2,3]. The undesired metabolites, namely acetate, acetaldehyde, and ethyl acetate, indirectly arise because of the increased synthesis of glycerol, the primary compound responsible for the adaptation to hyperosmotic stress [4]. Synthesis of glycerol is a two-step process that involves the reduction and dephosphorylation of the glycolytic intermediate, dihydroxyacetone phosphate (DHAP). Glycerol synthesis is not redox neutral, as one equivalent of NADH (nicotinamide adenine dinucleotide (NAD^+) + hydrogen (H)) is required per synthesized glycerol, resulting in the production of NAD^+ [5,6]. Therefore, a second reaction is required to reduce the generated NAD^+ back to NADH. The NADH generating step of glycolysis, the oxidation of glyceraldehyde 3-phosphate by glyceraldehyde 3-phosphate dehydrogenase, is balanced by coupling this step to alcoholic fermentation, where acetaldehyde is reduced to ethanol by alcohol dehydrogenase, regenerating NAD^+ and making this a redox neutral process. Excess oxidation of NADH through glycerol synthesis can lead to a significant shift in the NAD^+/NADH ratio [7]. Due to the importance of NADH for the reduction of acetaldehyde to ethanol, reduced levels of NADH relative to NAD^+ causes insufficient acetaldehyde reduction, thus leading to elevated levels of acetaldehyde, and subsequently acetate, due to acetaldehyde oxidation by aldehyde dehydrogenases [7–9].

In addition to glycerol synthesis, yeast can also actively uptake glycerol from the extracellular environment [10]. One such transporter, Stl1p, is a plasma membrane H^+/glycerol symporter. The proton-glycerol symport mechanism of Stl1p allows for extracellular glycerol to be imported against a concentration gradient, at the cost of H^+ influx into the cell [11]. As the name *STL* (sugar transporter like) suggests, the protein is one of 34 sugar permeases in yeast and is part of the major facilitator superfamily (MFS) [12]. In laboratory yeast strains of *S. cerevisiae*, *STL1* is reported as upregulated by salt-induced osmotic stress and repressed and inactivated by glucose, but glucose repression was overcome at high temperatures [11,13]. Previous gene expression analysis on the short-term hyperosmotic stress response in wine yeast to grape must inoculation revealed that *STL1* was highly expressed between the first 10 and 30 minutes, before rapid downregulation to near undetectable levels [14]. Another gene expression analysis study found that *STL1* expression upregulation was very low in *S. cerevisiae* wine yeast throughout table wine fermentations and in glucose-containing stress conditions [15]. Conversely, we have previously reported that *STL1* is the most highly upregulated gene in *S. cerevisiae* K1-V1116 during osmotically challenging Icewine fermentations, with a 25-fold increase relative to that measured in table wine fermentation being seen on the fifth day of fermentation [16].

While gene expression analysis is a useful tool for identifying potential gene targets associated with a given phenotype, further investigation is often required to confirm protein function and phenotypic impact. Genetic modification is one such tool that can be used for gene function elucidation. Given the hyperosmotic conditions of Icewine, and the differences between laboratory yeast strains and wine yeast strains, constructing engineered wine yeast strains would be optimal for studying Icewine hyperosmotic stress condition responses [17–19]. While *S. cerevisiae* is amendable with existing genome editing techniques, primarily through the chromosomal integration of selectable markers in haploid auxotrophic strains, applying these techniques to industrially relevant diploid and polyploid strains, like wine yeasts, can be challenging and time-consuming [20,21]. Luckily, with the advent of CRISPR-Cas9 genome editing methods, yeast strain engineering has become rapid, efficient, and multiplexed [22,23]. At the time of this work, there has only been one other published study on the use of CRISPR-Cas9 for genome editing in wine yeasts. The work of Vigentini et al. (2017) used a double plasmid CRISPR-Cas9 method to eliminate *CAN1*, an arginine permease, in the popular wine yeast strains, EC-1118 and AWR1796, thus lowering urea production [24].

In this work, we developed a rapid and effective CRISPR-Cas9 genome editing method for the wine yeast *S. cerevisiae* strain, K1-V1116, which is commonly used for Icewine production. Using a single linearized CRISPR-Cas9 plasmid, that is repaired in vivo with a linear DNA fragment containing the target sequence of the sgRNA, we were able to construct a K1-V1116 strain featuring the complete removal of the *STL1* open reading frame. Subsequent Icewine fermentations were conducted with K1-V1116 and K1-V1116 Δ*STL1* strain. The reduced fermentation performance, along with increased glycerol and acetic acid production per gram of sugar consumed, of the Δ*STL1* strain, provides further evidence that Stl1p contributes to the osmotolerance of K1-V1116 during Icewine fermentations, despite the high concentration of glucose.

2. Materials and Methods

2.1. Yeast Strains and Media

The commercial wine yeast used in this work was *S. cerevisiae* K1-V1116 (Lallemand, Montreal, QC, Canada). K1-V1116 strains were propagated using YPD (0.5% (*w*/*v*) Yeast Extract, 1% (*w*/*v*) Peptone, and 2% (*w*/*v*) Dextrose), unless otherwise stated. For solid media, 1% (*w*/*v*) agar was added. For selection media geneticin (G418) (Teknova, Hollister, CA, USA) was supplemented on YPD plates at 200 μg mL^{-1}.

2.2. Description of CRISPR-Cas9 Method

The CRISPR-Cas9 method used in this work was built upon the work by the Tom Ellis lab [25]. The plasmid used, pWS173, contains cassettes for the yeast expression of both Cas9 and sgRNA (Figure 1). The yeast-optimized Cas9 contains a nuclear localization sequence and is driven by the PGK1 promoter. The optimized sgRNA cassette contains a yeast tRNA and 5′ HDV ribozyme and is flanked by 500 bp junk DNA homology arms. In pWS173, the sgRNA cassette contains a BsmBI-flanked GFP dropout for the 20 bp protospacer sequence required for targeted endonuclease activity. The method relies on the repair of a BsmBI linearized pWS173 with a 1.1 kb linear DNA fragment, which has the sgRNA cassette, with a 20 bp protospacer sequence, inserted within. The sgRNA repair DNA is generated using pWS173 as a template, with two PCR amplicons containing primer overhangs that encode for the protospacer sequence and the terminal sequence of the adjacent fragment. The two protospacer encoding amplicons are then assembled using overlap extension PCR, as described in Section 2.3. When the linearized plasmid, sgRNA repair DNA, and donor DNA are transformed into yeast, the homology arms on the sgRNA repair DNA and pWS173 are recognized and the plasmid is circularized (Figure 2a). The circularization of the plasmid restores its functionality, offering G418 resistance and expression of CRISPR-Cas9 and the active sgRNA species (Figure 2b).

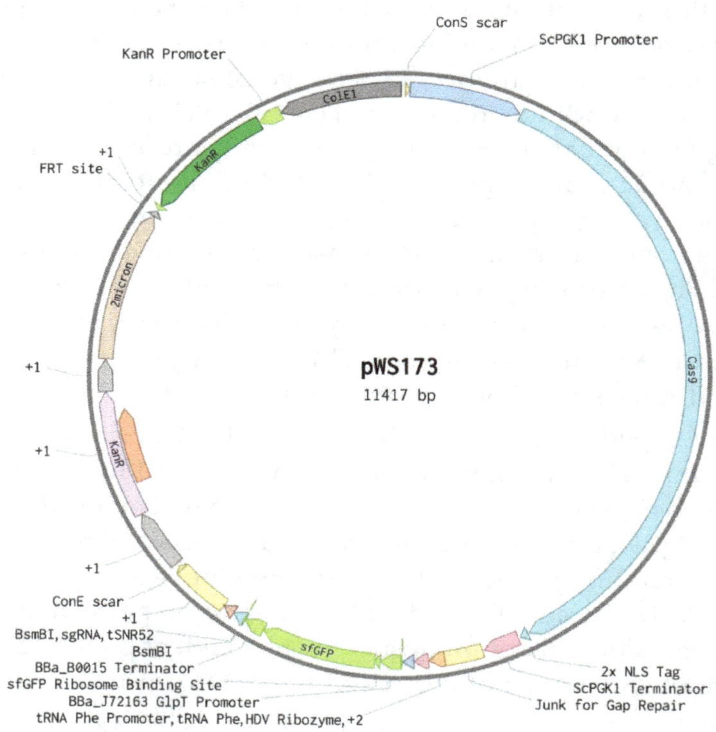

Figure 1. Representation of pWS173 plasmid (Benchling).

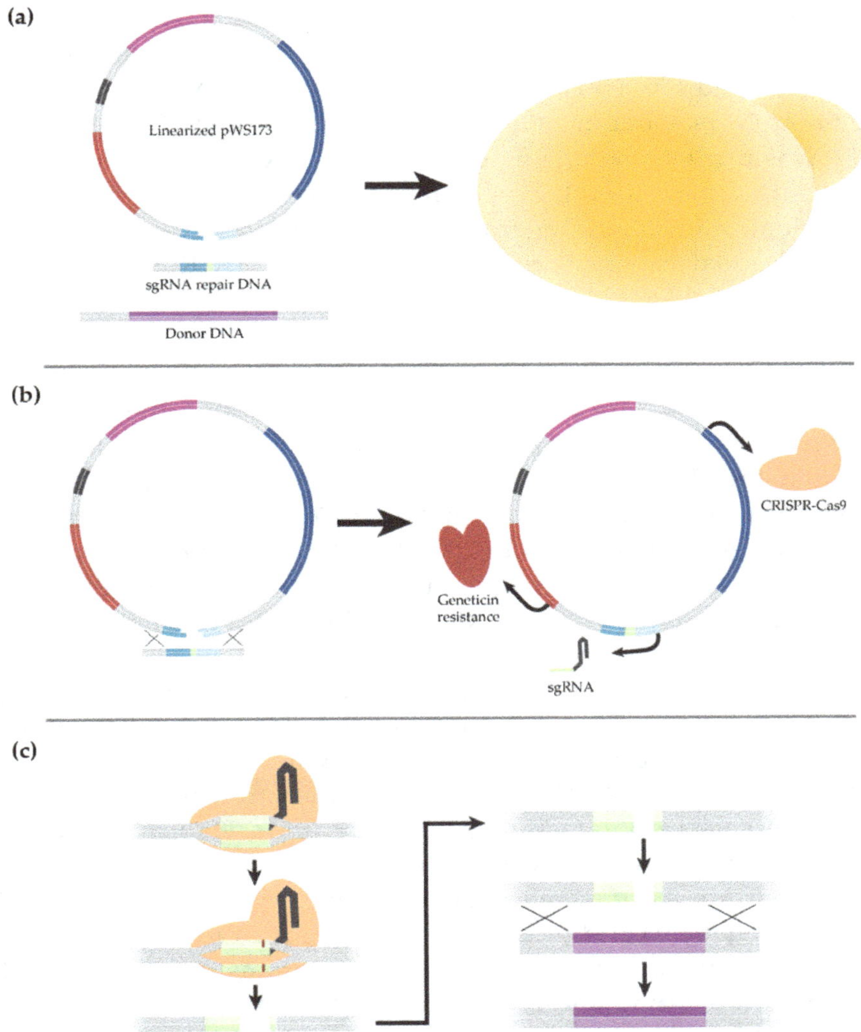

Figure 2. Schematic of DNA components which are transformed in yeast (**a**) and in vivo repair of pWS173, and subsequent expression of CRISPR-Cas9, sgRNA, and G418 resistance cassettes (**b**). The CRISPR-Cas9 ribonucleoprotein recognizes the genomic target sequence and cleaves the site, resulting in a double-stranded DNA break in the genomic DNA. Subsequent homology directed repair with the exogenous donor DNA results in successful genome editing (**c**).

2.3. Cloning

The plasmid used in this work, pWS173, was a gift from Tom Ellis (Addgene plasmid # 90960). Plasmids were propagated in *NEB Stable™ Competent Escherichia coli* (New England Biolabs), under the selection of LB (1% (*w*/*v*) Tryptone, 0.5% (*w*/*v*) Yeast Extract, 1% (*w*/*v*) Sodium Chloride) supplemented with 50 µg mL^{-1} kanamycin (Tocris Biosciences, Bristol, UK). The plasmid was isolated using QIAprep Spin Miniprep (Qiagen, Hilden, Germany), linearized using BsmBI (New England Biolands, Ipswich, MA, USA) and used without gel extraction. The sequences from EC-1118 [26] were used in place of K1-V1116 for all construct and primer design purposes, since the sequence of EC-1118 was available on the Ensembl genome browser. Benchling (San Francisco, CA, USA) was used for plasmid and construct design. The target sequence for CRISPR-Cas9 sgRNA, used in K1-V1116, was designed using the Benchling CRISPR tool, with the default parameters for the NGG PAM single guide RNA for Cas9. The R64-1-1 *S. cerevisiae* reference genome was used for off-target analysis. The target sequence used can be found in Table A1. Donor DNA and sgRNA repair DNA was synthesized using 2-part overlap extension PCR, using Q5 polymerase (New England Biolands, Ipswich, MA, USA) and the primers in Table A2. An initial annealing temperature of 55 °C was used for most PCR reactions, with the temperature adjusted empirically to resolve low yield reactions. An extension time of 15 second per kb and 35 cycles was used for all PCRs. For overlap extension PCR, 0.5 μL of the PCR mix each fragment was used as templates in a 50 μL reaction. Genomic DNA was isolated and purified from K1-V1116 using the Fungi/Yeast Genomic DNA Isolation Kit (Norgen Biotek, Thorold, ON, Canada). PCR amplified DNA was purified using QIAquick Gel Extraction Kit (Qiagen, Hilden, Germany). DNA was quantified using a NanoDrop Lite Spectrophotometer (ThermoFisher, Waltham, MA, USA). PCR amplicons were visualized using 1% agarose-TAE (40 mM Tris, 20 mM acetate, and 1 mM EDTA), containing 0.003% (*v*/*v*) RedSafe nucleic acid staining solution (iNtRON Biotechnology, Seongnam, Korea).

2.4. Strain Construction

The *S. cerevisiae* strain, K1-V1116, was transformed using a modified electroporation protocol, which was largely based off the protocol described by DiCarlo et al. (2013) [27]. A colony of K1-V1116 was inoculated into 5.5 mL of YPD. The culture was then incubated overnight at 30 °C in a roller drum. The following day, the culture was centrifuged at 8000 G for 1 min in a microcentrifuge. The media was removed, and the pellets were resuspended in 250 μL of room-temperature water and consolidated into a single tube. The cells were washed twice more, using 1 mL of water, and then resuspended in 1.5 mL of 500 mM lithium acetate with 5 mM of dithiothreitol (DTT) before being incubated for 30 min at 30 °C in a roller drum. Following conditioning, the cells were collected by centrifugation and washed twice with 1.5 mL of ice-cold electroporation buffer (1 M sorbitol and 1 mM calcium chloride). The final cell pellet was then resuspended in 950 μL of ice-cold electroporation buffer and kept on ice. For each transformation, 100 μL of cells were mixed with linearized pWS173 plasmid, sgRNA repair DNA, and donor DNA. The mixture was then transferred to an ice cold 0.2 cm gap electroporation cuvette and electroporated at 2.5 kV, 25 μF, and 200 Ω. The electroporated cells were recovered in 900 μL of room-temperature recovery media (0.5 M sorbitol and 0.5x YPD) and incubated for 4 h at 30 °C in a roller drum. After recovery, 50 μL of culture was plated onto YPD, supplemented with 200 μg mL^{-1} G418 (geneticin) and incubated at 30 °C for 24 to 48 h. The remaining recovered culture was saved at 4 °C and plated again if required. Integration events were confirmed with colony PCR using Q5 polymerase (New England Biolabs, Ipswich, MA, USA). To obtain colony PCR DNA, a single colony was mixed into 25 μL of 20 mM sodium hydroxide and boiled at 90 °C for 10 min. PCR products were visualized using agarose gel electrophoresis, as described in Section 2.2.

2.5. Icewine Juice

Vidal Icewine juice was previously acquired from Huebel Grape Estates in Niagara-on-the-Lake, Ontario, Canada and stored at −35 °C until use. The Icewine juice was thawed at 7 °C for 24 h, before being racked off and filtered through a series of coarse, medium, and fine pore size pad filters using the Bueno Vino Mini Jet filter (Vineco, St. Catharines, ON, Canada). The juice was then filter sterilized using 0.45 μm membrane cartridge (Millipore, Etobicoke, ON, Canada). The final sterile filtered Icewine juice (40.2 Brix) was used for Icewine fermentations. The sterile filtered Icewine juice contained 447 g L^{-1} sugar, pH 3.79, titratable acidity of 7.0 g L^{-1} tartaric acid, glycerol at 8.67 g L^{-1}, acetic acid at 0.06 g L^{-1} and yeast assimilable nitrogen (YAN) at 441 mg N L^{-1}.

2.6. Permissive Condition Growth Kinetics Characterization

Permissive condition growth kinetics characterization was conducted aerobically in YPD liquid media at 30 °C in an orbital shaker. In a sterile bottle with a stir bar, 250 mL of YPD was added, along with 450 mg of yeast cells from an agar plate, for each strain. This culture was stirred for 5 min on a stir plate and then was sampled into 75 mL aliquots in 250 mL Erlenmeyer flasks with sponge stoppers. The flasks were then placed in an orbital incubator set to 120 RPM and 30° C, and were sampled every two hours. Optical density at 600 nm (OD_{600}) values were obtained using a Genesys 10S UV-Vis Spectrometer (Thermo Scientific, Waltham, MA, USA) set to 600 nm in absorbance mode, with methacrylate cuvettes. Dilutions were made with sterile YPD as needed, to stay under a reading of 1.0. All readings were done in triplicate.

2.7. *Icewine Fermentations and Samplings*

The K1-V1116 strains used for fermentations were prepared as starter cultures. The starter culture consisted of filter sterilized Icewine juice diluted to 10 Brix, with an additional 2 g L^{-1} of diammonium phosphate (DAP). From YPD plates, four freshly cultured yeast colonies were collected and used to inoculate 100 mL of starter culture media. The culture was incubated until the cell concentration reached 2×10^8 cells mL^{-1}, or approximately 17 h, using a shaker table incubator set to 30 °C.

The starter culture was used to inoculate 1 L of Icewine juice to a final cell density of approximately 1×10^7 cells mL^{-1}. The fermentations were incubated at 17 °C for 34 days. Samples were taken every day between days 0 and 6, every other day between days 8 and 16, and every three days between days 19 and 34. Before sampling, the fermentations were stirred for 5 min to ensure homogeneity, after which 2 mL of the culture was taken. After centrifugation, the supernatant of the sample was transferred to a clean tube and stored at −35 °C for later metabolite analysis. The remaining cell pellet was used for total and viable cell counts using methylene blue on a hemocytometer. Larger 50 mL samples were taken from the initial juice and final wine and stored at −35 °C for later analysis. Colony PCR was conducted on the strains at the beginning and end of fermentation to verify the genotype. The fermentations were conducted in triplicate for each strain.

2.8. *Metabolite Analysis*

Soluble solids were determined by ABBE bench top refractometer (model 10450; American Optical, Buffalo, NY, USA). Acidity was determined by pH measurement using a sympHony pH meter (model B10P; VWR, Mississauga, ON, Canada), and titratable acidity by titration against 0.1 mol L^{-1} NaOH, to an endpoint of pH 8.2 [28]. Glucose, fructose, glycerol, acetic acid, amino nitrogen, and ammonia nitrogen were measured using Megazyme assay kits (K-FRUGL, K-ACET, K-GCROL, K-PANOPA, K-AMIAR; Megazyme International Ireland, Ltd., Bray, Co. Wicklow, Ireland). Ethanol was determined by gas chromatography (Model 6890; Agilent Technologies Inc., Palo Alto, CA, USA) equipped with a flame ionization detector (FID), split/split-less injector, and Chemstation software (version E.02.00.493). Separations were carried out with a DB®-WAX (30 m, 0.25 mm, 0.25 μm) GC column (122-7032 model; Agilent Technologies, Santa Clara, CA, USA) with helium as the carrier gas, at a flow rate of 1.5 mL min^{-1}. Metabolite production during fermentation was calculated by the difference in the respective metabolite concentration measured between the time zero point (immediately after inoculation) and at each sampling time point throughout fermentations. Normalized metabolite production was determined by dividing the final metabolite production by the final sugar consumed.

2.9. *Statistical Analysis*

XLSTAT-Pro by Addinsoft (New York, NY, USA) was used for statistical analysis. Analysis of variance (ANOVA) with mean separation by Tukey's Honest Significant Difference ($p < 0.05$) and Student's *t*-Test ($p < 0.05$, $p < 0.01$, $p < 0.001$) were used to evaluate differences between variables.

3. Results

3.1. *Development of a CRISPR-Cas9 Genome Editing Protocol for K1-V1116*

3.1.1. K1-V1116 Drug Sensitivity Testing

To assess the selection potential of Geneticin (G418) for the isolation of successful transformants, the wild type resistance of K1-V1116 needed to be determined. Given the proven effectiveness of G418 in previous studies [23,27], a rapid approach was used to determine wild type sensitivity: 200 μL of stationary phase K1-V1116 cultured in YPD was plated onto a series of YPD plates with G418 (140,

280, 560, 710, and 850 μg mL^{-1}) and incubated at 32 °C. After three days, only one colony appeared on the lowest concentration plate. As such, YPD with 200 μg mL^{-1} G418 was selected as the future selection media.

3.1.2. Initial K1-V1116 Transformations

The following step was to assess the G418 positive selection functionality of the intact and linearized pWS173, as well as validate the electroporation protocol developed. As the goal of this work was to develop a rapid and facile genome editing method, testing the use of stationary phase cultures as the source of cells for electroporation was desired. The use of stationary phase cultures would eliminate the need to monitor the OD_{600}, allow for more schedule flexibility and reduce the hands-on time by over 5 h on the day of transformation. While convenient, the previously reported cost of using stationary phase yeast is reduced transformation efficiency, thus we needed to determine if using stationary phase culture would be a viable strategy for future strain construction [29,30]. Three YPD cultures of K1-V1116 were incubated at 32 °C in a roller drum until log phase (OD_{600} of 1.8), 18 h of growth, and 40 h of growth, until prepared and transformed with either 300 ng of non-linearized pWS173 or 300 ng of linearized pWS173 plus 1 μg of sgRNA repair DNA, which would produce a truncated, non-functional sgRNA (Figure 3).

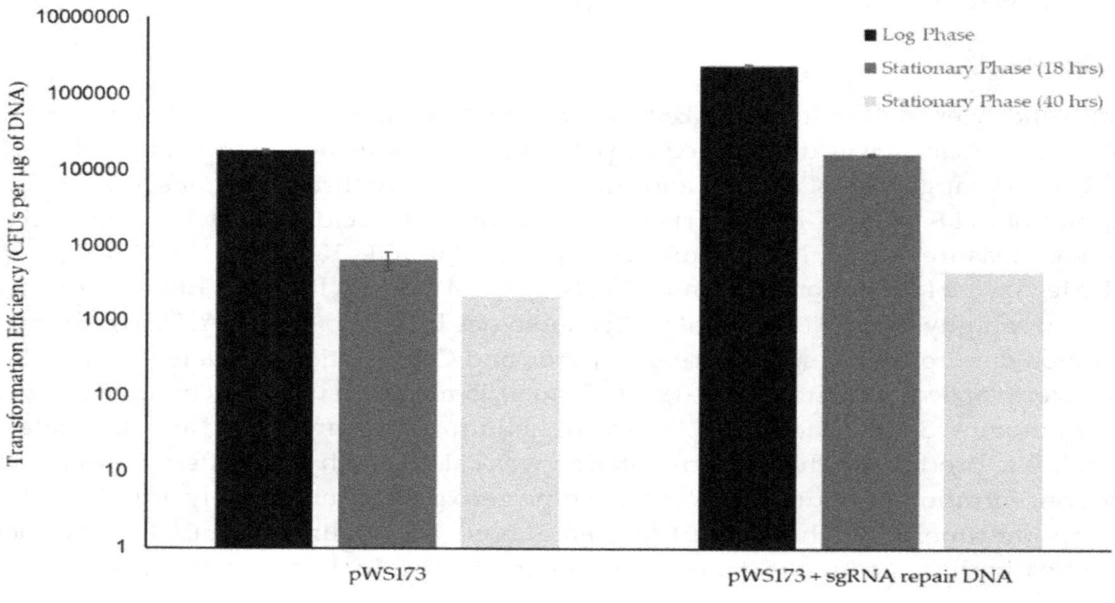

Figure 3. Log transformation efficiency of log phase culture, 18-hour culture, and 40-hour culture, with 300 ng of pWS173 or 300 ng of linearized pWS173, with 1 μg of non-functional sgRNA repair DNA. Single transformations were plated in triplicate on YPD with 200 μg mL^{-1} G418. Transformation efficiency represents the average ± standard deviation.

There was a difference in transformation efficiency between the log phase culture, 18-hour culture, and 40-hour culture. For the repaired pWS173 plasmid, the transformation efficiency of the log phase culture was 14x greater than the 18-hour culture, and 541x greater than the 40-culture. Despite this difference, the 18-hour culture still achieved a transformation efficiency of $1.6 \times 10^5 \pm 5.1 \times 10^3$ CFUs per μg of DNA. The same trend of transformation efficiency between cell culture phases was shown using the intact pWS173. The transformation efficiency measured at each growth phase of the repaired pWS173 appears greater than that observed with the intact pWS173, although we cannot say if this difference is significant due to the experimental design. The transformation efficiency seen in the 18-hour culture was deemed acceptable for future strain construction given the isogenic goal of this work.

3.1.3. Construction of a K1-V1116 ΔSTL1 Strain

Upon validation of the electroporation protocol and the G418 positive selection of the pWS173, K1-V1116 Δ*STL1* was constructed. For this, the sgRNA repair DNA that contained a target sequence within the *STL1* open reading frame was synthesized (Figure 4a). The donor DNA for homology-directed repair contained 500 bp homology arms, and would result in the deletion of the entire *STL1* open reading frame (Figure 4b). K1-V1116 was transformed with 525 ng of linearized pWS173, 1 μg of sgRNA repair DNA, containing a target in *STL1*, and 1.3 μg of donor DNA. The transformation efficiency was 932 ± 62 CFUs per μg of DNA. Of four selected transformants, all four had the genotype which corresponded with the intended integration event (Figure 4c).

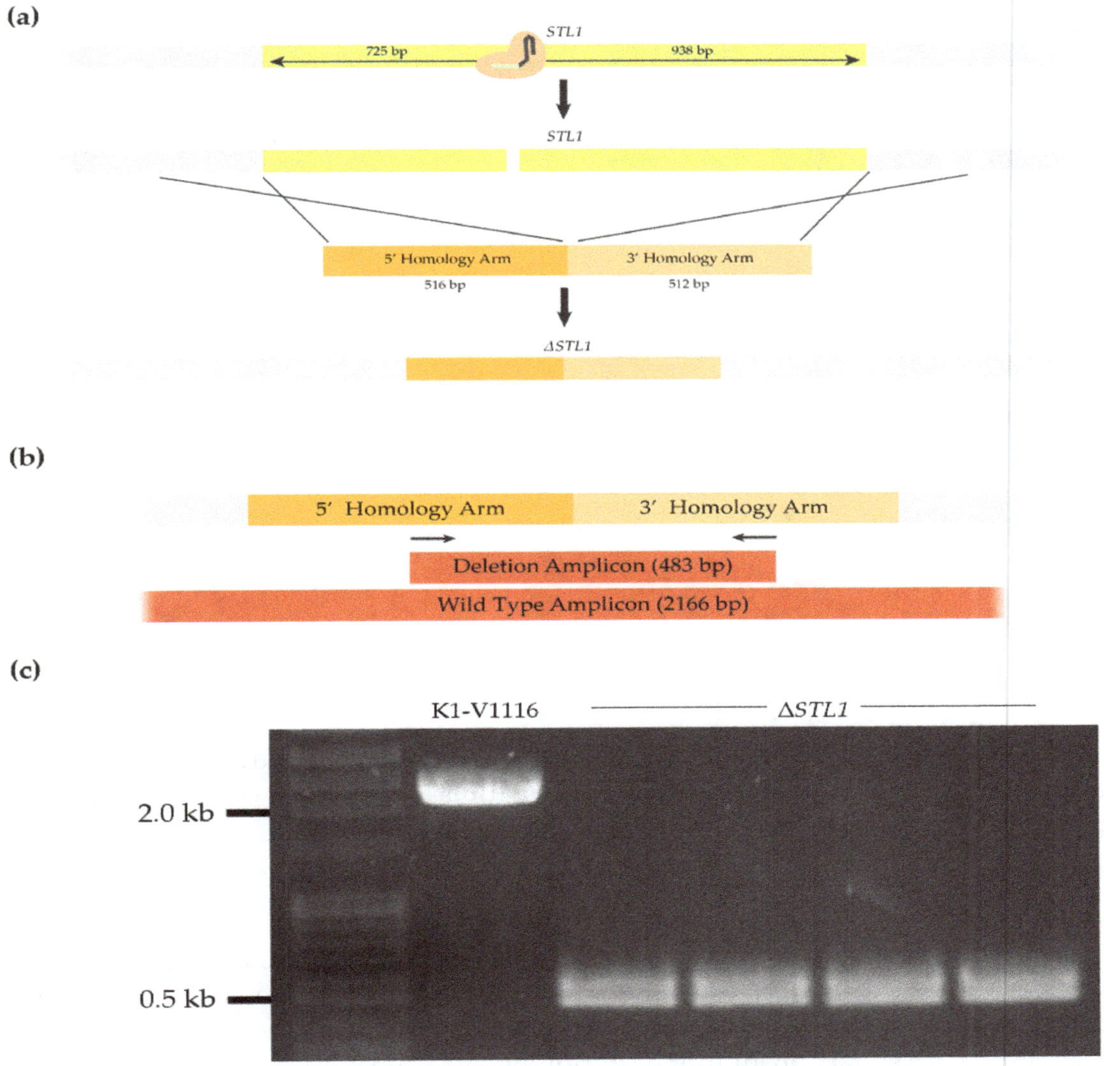

Figure 4. Schematic of *STL1* and mechanism of CRISPR-Cas9 genome edit. CRISPR-Cas9 cleaves at the depicted location, resulting in homology directed repair with the Δ*STL1* donor DNA (**a**). PCR genotype scheme used to validate the integration (**b**). Gel image of *STL1* genotype of K1-V1116 in lane 2 and K1-V1116 Δ*STL1* transformants in lanes 3 to 6 (**c**).

Further testing of an additional 15 transformants revealed that one isolate was a Δ*STL1* heterozygote with an additional two transformants maintaining a wild type genotype, thus resulting in an overall homozygous editing efficiency of 84%. Eight transformants were streaked to single colonies on YPD with G418, followed by two more rounds on YPD, in order to eliminate pWS173. The final K1-V1116 Δ*STL1* strain used for further testing was negative for pWS173.

3.2. ΔSTL1 Growth Characterization in Permissive Conditions and Icewine Fermentations

3.2.1. ΔSTL1 Growth Kinetics in Permissive Conditions

To ensure that the modification to the Δ*STL1* strain did not cause any unintended permissive growth defects, its growth kinetics were characterized in permissive conditions. As *STL1* is reported as having remarkably low expression during conditions containing glucose, especially in absence of significant osmotic stress, this strain was expected to have a similar growth kinetic profile to the parent strain in YPD [11]. The YPD OD_{600} growth curve of K1-V1116 and K1-V1116 Δ*STL1* did reveal similar growth kinetics between the two strains (Figure 5). Interestingly, K1-V1116 Δ*STL1* consistently reached a slightly higher stationary phase OD_{600} value compared to K1-V1116.

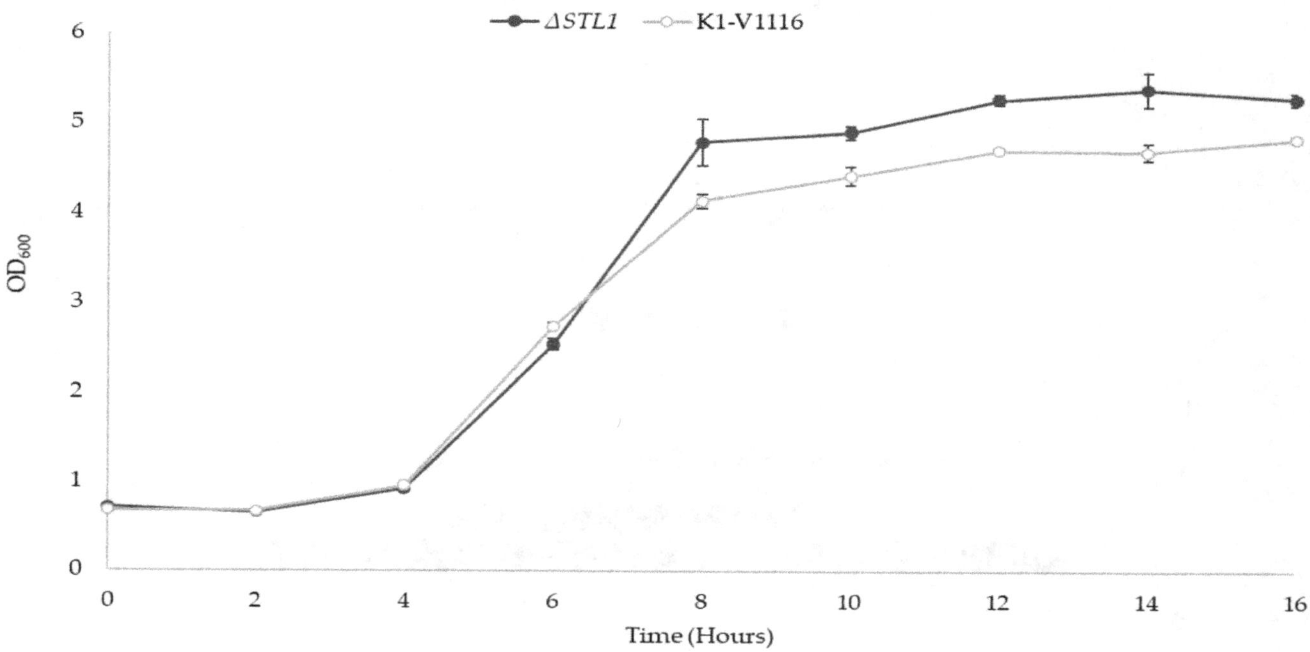

Figure 5. Growth kinetics of K1-V1116 Δ*STL1* (●) and K1-V1116 (○) cultured aerobically in YPD liquid media at 30 °C. Cultures were setup and measured in triplicate. Samples were taken every 2 h, with each point representing a corrected OD_{600} average ± standard deviation.

3.2.2. ΔSTL1 Icewine Fermentation

The fermentation performance of the K1-V1116 Δ*STL1* compared to K1-V1116 was compared in Icewine juice. Within the first 10 days of the fermentation, K1-V1116 Δ*STL1* began to lag in sugar consumption compared to K1-V1116 (Figure 6). The sluggish performance continued for the remainder of the fermentation, with sugar consumption continuing even as K1-V1116 reached a plateau at around day 25.

The total cell concentrations of strains on day 7 revealed that K1-V1116 Δ*STL1* had only reached around 3.5×10^7 cells mL^{-1}, versus the 5.5×10^7 cells mL^{-1} of K1-V1116 (Figure 7a). The reduction in viable cells seen in K1-V1116 later in the fermentation was not as prominent with K1-V1116 Δ*STL1* (Figure 7b).

The average total amount of glycerol and acetic acid produced throughout fermentation by K1-V1116 Δ*STL1* is lower than K1-V1116 but K1-V1116 Δ*STL1* produces more glycerol and acetic acid per gram of sugar consumed (Figure 8). The differences in glycerol production appear in the first few days of fermentation, while the differences in acetic acid production appear after day 5.

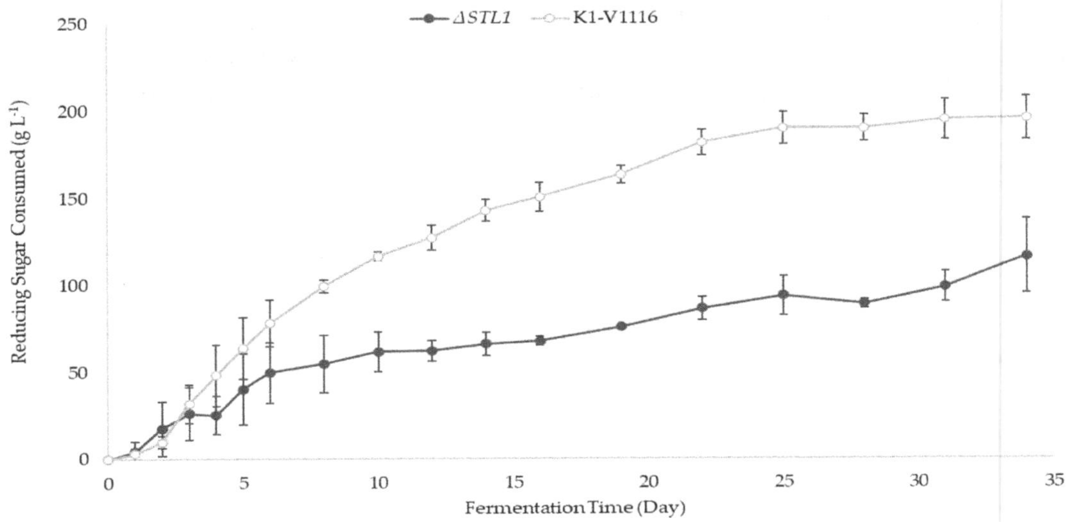

Figure 6. Reducing sugar consumed during fermentation. Fermentations of K1-V1116 Δ*STL1* (●) and K1-V1116 (○) were performed in triplicate, with samples from each trial being analyzed in duplicate. Sugar values represent the average ± standard deviation.

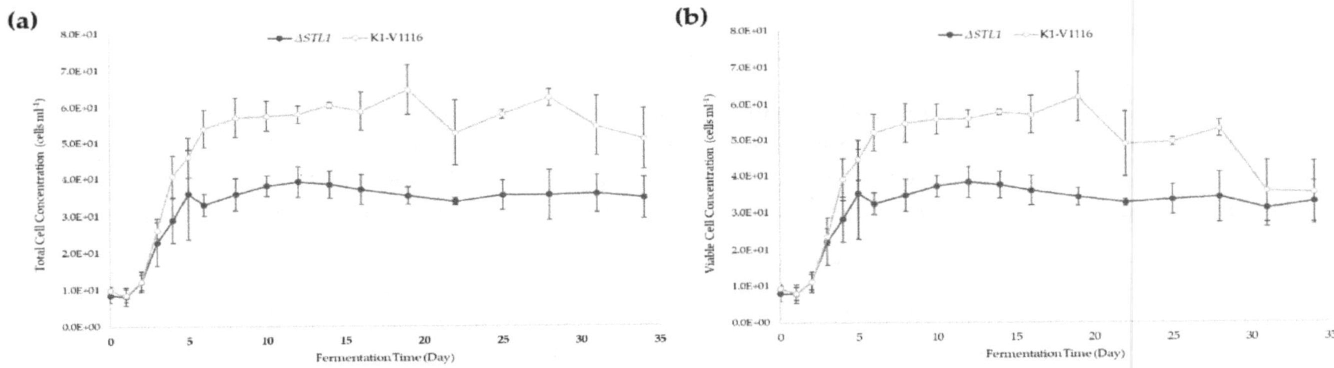

Figure 7. Fermentation total cell concentration (**a**) and viable cell concentration (**b**) of K1-V1116 Δ*STL1* (●) and K1-V1116 (○). Fermentations were performed in triplicate, with counts being determined in duplicate. Cell concentrations represent the average ± standard deviation.

The final wine parameters (Table 1) demonstrate the significant differences between K1-V1116 and K1-V1116 Δ*STL1*. As a result of the difference in reducing sugar consumption, K1-V1116 Icewine had 80% more alcohol than that of K1-V1116 Δ*STL1*. The differences in total glycerol and acetic acid were not statistically significant, despite the differences in the final alcohol and reducing sugar levels.

Table 1. Icewine juice and final wine chemical composition. Data represent the mean value ± standard deviation from triplicate fermentations, with each sample tested in duplicate. Lowercase letters within the same parameter indicate differences between treatments as tested by ANOVA, with mean separation by Tukey's Honest Significant Difference ($HSD_{0.05}$).

Parameter	Icewine Juice	Wine from K1-V1116	Wine from K1-V1116 Δ*STL1*
Reducing Sugars (g L^{-1})	447 ± 19 a	251 ± 11 c	333 ± 19 b
pH	3.79 ± 0.04 b	3.90 ± 0.02 a	3.84 ± 0.08 a,b
Titratable Acidity (g L^{-1})	7.0 ± 0.1 b	8.6 ± 0.2 a	8.6 ± 0.3 a
Yeast Assimilable Nitrogen (mg N L^{-1})	441 ± 2 a	345 ± 13 c	396 ± 14 b
Glycerol (g L^{-1})	8.67 ± 0.32 b	17.42 ± 1.12 a	16.36 ± 0.32 a
Acetic Acid (g L^{-1})	0.06 ± 0.01 b	1.48 ± 0.07 a	1.39 ± 0.05 a
Ethanol (% *v*/*v*)	1.3 ± 0.0 c	12.3 ± 0.6 a	6.7 ± 0.6 b

The final differences in normalized glycerol and acetic acid production were significant, despite the insignificant differences in the total glycerol and acetic acid levels (Table 2). K1-V1116 Δ*STL1* produced nearly 49% more glycerol and 71% more acetic acid per gram of sugar consumed.

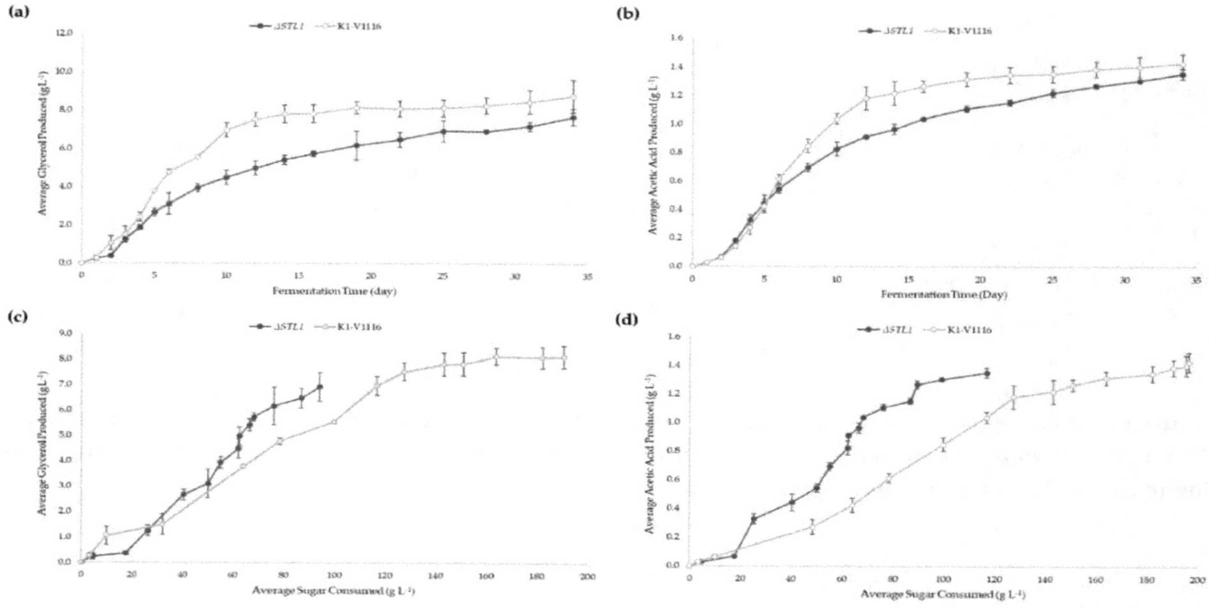

Figure 8. Average glycerol produced (**a**) and average acetic acid produced (**b**) over the time course of the fermentation. Average glycerol produced (**c**) and average acetic acid produced (**d**) versus sugar consumed over the course of the fermentation. Includes K1-V1116 Δ*STL1* (●) and K1-V1116 (○). Values are the averages of triplicate fermentations, with each sample tested in duplicate for metabolite analysis.

Table 2. Consumption of juice components and production of yeast metabolites.

Parameter	K1-V1116	K1-V1116 Δ*STL1*
Reducing Sugar consumed (g L^{-1})	196 ± 13 **	117 ± 21 **
Yeast Assimilable Nitrogen consumed (mg N L^{-1})	92 ± 17 *	44 ± 12 *
Glycerol produced (g L^{-1})	8.76 ± 0.84	7.68 ± 0.44
Acetic Acid produced (g L^{-1})	1.42 ± 0.07	1.35 ± 0.07
Normalized Glycerol production (mg glycerol per g sugar consumed)	45 ± 3 *	67 ± 12 *
Normalized Acetic Acid production (mg acetic acid per g sugar consumed)	7 ± 0 **	12 ± 2 **

Production refers to the difference between the initial Icewine juice and the final wine of each strain. Data represent the mean value ± standard deviation of duplicate measurements per sample (three winemaking replicates per treatment). Asterisks (* $p < 0.05$, ** $p < 0.01$) indicate significant differences between yeast strains (Student's *t*-Test).

4. Discussion

The CRISPR-Cas9 genome editing method developed was effective in modifying the commercial wine yeast strain, K1-V1116, with various protocol optimization strategies. The use of a single plasmid which contains all required elements, except for the protospacer sequence, in conjunction with overlap extension PCR generated sgRNA repair DNA, allows for rapid testing of other loci. While the use of primers and in vitro DNA assembly can lead to mutations, the resulting 1.1 kb sgRNA repair DNA can be rapidly sequenced with Sanger sequencing. Alternatively, commonly used sgRNA repair DNA fragments could be cloned into an *E. coli* vector for sequence verification and maintenance. The use of overlap extension PCR to generate the Δ*STL1* donor DNA was fitting for an edit which is not sensitive to single-nucleotide polymorphisms and indels, but such a cloning method is not appropriate for a sequence sensitive edit, like a gene overexpression cassette. Future work that ventures into sensitive edit types will need to rely on other high-fidelity and verifiable cloning methods.

The initial antibiotic resistance testing of K1-V1116 demonstrated that the strain is no more resistant than previously used strains. We believe the rapid test was sufficient, as notable background growth was not encountered for the remaining transformations. The initial transformation trials, with various K1-V1116 cultures as the source for electrocompetent cells, reinforced the previous findings that use of stationary phase cells was many orders of magnitude less efficient than log phase cells [29,30]. For the isogenic goal of this work, the use of stationary phase yeast for electroporation resulted in acceptable transformation efficiencies and greatly reduced the work required for transformation. Whether future work utilizes stationary phase cells will depend on the goal of the project. Conveniently, the use of the repaired pWS173 and sgRNA repair DNA appears to be more effective than a larger circularized plasmid. This observation indicates that the homologous recombination in K1-V1116 is active and potentially comparable to that of many laboratory *S. cerevisiae* strains [31]. The editing efficiency seen with the Δ*STL1* edit, 84%, was high enough for practical uses, and is within the ranges seen across many CRISPR-Cas9 methods [22].

The Icewine fermentation trials conducted with K1-V1116 Δ*STL1* demonstrated the serious consequences that removing the Stl1p transporter has on fermentation performance. The sluggish fermentation, as indicated by the drastically reduced sugar consumption and total cell count, indicates that cell growth was inhibited by the deletion. The higher level of normalized glycerol production suggests that, since glycerol uptake is disrupted, the cell has an increased reliance on endogenous glycerol synthesis to satisfy demand under the hyperosmotic conditions. With the increased glycerol production per sugar consumed, the normalized acetic acid levels increased as well. Given the current understanding of glycerol synthesis and the NAD^+/NADH ratio during Icewine fermentations, the increased glycerol demand correlates with higher acetic acid levels [4,8,16]. The exact cause of the reduced fermentation performance still needs to be determined. The increased demand on glycerol synthesis may further increase the NAD^+/NADH ratio and disrupt the metabolic flux of the many reactions which rely on NAD^+/NADH [32,33]. The inability to meet the glycerol demand could result in cell shrinkage, thus leading to the crowding of macromolecules and the slowing of cellular processes [34]. The accumulation of the toxic fermentation intermediate, acetaldehyde, may also contribute to the sluggish fermentation [35]. The exact cause of this phenomenon appears to not be acutely toxic to the cells, as the viable cell concentration of K1-V1116 Δ*STL1* was relatively stable throughout stationary phase. The maintained viability, despite the reduced performance, is not abnormal for other known causes of sluggish fermentations [36].

These results, in conjunction with previous findings that *STL1* was the most highly upregulated gene during Icewine fermentations, strongly suggest that Stl1p has a role in maintaining internal glycerol levels during the hyperosmotic stress conditions of an Icewine fermentation [4,16]. The previous studies on *STL1* expression reveal that expression is low when glucose is present, even in the relatively stressful environments of table wine fermentations [11,13–15]. Our findings are similar to recent gene expression analysis research in sugarcane bioethanol strains, which showed that *STL1* glucose repression was overcome in 30 and 35 °Brix conditions [37]. Future work should investigate whether overcoming glucose repression and inactivation at high sugar concentrations is a shared attribute between strains, or if certain strains have evolved the ability to overcome it. Considering many other yeast species rely on glycerol uptake to satisfy osmotic stress glycerol demand, it appears natural for *S. cerevisiae* strains like K1-V1116 to rely more on glycerol uptake during hyperosmotic stress conditions [38,39]. Furthermore, the role of initial glycerol levels in Icewine juice needs to be determined, as reduced starting glycerol levels may negatively impact the effectiveness of Stl1p-mediated osmotolerance.

5. Conclusions

Our work demonstrates that CRISPR-Cas9 genome editing is an effective tool for studying the role of glycerol uptake by Stl1p in the commercial wine yeast, K1-V1116. The CRISPR-Cas9 genome editing method developed allows for rapid strain engineering method by employing time- and cost-saving strategies typically only used with laboratory yeast strains. We found that K1-V1116 is not challenging to engineer despite the lack of genome sequence information and oenological nature. The use of the engineered K1-V1116-lacking *STL1* in Icewine fermentations generated significant data, which support the idea that Stl1p contributes to osmotolerance during the harsh hyperosmotic conditions.

Author Contributions: J.M. and D.L.I. conceived and designed the experiments; J.M. performed the genome editing experiments; R.A. performed the yeast growth curves in YPD, filter sterilized the juice and determined the chemical analysis of the sterile-filtered Icewine juice; L.M. conducted the Icewine fermentations and measured the yeast metabolites and final wine composition; J.M. and D.L.I. contributed to the writing of the manuscript and J.M., D.L.I. and R.A. contributed to the review and editing of the manuscript.

Acknowledgments: We would like to thank the Tom Ellis lab for sharing pWS173 and for assembling an in-depth guide on using their CRISPR/Cas9 method. We appreciate the additional technical support provided by Sudarsana Poojari.

Appendix A

Table A1. CRISPR-Cas9 Target sequence used for STL1 deletion.

Protospacer Sequence	Protospacer Adjacent Motif
GGTAGGAACACTAGACGACG	CGG

Table A2. Primers used in this work.

Primer Name	Sequence
pWS173 sgRNA FW	ATCAACAACAGAGGACATATGCCCTACCTCCATG
pWS173 sgRNA RV	ATCCTGCACTCATCTACTACCCGCATCCC
STL1 sgRNA FW	GCTATTTCTAGCTCTAAAACCGTCGTCTAGTGTTCCTACC AAAGTCCCATTCGCCACCCG
STL1 sgRNA RV	CGGGTGGCGAATGGGACTTTGGTAGGAACACTAGA CGACGGTTTTAGAGCTAGAAATAGC
No target sgRNA FW	GCCTTATTTTAACTTGCTATTTCTAGCTCTAAAACAAAGT CCCATTCGCCACCCG
No target sgRNA RV	GCTGGGCAACACCTTCGGGTGGCGAATGGGACTTTGTTTT AGAGCTAGAAATAGC
STL1 5′ homology arm FW	CGCGGGGTATTCGGCGCTACG
STL1 5′ homology arm RV	GTGTATGAATCAACCCTCAAAATTTGCTTTATCGTGGTCTAA AACTTTCTATGTTCTAC
STL1 3′ homology arm FW	TAAATAAAAAATAGAAAAGTAGAACATAGAAAGTTTTAGACC ACGATAAAGCAAATTTTG
STL1 3′ homology arm RV	GCAGTGTTCGAAAGTTGGTGGCTTTTTGTATGTTGC
STL1 genotyping FW	CAGAATTGCAGTTCAGGAGTAGTCACAC
STL1 genotyping RV	CGCTCCAATGTAATAACACGTCCACGTTATC
pWS173 KanR genotyping FW	GGATGGCAAGATCCTGGTATCGG
pWS173 KanR genotyping RV	ACTGAATCGGGAGGATCTGCTG

References

1. Inglis, D.L.; Pickering, G.P. Vintning on thin ice—The making of Canada's iconic dessert wine. In *The World of Niagara Wine*; Ripmeester, M., Mackintosh, P., Fullerton, C., Eds.; Wilfrid Laurier University Press: Waterloo, ON, Canada, 2013; pp. 229–248.
2. Kontkanen, D.; Inglis, D.L.; Pickering, G.J.; Reynolds, A. Effect of Yeast Inoculation Rate, Acclimatization, and Nutrient Addition on Icewine Fermentation. *Am. J. Enol. Vitic.* **2004**, *55*, 363–370.
3. Pigeau, G.M.; Inglis, D.L. Upregulation of ALD3 and GPD1 in *Saccharomyces cerevisiae* during Icewine fermentation. *J. Appl. Microbiol.* **2005**, *99*, 112–125. [CrossRef] [PubMed]

4. Yang, F.; Heit, C.; Inglis, D.L. Cytosolic Redox Status of Wine Yeast (*Saccharomyces cerevisiae*) under Hyperosmotic Stress during Icewine Fermentation. *Fermentation* **2017**, *3*, 61. [CrossRef]
5. Björkqvist, S.; Ansell, R.; Adler, L.; Lidén, G. Physiological response to anaerobicity of glycerol-3-phosphate dehydrogenase mutants of *Saccharomyces cerevisiae*. *Appl. Environ. Microbiol.* **1997**, *63*, 128–132.
6. Bakker, B.M.; Overkamp, K.M.; van Maris, A.J.A.; Kötter, P.; Luttik, M.A.H.; van Dijken, J.P.; Pronk, J.T. Stoichiometry and compartmentation of NADH metabolism in *Saccharomyces cerevisiae*. *FEMS Microbiol. Rev.* **2001**, *25*, 15–37. [CrossRef]
7. Cronwright, G.R.; Rohwer, J.M.; Prior, B.A. Metabolic Control Analysis of Glycerol Synthesis in *Saccharomyces cerevisiae*. *Appl. Environ. Microbiol.* **2002**, *68*, 4448–4456. [CrossRef]
8. Blomberg, A.; Adler, L. Roles of glycerol and glycerol-3-phosphate dehydrogenase (NAD +) in acquired osmotolerance of *Saccharomyces cerevisiae*. *J. Bacteriol.* **1989**, *171*, 1087–1092. [CrossRef]
9. Aranda, A.; del Olmo, M.L. Response to acetaldehyde stress in the yeast *Saccharomyces cerevisiae* involves a strain-dependent regulation of several ALD genes and is mediated by the general stress response pathway. *Yeast* **2003**, *20*, 747–759. [CrossRef]
10. Hohmann, S. An integrated view on a eukaryotic osmoregulation system. *Curr. Genet.* **2015**, *61*, 373–382. [CrossRef]
11. Ferreira, C.; van Voorst, F.; Martins, A.; Neves, L.; Oliveira, R.; Kielland-Brandt, M.C.; Lucas, C.; Brandt, A. A Member of the Sugar Transporter Family, Stl1p Is the Glycerol/H+ Symporter in *Saccharomyces cerevisiae*. *Mol. Biol. Cell* **2005**, *16*, 2068–2076. [CrossRef]
12. Nelissen, B.; De Wachter, R.; Goffeau, A. Classification of all putative permeases and other membrane plurispanners of the major facilitator superfamily encoded by the complete genome of *Saccharomyces cerevisiae*. *FEMS Microbiol. Rev.* **1997**, *21*, 113–134. [CrossRef] [PubMed]
13. Ferreira, C.; Lucas, C. Glucose repression over *Saccharomyces cerevisiae* glycerol/H+ symporter gene STL1 is overcome by high temperature. *FEBS Lett.* **2007**, *581*, 1923–1927. [CrossRef] [PubMed]
14. Noti, O.; Vaudano, E.; Pessione, E.; Garcia-Moruno, E. Short-term response of different *Saccharomyces cerevisiae* strains to hyperosmotic stress caused by inoculation in grape must: RT-qPCR study and metabolite analysis. *Food Microbiol.* **2015**, *52*, 49–58. [CrossRef] [PubMed]
15. Pérez-Torrado, R.; Oliveira, B.M.; Zemančíková, J.; Sychrová, H.; Querol, A. Alternative Glycerol Balance Strategies among *Saccharomyces* Species in Response to Winemaking Stress. *Front. Microbiol.* **2016**, *7*. [CrossRef]
16. Heit, C.; Martin, S.J.; Yang, F.; Inglis, D.L. Osmoadaptation of wine yeast (*Saccharomyces cerevisiae*) during Icewine fermentation leads to high levels of acetic acid. *J. Appl. Microbiol.* **2018**, *124*, 1506–1520. [CrossRef] [PubMed]
17. Hauser, N.C.; Fellenberg, K.; Gil, R.; Bastuck, S.; Hoheisel, J.D.; Pérez-Ortín, J.E. Whole Genome Analysis of a Wine Yeast Strain. *Comp. Funct. Genomics* **2001**, *2*, 69–79. [CrossRef] [PubMed]
18. Jiménez-Martí, E.; Gomar-Alba, M.; Palacios, A.; Ortiz-Julien, A.; del Olmo, M. Towards an understanding of the adaptation of wine yeasts to must: Relevance of the osmotic stress response. *Appl. Microbiol. Biotechnol.* **2011**, *89*, 1551–1561. [CrossRef]
19. Jiménez-Martí, E.; Zuzuarregui, A.; Gomar-Alba, M.; Gutiérrez, D.; Gil, C.; del Olmo, M. Molecular response of *Saccharomyces cerevisiae* wine and laboratory strains to high sugar stress conditions. *Int. J. Food Microbiol.* **2011**, *145*, 211–220. [CrossRef]
20. Pronk, J.T. Auxotrophic Yeast Strains in Fundamental and Applied Research. *Appl. Environ. Microbiol.* **2002**, *68*, 2095–2100. [CrossRef]
21. Le Borgne, S. Genetic Engineering of Industrial Strains of *Saccharomyces cerevisiae*. In *Recombinant Gene Expression*; Lorence, A., Ed.; Springer: Berlin/Heidelberg, Germany, 2012; pp. 451–465. [CrossRef]
22. Stovicek, V.; Holkenbrink, C.; Borodina, I. CRISPR/Cas system for yeast genome engineering: Advances and applications. *FEMS Yeast Res.* **2017**, *17*. [CrossRef]
23. Zhang, Y.; Wang, J.; Wang, Z.; Zhang, Y.; Shi, S.; Nielsen, J.; Liu, Z. A gRNA-tRNA array for CRISPR-Cas9 based rapid multiplexed genome editing in *Saccharomyces cerevisiae*. *Nat. Commun.* **2019**, *10*, 1–10. [CrossRef] [PubMed]
24. Vigentini, I.; Gebbia, M.; Belotti, A.; Foschino, R.; Roth, F.P. CRISPR/Cas9 System as a Valuable Genome Editing Tool for Wine Yeasts with Application to Decrease Urea Production. *Front. Microbiol.* **2017**, *8*. [CrossRef] [PubMed]

25. Shaw, W. Quick and Easy CRISPR Engineering in Saccharomyces cerevisiae. 2018. Available online: https://www.benchling.com/2018/11/06/quick-and-easy-crispr-engineering-in-saccharomyces-cerevisiae/ (accessed on 10 November 2018).
26. Novo, M.; Bigey, F.; Beyne, E.; Galeote, V.; Gavory, F.; Mallet, S.; Cambon, B.; Legras, J.-L.; Wincker, P.; Casaregola, S.; et al. Eukaryote-to-eukaryote gene transfer events revealed by the genome sequence of the wine yeast *Saccharomyces cerevisiae* EC1118. *Proc. Natl. Acad. Sci. USA* **2009**, *106*, 16333–16338. [CrossRef]
27. DiCarlo, J.E.; Norville, J.E.; Mali, P.; Rios, X.; Aach, J.; Church, G.M. Genome engineering in *Saccharomyces cerevisiae* using CRISPR-Cas systems. *Nucleic Acids Res.* **2013**, *41*, 4336–4343. [CrossRef]
28. Zoecklein, B.W.; Fugelsang, K.C.; Gump, B.H. Laboratory Procedures. In *Wine Analysis and Production*, 1st ed.; Chapman and Hall, International Thomson Publishing: New York, NY, USA, 1995; pp. 374–378.
29. Grey, M.; Brendel, M. A ten-minute protocol for transforming *Saccharomyces cerevisiae* by electroporation. *Curr. Genet.* **1992**, *22*, 335–336. [CrossRef]
30. Tripp, J.D.; Lilley, J.L.; Wood, W.N.; Lewis, L.K. Enhancement of plasmid DNA transformation efficiencies in early stationary phase yeast cell cultures. *Yeast* **2013**, *30*, 191–200. [CrossRef]
31. Kuijpers, N.G.; Solis-Escalante, D.; Bosman, L.; van den Broek, M.; Pronk, J.T.; Daran, J.-M.; Daran-Lapujade, P. A versatile, efficient strategy for assembly of multi-fragment expression vectors in *Saccharomyces cerevisiae* using 60 bp synthetic recombination sequences. *Microb. Cell Factories* **2013**, *12*, 47. [CrossRef]
32. Förster, J.; Famili, I.; Fu, P.; Palsson, B.Ø.; Nielsen, J. Genome-Scale Reconstruction of the *Saccharomyces cerevisiae* Metabolic Network. *Genome Res.* **2003**, *13*, 244–253. [CrossRef]
33. Nielsen, J. It Is All about Metabolic Fluxes. *J. Bacteriol.* **2003**, *185*, 7031–7035. [CrossRef]
34. Miermont, A.; Waharte, F.; Hu, S.; McClean, M.N.; Bottani, S.; Léon, S.; Hersen, P. Severe osmotic compression triggers a slowdown of intracellular signaling, which can be explained by molecular crowding. *Proc. Natl. Acad. Sci. USA* **2013**, *110*, 5725–5730. [CrossRef]
35. Jones, R.P. Biological principles for the effects of ethanol. *Enzym. Microb. Technol.* **1989**, *11*, 130–153. [CrossRef]
36. Bisson, L.F. Stuck and Sluggish Fermentations. *Am. J. Enol. Vitic.* **1999**, *50*, 107–119.
37. Monteiro, B.; Ferraz, P.; Barroca, M.; da Cruz, S.H.; Collins, T.; Lucas, C. Conditions promoting effective very high gravity sugarcane juice fermentation. *Biotechnol. Biofuels* **2018**, *11*, 251. [CrossRef] [PubMed]
38. Lages, F.; Silva-Graça, M.; Lucas, C. Active glycerol uptake is a mechanism underlying halotolerance in yeasts: A study of 42 species. *Microbiology* **1999**, *145*, 2577–2585. [CrossRef]
39. Dušková, M.; Ferreira, C.; Lucas, C.; Sychrová, H. Two glycerol uptake systems contribute to the high osmotolerance of *Zygosaccharomyces rouxii*. *Mol. Microbiol.* **2015**, *97*, 541–559. [CrossRef]

8

Robotic Cell Printing for Constructing Living Yeast Cell Microarrays in Microfluidic Chips

Charlotte Yvanoff [1,2,*], Stefania Torino [1,3] and Ronnie G. Willaert [1,2,4]

1 Research Group Structural Biology Brussels, Vrije Universiteit Brussel, 1050 Brussels, Belgium; Stefania.Torino@vub.vib.be (S.T.); Ronnie.Willaert@uantwerpen.be (R.G.W.)
2 Alliance Research Group VUB-UGent NanoMicrobiology (NAMI), International Joint Research Group VUB-EPFL NanoBiotechnology & NanoMedicine (NANO), Vrije Universiteit Brussel, 1050 Brussels, Belgium
3 Center for Structural Biology, Vlaams Instituut voor Biotechnologie (VIB), 1050 Brussels, Belgium
4 Department Bioscience Engineering, University Antwerp, 2000 Antwerp, Belgium
* Correspondence: Charlotte.Grabielle.Yvanoff@vub.be

Abstract: Living cell microarrays in microfluidic chips allow the non-invasive multiplexed molecular analysis of single cells. Here, we developed a simple and affordable perfusion microfluidic chip containing a living yeast cell array composed of a population of cell variants (green fluorescent protein (GFP)-tagged *Saccharomyces cerevisiae* clones). We combined mechanical patterning in 102 microwells and robotic piezoelectric cell dispensing in the microwells to construct the cell arrays. Robotic yeast cell dispensing of a yeast collection from a multiwell plate to the microfluidic chip microwells was optimized. The developed microfluidic chip and procedure were validated by observing the growth of GFP-tagged yeast clones that are linked to the cell cycle by time-lapse fluorescence microscopy over a few generations. The developed microfluidic technology has the potential to be easily upscaled to a high-density cell array allowing us to perform dynamic proteomics and localizomics experiments.

Keywords: cell printing; piezoelectric dispensing; *Saccharomyces cerevisiae*; GFP-tagged yeast clone collection; living cell microarrays; microfluidic chip; dynamic single-cell analysis

---

1. Introduction

Cell assays have been miniaturized by growing cells in multiwell plates with increasing well number from 96 to 384 and 1536 wells and decreasing well volume from 280 μL down to 3 μL. These experiments are typically integrated in a robotic analysis platform. Major drawbacks of robotic platforms are the expense of the instrumentation, the cost of experimental consumables, closed systems (batch growth), and still relative medium throughput compared to recently developed microfluidic chips. As cell collections grow, further miniaturization of cell assays is needed to increase parallelism of the analyses. Cell microarrays provide an attractive solution, as they could increase the throughput significantly [1–3]. A cellular microarray consists of a solid support wherein small volumes of different biomolecules and cells can be displayed in defined locations, allowing the multiplexed interrogation of living cells and the analysis of cellular responses [4,5]. Living cell microarrays have been combined with microfluidic bioreactors, which provide multiple advantages for multiplex dynamic analyses and high-throughput screening [4,6]. Cellular arrays are emerging as important tools for functional genomics, drug discovery, toxicology, and stem cell research [4,7,8]. A major advantage of cell microarrays over microtiter plates is the opportunity to measure parameters on hundreds of individual single cells and average them, instead of measuring the parameters of a whole cell population. An interesting application where living yeast cell arrays are used is "dynamic proteomics" in which living cell microarrays using a fluorescent protein (e.g. GFP)-tagged yeast clone collection and automated time-lapse microscopy is used for the rapid acquisition of *in vivo* quantitative data about the dynamic proteome and protein subcellular localization ("localizomics"). This enables the identification of members of protein complexes and coregulated proteins, and the unravelling of signaling pathways such as exposure to stress compounds, antimicrobials, or mating conditions [9–13].

Mechanical cell patterning, where mechanical barriers capture the cells at specified spots, has been frequently used to create cellular microarrays [14] in microfluidic chips. Yeast cells have been mechanically patterned in (single-cell) microwells [15,16], microchambers [17,18], and mechanical trap barriers [19–21]. As for the creation of classical DNA microarrays, a fluid-dispensing device can be used to spot or "print" living cells in an array format [4]. Dispensing techniques are categorized as contact and non-contact dispensing [22,23]. Robotic contact printing (e.g. printing DNA microarrays) was initially used to print cells on a semi-solid agar growth medium [24]. Today, mostly non-contact-based devices are used to produce living cellular arrays [25–28]. Here, the fluid is ejected as a flying droplet or jet toward the surface from a short distance. One concept of non-contact printing is based on syringe–solenoid-driven printers, where a reservoir and a high-speed microsolenoid valve are connected to a high-resolution syringe [29,30]. Typically, droplet volumes of 10 to 20 nanoliter are the lower dispensing limit. Another concept is piezoelectrical dispensing, where a technology similar to the one used in an ink-jet printer is used [31,32]. A piezo actuator is fixed around a glass capillary close to the end of the tip. The squeezing of the capillary forced by the piezo actuation induces droplet ejection out of the capillary. The fast response time of the piezoelectric crystal permits fast dispensing rates (kHz range), and the small deflection of the crystal generates droplets from tens of picoliters to a few nanoliters. Robotic cell printing can be used to easily create a living cell microarray composed of a clone collection and at a positional xy accuracy of a few micrometer.

Here, we developed a perfusion microfluidic chip containing a living GFP-clone collection that can be used to perform dynamic proteomics and localizomics experiments and demonstrate its performance. We combined mechanical patterning and robotic cell printing to produce living cell arrays. Soft lithography was used to create a 102-microwell array in the epoxy resin SU-8 (Structured by UV-8) on top of a glass coverslip substrate. A commercial microchannel top plate was used to close the microfluidic chip. Piezo dispensing was optimized for the development of living yeast cell arrays in microfluidic chips. We demonstrated that a clone collection can successfully be printed into the microwell array and yeast cells can be grown in the microwells in continuous mode. Finally, time-lapse fluorescence imaging was performed using 6 selected GFP-tagged clones demonstrating that the approach is suitable to perform dynamic analyses of protein expression and protein localization in living cells.

2. Materials and Methods

2.1. Yeast Strain and Media

Clones from the *S. cerevisiae* Yeast GFP Clone Collection (ThermoFisher Scientific, Waltham, MA, USA) were used [33]. We selected 34 clones linked to the cell cycle (Table S1) and revived them from the cryo-stock on YPD-agar plates (yeast extract 10 g/L, peptone 20 g/L, dextrose 20 g/L, agar 15 g/L). Single colonies were then transferred to liquid cultures in Synthetic Complete (SC) medium supplemented with 2% (*m/v*) glucose. The yeast cells were grown overnight in a shaking incubator at 170 rpm and 30 °C. Prior to cell printing, the cultures were diluted in Phosphate Buffered Saline (PBS) to an optical density at 600 nm (OD_{600}) of 0.5.

2.2. Fabrication of the Microwell Substrate and the Microfluidic Chip

The microfluidic chip consists of two parts: a bottom glass slide with SU-8 microwells and a top plate containing a microfluidic channel with in- and outlet ports (Figure 1). The microwell substrate was produced in-house whereas the top plate was obtained commercially (sticky-Slide I 0.4 Luer, Ibidi, Gräfelfing, Germany). Microwells were fabricated by SU-8 photolithography on glass coverslips. Glass coverslips (75 × 25 mm, thickness of 170 μm; Ibidi, Gräfelfing, Germany) were cleaned in acetone

(Carl Roth, Karlsruhe, Germany) and 2-propanol (Carl Roth, Karlsruhe, Germany) for 15 minutes each, then rinsed with ultrapure water and finally blow-dried. The glass slides were then exposed to air plasma at 100 W, 50 kHz, for 5 minutes (Plasma System Cute, Femto Science, Dongtangiheung-Ro, Korea). The negative photoresist SU-8 2050 (Kayaku Advanced Materials, Westborough, MA, USA) was spin-coated onto the glass slides at 3500 rpm for 30 seconds in order to reach an approximative thickness of 50 μm. Next, the SU-8 was soft-baked for six minutes at 95 °C on a hot plate. The SU-8 slides were then aligned with the photomask (film photomask, Selba, Versoix, Switzerland) in the mask aligner UV-KUB3 (Kloé, Saint-Mathieu-de-Tréviers, France) and illuminated with 365 nm ultraviolet (UV) light (intensity of 35 mW/cm^2) for 10 seconds. Following UV exposure, the SU-8 slides were post-baked for six minutes at 95 °C on a hot plate. Finally, the slides were treated with SU-8 developer for seven minutes and next washed with 2-propanol and blow-dried. The dimensions of the wells were then evaluated by optical microscopy (Nikon Eclipse Ti2, Nikon, Tokyo, Japan) with a 10x objective (for the diameter and pitch distance) and with a 3D profilometer (Profilm 3D, Filmetrics, San Diego, CA, USA) for the well's depth. The microfluidic chip was assembled by pressing the sticky top plate (sticky-Slide I 0.4 Luer, Ibidi, Gräfelfing, Germany) to the microwell substrate.

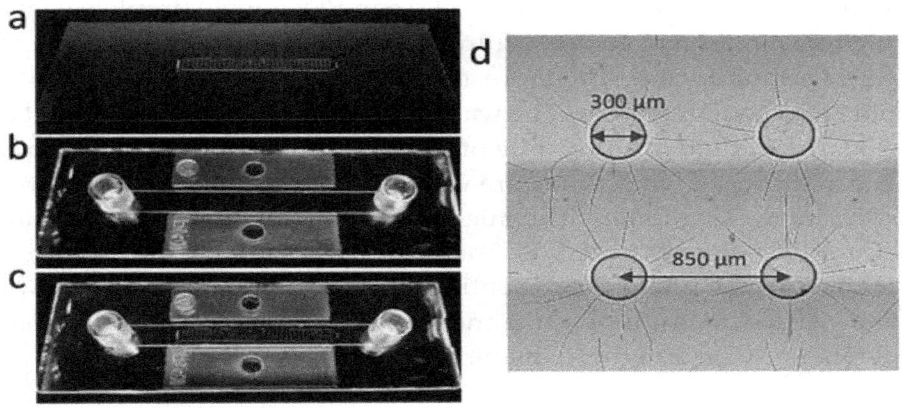

Figure 1. Construction of the microfluidic chip containing the cell microarray. (**a**) The bottom substrate: the SU-8 microwell array on the glass coverslip. (**b**) The top plate containing the channel, inlet and outlet (Ibidi, Gräfelfing, Germany). (**c**) The integrated microfluidic chip obtained by sticking the top plate to the bottom plate using double-sided sticky tape. (**d**) The microwells constructed by SU-8 UV-photolithography.

2.3. Cell Printing

Cell printing was performed using a non-contact iTWO-400 dispenser (M2 Automation, Berlin, Germany) (Figure S2a). The printer is established in an environmental enclosure for live cell printing that contains a HEPA filter and recirculating air is sterilized with a UV lamp. The environmental temperature can be controlled as well as the relative humidity. Moreover, the instrument deck contains a cooling system for source plates and target plates with dew point control, which prevents evaporation during spotting. Samples to be printed are manually dispensed in 384-multiwell plates (ShallowWell 384-multiwell plate, Thermo Fischer Scientific, Waltham, MA, USA), which are then mounted onto the "source plate" locations of the robot (Figure S2a). Likewise, the substrate to be printed is mounted onto the "target plate" locations of the robot (Figure S2a). The actual printing is performed with a piezo dispenser made of a borosilicate glass capillary surrounded by a piezo ceramic actuator (PDMD, M2 automation, Berlin, Germany), which is able to shoot pico- to nanoliter droplets at high frequency with high volume and position accuracies. Finally, the instrument deck is also equipped with a wash station, which enables to clean the tip of the piezo dispenser after sample aspiration and sample printing, in order to avoid cross-contamination. A typical printing run consists in aspirating the sample to be printed from the "source plate", washing the outside of the tip to get rid of contaminants that could impair the shooting, shooting the sample at its desired position on the "target plate" and finally dispensing the remaining sample at the washing station and washing the outside of the tip so that it is cleaned for the next sample. Each step of this procedure can be specified in the software controlling the robot. Additionally, the robot is equipped with a camera annexed to the piezo dispenser, which enables us to verify the successful printing of the samples.

Ultrapure water was printed with a pulse duration of 15 µs and an amplitude voltage of 75 V. The environmental temperature was around 28 °C, the relative humidity was controlled at 50% and the temperature of the target plate was at 15 °C. These parameters were maintained throughout each of the cell printing experiments. For optimization purposes, ultrapure water was printed on standard glass coverslips (24 × 24 × 0.17 mm) and in microwells of commercial nanotiter plates (Microfluidic ChipShop, Jena, Germany) made of the cyclo-olefin copolymer material Topas.

Prior to cell printing, the SU-8 microwells were coated with a solution of concanavalin A (Con A) (Con A from *Canavalia ensiformis*, Sigma Aldrich, Overijse, Belgium) at 2 mg/mL in H_2O, with 5 mM $CaCl_2$ and 5 mM $MnSO_4$. Sixty drops of Con A were printed per well with the parameters mentioned above and were left to incubate for 15 minutes before being allowed to dry.

Yeast suspensions were prepared at an OD_{600} of 0.5 in PBS and 40 droplets were printed into each well with the parameters mentioned above. Multiple yeast suspensions were successively printed in different wells of the microfluidic chip. Therefore, the piezo dispenser was thoroughly washed between each sample to avoid cross-contamination. The washing procedure consisted in first discarding the old sample by ejecting 30 µL of liquid at a flow rate of 30 µL/s using the syringe pump and next, flushing the outside of the piezo dispenser with ultrapure water for five seconds at the wash station. The yeast cells were left to sediment for 10 minutes. The microfluidic chip was closed and connected to a syringe filled with SC medium via silicone tubing. SC medium was gently perfused into the channel of the microfluidic chip with a syringe pump (KD Scientific, Holliston, MA, USA) at a flow rate of 25 µL/min. Finally, the complete set-up (microfluidic chip and syringe pump) was installed on a microscope for direct imaging or kept at 4 °C overnight to image the next day.

2.4. Microscopy

The microfluidic chip and syringe pump set-up were installed on a Nikon Eclipse Ti2 epifluorescence microscope (Nikon, Tokyo, Japan) for time-lapse imaging. The microfluidic chip was inserted into a temperature-controlled chamber (Ibidi, Gräfelfing, Germany), which was mounted onto an automated scanning stage (ProScan III, Prior Scientific Instruments, Cambridge, United Kingdom). The syringe pump was placed next to the microscope and the syringe was covered with a syringe heater (New Era Pump Systems, Farmingdale, NY, USA). Both the temperature controller and the syringe heater were set at 30 °C for the duration of the time-lapse experiments.

For the growth experiment with 34 clones, bright field images of yeasts were acquired every 20 minutes, for 18 hours with a 20x objective. At the final time point (18 h), the yeasts were also imaged with a 60x objective in bright field and fluorescence. The GFP fluorescence was observed by exciting the sample with a LED light source (pE-300^{white}, CoolLED, Andover, United Kingdom) and detecting it through a FITC filter. For the growth experiment with six clones, the yeasts were imaged in bright field and fluorescence and images were recorded every 30 minutes for three hours using a 60x objective.

2.5. Image Processing

Images acquired by the camera of the iTWO-400 dispenser were post-processed using Fiji [34]. More precisely, they were stitched together to form the pictures shown in Figure 2, Figure 3, and Figure 4 using the "Grid/Collection Stitching" plugin [35]. The bright field and GFP-fluorescent pictures were also processed with Fiji for background correction and manual stack alignment.

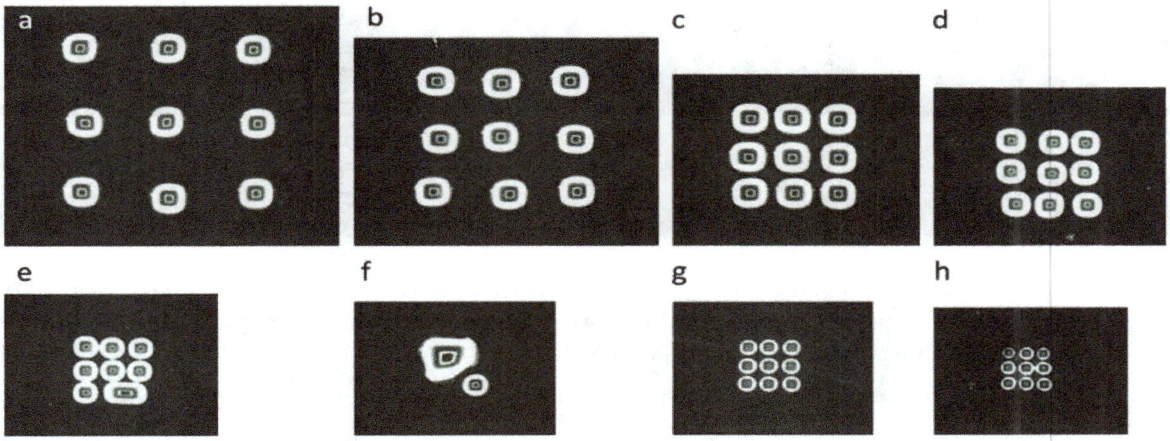

Figure 2. Optimization of the printing on a glass substrate. Water droplets were piezo dispensed (pulse duration of 15 μs and voltage of 75 V) as a 3 × 3 array on a glass coverslip: (**a**) 100 droplets at a frequency of 50 Hz with a pitch of 1000 μm; (**b**) 100 droplets at a frequency of 50 Hz with a pitch of 750 μm; (**c**) 100 droplets at a frequency of 50 Hz with a pitch of 500 μm; (**d**) 50 droplets at a frequency of 25 Hz with a pitch of 400 μm; (**e**) 25 droplets at a frequency of 25 Hz with a pitch of 300 μm; (**f**) 25 droplets at a frequency of 25 Hz with a pitch of 250 μm; (**g**) 10 droplets at a frequency of 5 Hz with a pitch of 250 μm; (**h**) five droplets at a frequency of 4 Hz with a pitch of 200 μm.

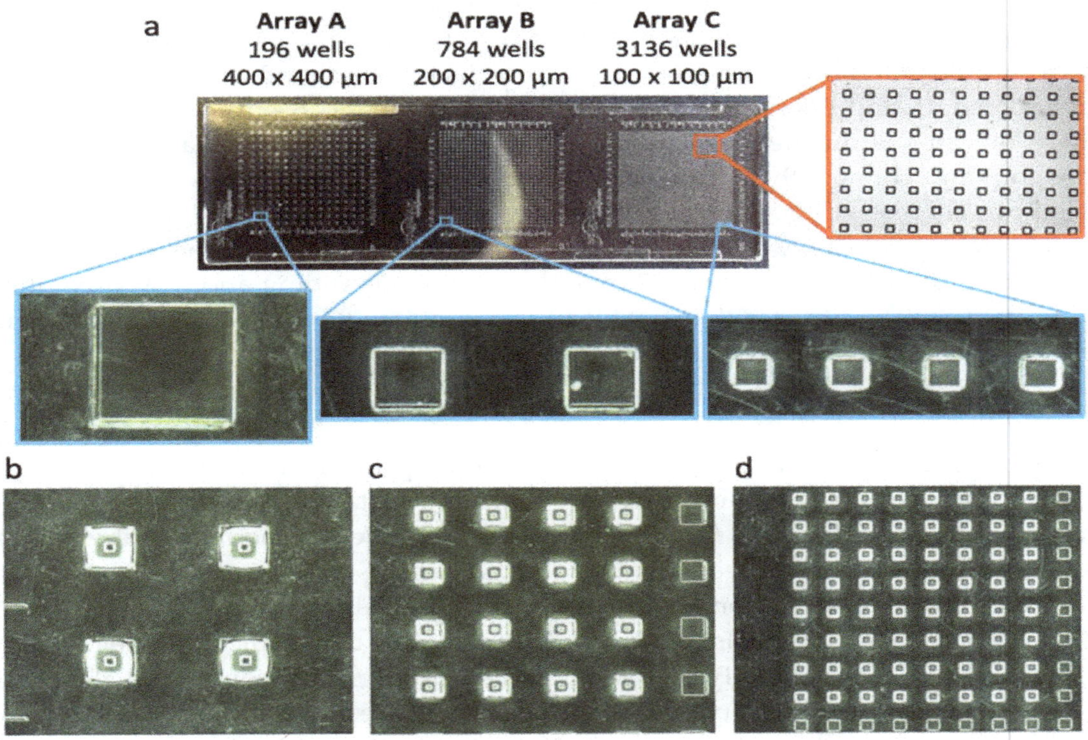

Figure 3. Optimization of piezo printing into microwells. (**a**) Commercial microtiter plate (Microfluidic ChipShop, Jena, Germany) containing three microwell arrays with square microwells of varying width; array A: width of 400 μm and pitch of 1125 μm, array B: width of 200 μm and pitch of 563 μm, array C: width of 100 μm and pitch of 281 μm. Water droplets were piezo dispensed (pulse duration of 15 μs, amplitude voltage of 75 V and frequency of 50 Hz) as (**b**) a 2 × 2 array in array A wells at 200 droplets/well, (**c**) a 4 × 4 array in array B wells at 25 droplets/well, (**d**) a 8 × 8 array in array C wells at five droplets/well.

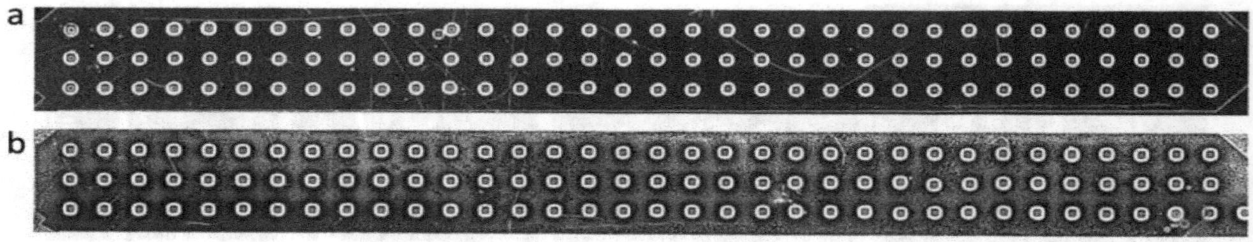

Figure 4. Con A and yeast cells dispensing into the microwells of the microfluidic ship (pulse duration of 15 μs, voltage of 75 V and frequency of 50 Hz): (**a**) 60 droplets of Con A were printed, (**b**) 40 droplets were printed from the yeast solution (OD_{600} of 0.5).

3. Results

3.1. Construction of the Microfluidic Chip

The microfluidic chip (Figure 1c) was made of two parts: the microwell array on a glass slide (Figure 1a) and the top plate with a microfluidic channel (Figure 1b). The microwell array was produced in the epoxy-based photoresist SU-8 through a standard photolithography protocol. The mask design of the microwell array consisted of three rows of 34 wells with a well diameter of 300 μm and a pitch distance of 850 μm (Figure S1). Visual inspection of the microwells after photolithography showed that the SU-8 wells matched the mask's dimensions (Figure 1d). Furthermore, the measured well's depth was approximatively 50 μm (Figure S1b), which fits the expected thickness of the SU-8 layer, according to the protocol mentioned earlier. Altogether, these results prove the photolithography process to be successful and a quick method to produce the microwells in the microfluidic chip.

The overall length and thickness of the microwell array were 28.4 mm and 2.0 mm, respectively. These dimensions fit within the microfluidic channel of the top plate that was used to seal the microfluidic chip and connect the inlet and outlet tubes. The top plate was commercially available and consisted of a channel of 50 × 5 × 0.4 mm (l × w × h) surrounded by a double-sided tape, which enabled us to quickly and easily stick it onto the bottom substrate. Once the microfluidic chip was sealed (after filling with the cells), it was connected to silicone tubing and a syringe filled with SC growth medium. The medium was then perfused at a very low flow rate (25 μL/min) with a syringe pump in order to ensure nutrient renewal for cell growth.

3.2. Living-Cell Microarray Development

First, piezo dispensing parameters were optimized by printing a water droplet array on a glass substrate. Piezo dispensing is accomplished by applying rectangular voltage pulses to the piezo ceramic actuator (Figure S2c). For the duration of each pulse, the actuator tube contracts a few micrometers, thereby initiating a pressure wave that causes the ejection of a droplet. The amplitude and the duration of the pulse influence the volume and the velocity of the droplet. The higher the amplitude (applied voltage) and the longer the pulse, the bigger the droplet and the higher its velocity. The frequency of the pulses is directly correlated to the duration of the pulse and also influences the droplet's volume. Finally, the viscosity of the sample to dispense also has an influence on the printing parameters. Therefore, piezo dispensing needs to be optimized for each solution that has to be printed. In this work, yeast suspensions were prepared in PBS prior to printing.

Considering that PBS solution and ultrapure water have similar viscosities; we first optimized the piezo dispensing parameters with ultrapure water for simplicity reason. Typically, we started by determining the amplitude and duration of the voltage pulses that were giving a reliable ejection of water droplets. We obtained a stable shooting of ultrapure water with an amplitude of 75 V and a pulse duration of 15 μs. These parameters resulted in a droplet volume ranging between 70 to 80 pL (Figure S2b). Next, we created 3 × 3 arrays of water droplets with decreasing pitch distance on a glass substrate (Figure 2). We started by shooting 100 drops/spot at 50 Hz with a pitch distance of 1000 μm, 750 μm and 500 μm. We could not print arrays with pitch distances lower than 500 μm without merging of neighboring spots. Hence, we reduced the droplet number and printed 50 drops/spot at 25 Hz with a decreasing pitch distance from 400 to 300 and 250 μm. Surprisingly, printing half the droplet number did not allow halving the pitch distance as expected. Additional testing established that five drops/spot at 4 Hz with a pitch distance of 200 μm were the limit parameters at which a 3×3 array of water droplets could be stably printed on a glass slide. This optimization process aimed at determining the minimal spot sizes and pitch distances that can be stably printed in order to increase the throughput of the platform.

Secondly, we evaluated the filling of microwells with water droplets using piezo dispensing. Therefore, we initially used a commercially available microtiter plate (size of a microscope slide) that contains squared microwells with dimensions of 400×400 μm (array A), 200×200 μm (array B), 100×100 μm (array C), and height of 20 μm (Figure 3). The pulse duration (15 μs) and amplitude voltage (75 V) were as for printing on the glass substrate. With a frequency of 50 Hz, the number of droplets decreased from 200 drops per well for array A to 25 drops per well for array B and five drops per well for array C.

Once the piezo-dispensing parameters were established with ultrapure water, the printing protocol was optimized for living cells. First, the microwells were filled with 60 droplets of Con A at a pulse duration of 15 μs, amplitude voltage of 75 V and frequency of 50 Hz (Figure 4a). Con A is a lectin, which binds to the mannose glycans at the yeast cell wall [36]. Hence, Con A was used as a coating, which anchored the yeast cells to the bottom of the microwells so that they were not flushed away when SC medium was perfused during continuous cultivation. Next, the density of printed yeast cells per well was optimized. The final objective of this study consisted in developing a method that allowed monitoring the growth of single yeast cells as well as their expression of GFP-tagged proteins by time-lapse fluorescence microscopy. Therefore, the printing process had to deliver only a few cells per well. To do so, two parameters were adapted: the cell density of the yeast suspensions to print and the number of droplets/well. The latter was evaluated first and 40 drops/well provided optimal filling of the microwell without overflowing (Figure 4b). Since this parameter was kept constant for all the living cell experiments, the cell density of the yeast suspensions was then optimized. Yeasts suspensions at OD_{600} of 0.8, 0.5, and 0.2 were tested (Figure S2d). The suspension at an OD_{600} of 0.8 proved difficult to print since this high cell concentration increased the viscosity of the sample, thereby preventing the droplet ejection. On the other hand, cell suspensions at an OD_{600} of 0.2 were possible to shoot but did not deliver yeast cells in every well. The suspension at an OD_{600} of 0.5 was then selected, as it could be reproducibly printed and guaranteed the presence of a significant number of single cells per well. All the yeast printing experiments were performed with the following optimized parameters: pulse duration of 15 μs, amplitude voltage of 75 V, and a frequency of 50 Hz.

Finally, since the objective was to print multiple clones from the GFP-tagged yeast collection in different wells of the microfluidic chip, a last printing procedure had to be optimized, namely the washing step between samples in order to avoid cross-contamination. The washing step includes the disposal of the printed sample followed by the cleaning of the outside of the piezo dispenser capillary. Discarding the sample can be done either by applying pressure to the liquid path of the robot or by ejecting it with a syringe pump. Indeed, the iTWO-400 robot is equipped with both a pressure unit and a syringe pump, which enable to flow liquid through the piezo dispenser on a passive or active basis, respectively. With the pressure unit, the liquid continuously flows at a fixed fluid flow rate for as long as the pressure is applied. With the syringe pump, a specified volume of liquid flows at a specified flow rate. Cleaning the outside of the piezo dispenser is performed at the washing station of the robot (Figure S2a), and only the washing time can be adapted. To evaluate the efficiency of the washing step on avoiding cross-contamination, wells from a 384-multiwell plate (MTP) were filled with both a yeast suspension in PBS (at OD_{600} of 0.5) and a colored ink according to a designed pattern (Figure 5a). The samples were then printed on solid agar medium in a petri dish and left to grow at 30 °C for 24 hours before evaluation. In the first experiment, the printed samples were discarded by means of pressure (at 450 mbar) for 10 seconds, and the outside of the piezo dispenser was cleaned for five seconds. Although most of the design was correctly printed, one cross-contaminated spot was visible (Figure 5b). The experiment was then repeated; however, the samples were discarded in the wash station by means of the syringe pump. The cleaning of the piezo dispenser was kept at five seconds. Using this procedure, no cross-contamination was observed (Figure 5c).

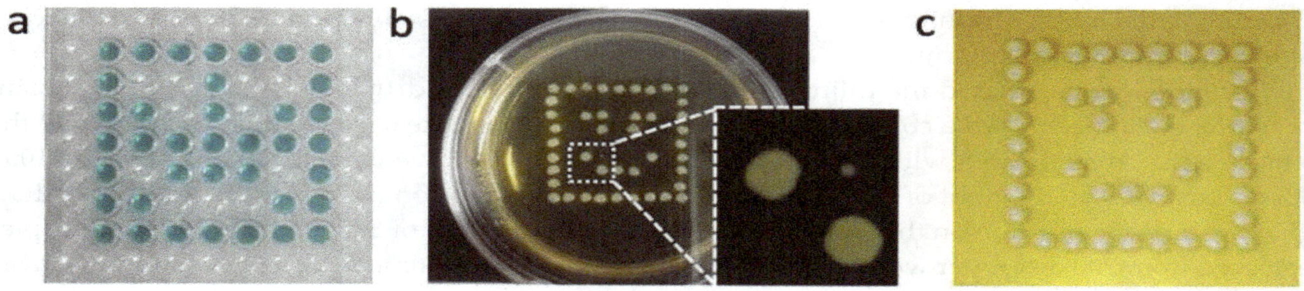

Figure 5. Evaluation of the washing protocol to avoid cross-contamination. (**a**) Printing pattern to evaluate the washing protocol of the piezo dispenser. Transparent wells contain a yeast solution in PBS, and the colored wells contain a blue dye (no cells). (**b**) Washing protocol based on pressure resulted in 1 cross-contamination colony. (**c**) Washing protocol based on the syringe pump resulted in spotting without cross-contamination.

3.3. *Growth of Yeast Cells in the Microfluidic Chip*

A selection of 34 GFP-tagged *S. cerevisiae* clones that are related to the cell cycle were printed as triplicates into the well array (Figure 4, S2e), and the chip was closed by sticking the microfluidic channel to the microwell substrate (Figure 1). Time-lapse microscopy was used to follow the growth during 18 h. An overview of all wells is presented in Figure S3 and some selected wells in Figure 6. At the end of the growth experiment, the GFP-tagged proteins were visualized.

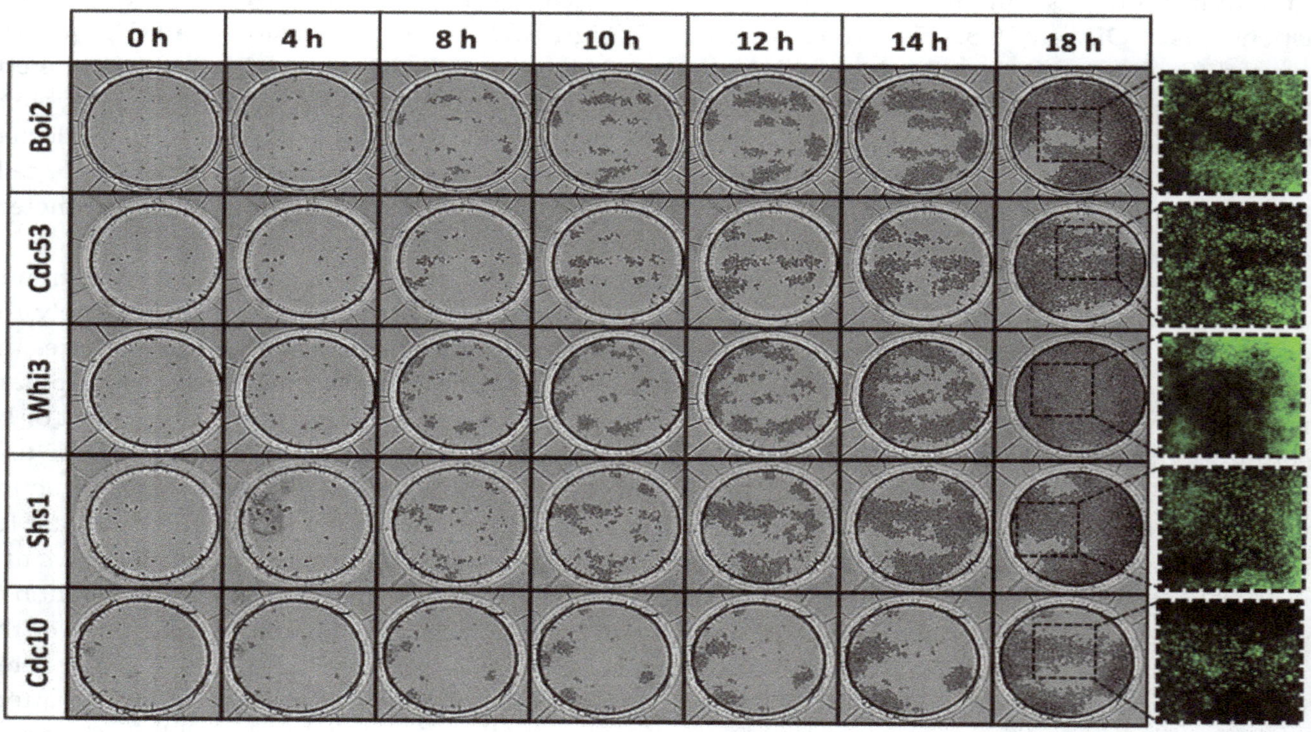

Figure 6. Yeast growth in some selected wells (see Figure S3 for the full overview).

To evaluate the suitability of the developed yeast chips to perform dynamic analyses of protein expression and protein localization in single yeast cells, a small set of 6 clones were selected, i.e. Cdc39, Cla4, Bem1, Shs1, Cdc14 and Cdc28. A time-lapse experiment was performed where these clones were observed at higher resolution (600× magnification) during 3 h (Figure 7).

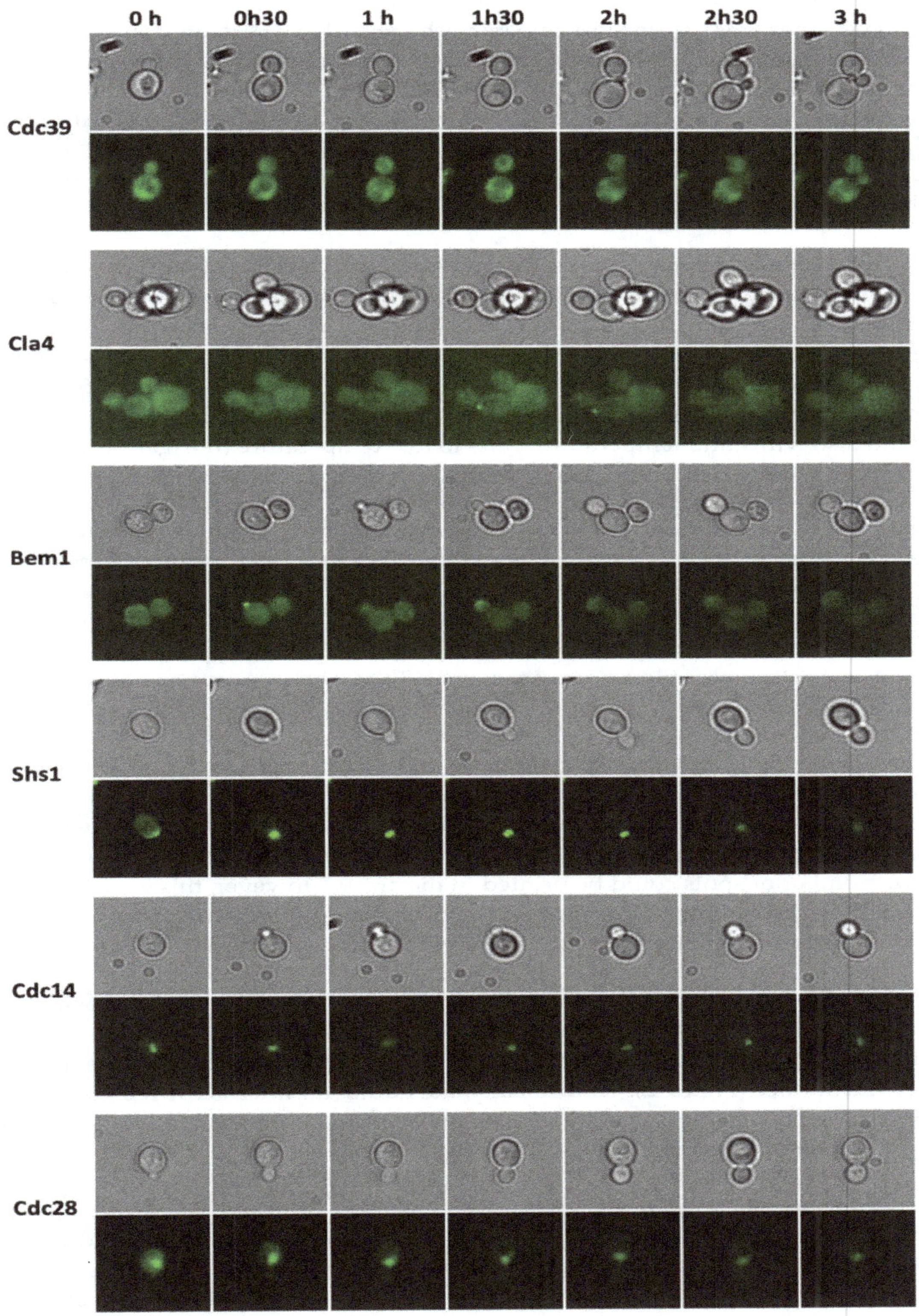

Figure 7. Time-lapse fluorescence microscopy of selected GFP-tagged clones related to the cell cycle.

4. Discussion

To create a living GFP-tagged yeast array, we combined mechanical patterning by constructing an array of microwells with cell printing (robotic cell patterning), which allows the controlled placement of the GFP-tagged clones in the selected wells. The cells were trapped in the wells by sticking them to the glass bottom of the microwells using the lectin Con A. Microfluidic chips where a GFP-tagged yeast clone collection was patterned as an array into microchambers have been previously developed [10,37]. Also, a microfluidic perfusion system where a robotic printed yeast array on agar and sandwiched with a track-etched membrane has been described [38]. These designs and fabrication methods are much more complex and difficult to construct. Due to the open design, robotic patterning of living cells in microwells is much more flexible in creating different filling designs. Additionally, it could be used to create high-density arrays without increasing much the complexity of the microfluidic chip design and construction.

A hybrid SU-8 on glass microfluidic chip containing a living cell array was developed. The direct fabrication of SU-8 microwells on the glass coverslip substrate is a simple, low cost and a high precision method that is suitable to construct disposable biochips [39]. SU-8 is biocompatible and has also the advantage that high density well arrays at high aspect ratios could be created [40]. A microwell array of 3 × 34 microwells containing microwells with a diameter of 300 μm and depth of 50 μm were created. The SU-8 microwell layer stuck well to the glass substrate. The bonding of the SU-8 microwell array to the glass could also withstand temperature shifts to low temperature (refrigerator). This allowed to store the chip filled with yeast cells for a few days before the growth experiment was performed.

Piezo printing was selected as the method to dispense a Con A protein solution and the GFP-tagged yeast collection into the microwells. We performed these dispensing steps with a piezoelectric dispenser (also called drop-on-demand ink-jet printing) [41,42]. Piezo printing of proteins and cells has been used previously for various applications including enzyme printing for glucose biosensors [43], protein arrays [44], netrin-1 adhesive micropattern construction [45], bacterial *Escherichia coli* arrays [46,47], bacterial and yeast cells on cantilever array sensors [48], *S. cerevisiae* cells on an agar layer [38], and mammalian cells [49].

We demonstrated that printing a small array on a flat glass substrate is possible and this method could be used to create cell arrays on glass substrate in a cheap and easy way. However, the droplet size was not proportional to the droplet volume which is a limitation to increase the throughput. We expected that half the number of droplets would result in half the spot volume, half the spot size and that twice the number of spots could be printed in the array. However, this was not the case. Also, significant variations in the droplet locations occurred, which resulted in arrays that were not perfectly arranged, which can result in merging of the spots in case of a small pitch distance (Figure 2e,f).

Printing of cells into microwells compared to on a flat surface has many advantages. Spatial confinement in the microwell results in higher resolution printing: smaller well sizes than droplet spot sizes and smaller pitch distances can be obtained. Droplet position in the wells is more accurate than droplets on a flat surface (Figure 2). Additionally, the liquid–air interface is reduced for microwells compared to droplets on a surface, resulting in a reduced evaporation rate. All these benefits result in higher throughput and stable printing.

The printer setup allowed to aspirate a yeast solution from an MTP (multiwell plate) well and deposit picoliter droplets containing living cells at the target microwell, and to perform this consecutively for all filled MTP wells. A tip dispenser washing protocol based on syringe pump cell solution ejection was successful to avoid cross contamination between different wells since syringe pump ejection occurred at a much higher flow rate than pressure-based ejection and could remove all yeast cells from the piezo tip. A GFP-tagged clone collection in a 3 × 34 cell array was cultivated during 18 h (Figure 6 and S3). This experiment demonstrated that a clone collection of 34 clones in triplicate could be successfully printed at a cell concentration that allows single cell observation during several generations.

As a proof-of-concept experiment to observe changes in GFP-tagged protein expression and cellular location, we selected GFP-tagged proteins that play a role during the yeast growth cycle. The Cdc39 protein is a subunit of the CCR4-NOT1 core complex that has multiple roles in the regulation of mRNA levels [50,51]. A high fluorescence intensity covering the cytoplasm in the mother and daughter cell can be observed (Figure 7) since it is present at a large number of protein molecules per cell (4300) [52]. Cla4p is a Cdc42p-activated signal transducing kinase that is involved in septin ring assembly, vacuole inheritance, cytokinesis, and sterol uptake regulation [53]. It is distributed in the cytoplasm and the bud. Higher intensity spots can be observed at the site where the bud appears (Figure 7). Bem1p is involved in establishing cell polarity and morphogenesis and functions as a scaffold protein for complexes that include Cdc24p, Ste5p, Ste20p, and Rsr1p [54]. A high intensity spot can be observed at the bud site and the growing bud (Figure 7). Shs1p is a component of the septin ring that is required for cytokinesis [55]. Initially, it is present at the cell periphery and moves to the bud site to create the septin ring (Figure 7). Cdc14p is a phosphatase required for the mitotic exit [56]. It is present in the nucleus and the nucleolus (Figure 7). Cdc28p is a cyclin-dependent kinase (CDK) catalytic subunit and master regulator of mitotic and meiotic cell cycles [57,58]. It alternately associates with G1, S, and G2/M phase cyclins. It is observed initially in the cytoplasm and next in the nucleus (Figure 7). These results demonstrate that the microfluidic chip can be used to perform dynamic experiments with subcellular resolution of fluorescently-tagged proteins.

In the future, the number of microwells in the array could be upscaled allowing to perform dynamic proteomics and localizomics experiments. This cell microarray-based systems-biology platform could be used to detect directly chemical disturbances in a small-molecule compound screen of the proteome in contrast to genetic approaches that are based on chemical-genetic interactions and necessitates a multistep indirect approach [59]. The genome-wide tagging of proteins of *S. cerevisiae*, including with GFP, has already provided a vast resource of such information [9,11,33,60]. Analysis of this high-resolution, high coverage localisation data set in the context of transcriptional, genetic, and protein-protein interaction (PPI) data revealed the combinatorial logic of transcriptional co-regulation and spatial-temporal regulation of proteins, and provided for example a comprehensive view of trafficking and signalling regulatory interactions within and between organelles in eukaryotic cells. As demonstrated here, dynamic movements from one location to another can also be followed as the cell proceeds through the cell cycle. Dynamic movements of proteins have also been described in yeast cells that respond to environmental stresses such as dithiothreitol (DTT) stress [9], hydrogen peroxide stress [9], osmotic stress by potassium chloride [37], and nitrogen starvation [9] or chemical perturbations by rapamycin [11], hydroxyurea [11], or methyl methane sulfonate [10].

5. Conclusions

We developed a perfusion microfluidic chip containing living yeast cell arrays that allows long term cultivation of the yeast cells and high-resolution time-lapse fluorescence microscopy. The creation of the cell array was based on mechanical patterning in SU-8 microwells and piezoelectric filling of the microwells. The microfluidic chip was closed by sticking a top plate that contained the microfluidic channel and inlet and outlet to the microwell substrate. The developed technology and method were validated by a growth experiment of a clone collection of 34 clones (in triplicate) that are linked to the cell cycle. Additionally, a set of six selected GFP-tagged clones were observed by time-lapse fluorescence microscopy at high resolution to observe single cell protein expression and subcellular location of the GFP-tagged proteins.

Future technological challenges lie in the further upscaling of the technology and procedures to construct genome/proteome-wide cell microarrays and analyzing cells dynamically on a whole proteome level. Dynamic proteomics profiling information based on chemical compound (such as which are highly desired due to the limited number of available antifungal drugs and the fast emergence of multiresistant pathogens [62,63].

Author Contributions: Conceptualization, C.Y. and R.G.W.; methodology, C.Y., S.T., and R.G.W.; formal analysis, C.Y.; investigation, C.Y.; resources, R.G.W.; writing—original draft preparation, C.Y. and R.G.W.; writing—review and editing, C.Y., S.T., and R.G.W.; supervision, R.G.W.; project administration, R.G.W.; funding acquisition, C.Y. and R.G.W. All authors have read and agreed to the published version of the manuscript.

Acknowledgments: We acknowledge Rouslan Efremov for the use of lithography equipment and Gangadhar Eluru for allowing us to use his design of the microwell mask. The Hercules Foundation (FWO Flanders) is acknowledged for the funding of the Raith Voyager e-beam equipment (AUGE/13/19).

References

1. Castel, D.; Pitaval, A.; Debily, M.A.; Gidrol, X. Cell microarrays in drug discovery. *Drug Discov. Today* **2006**, *11*, 616–622. [CrossRef] [PubMed]
2. Berthuy, O.I.; Muldur, S.K.; Rossi, F.; Colpo, P.; Blum, L.J.; Marquette, C.A. Multiplex cell microarrays for high-throughput screening. *Lab Chip* **2016**, *16*, 4248–4262. [CrossRef] [PubMed]
3. Hong, H.; Koom, W.; Koh, W.G. Cell Microarray Technologies for High-Throughput Cell-Based Biosensors. *Sensors* **2017**, *17*, 1293. [CrossRef] [PubMed]
4. Willaert, R.; Goossens, K. Microfluidic Bioreactors for Cellular Microarrays. *Fermentation* **2015**, *1*, 38–78. [CrossRef]
5. Jonczyk, R.; Kurth, T.; Lavrentieva, A.; Walter, J.-G.; Scheper, T.; Stahl, F. Living Cell Microarrays: An Overview of Concepts. *Microarrays* **2016**, *5*, 11. [CrossRef]
6. Willaert, R.; Sahli, H. On-Chip Living-Cell Microarrays for Network Biology. In *Bioinformatics—Trends and Methodologies*; InTechOpen: London, UK, 2011.
7. Fernandes, T.G.; Diogo, M.M.; Clark, D.S.; Dordick, J.S.; Cabral, J.M.S. High-throughput cellular microarray platforms: Applications in drug discovery, toxicology and stem cell research. *Trends Biotechnol.* **2009**, *27*, 342–349. [CrossRef]
8. Chen, D.; Davis, M. Molecular and functional analysis using live cell microarrays. *Curr. Opin. Chem. Biol.* **2006**, *10*, 28–34. [CrossRef]
9. Breker, M.; Gymrek, M.; Schuldiner, M. A novel single-cell screening platform reveals proteome plasticity during yeast stress responses. *J. Cell Biol.* **2013**, *200*, 839–850. [CrossRef]
10. Dénervaud, N.; Becker, J.; Delgado-Gonzalo, R.; Damay, P.; Rajkumar, A.S.; Unser, M.; Shore, D.; Naef, F.; Maerkl, S.J. A chemostat array enables the spatio-temporal analysis of the yeast proteome. *Proc. Natl. Acad. Sci. USA* **2013**, *110*, 15842–15847. [CrossRef]
11. Chong, Y.T.; Koh, J.L.Y.; Friesen, H.; Duffy, K.; Cox, M.J.; Moses, A.; Moffat, J.; Boone, C.; Andrews, B.J. Yeast proteome dynamics from single cell imaging and automated analysis. *Cell* **2015**, *161*, 1413–1424. [CrossRef]
12. Yuan, H.; Zhang, R.; Shao, B.; Wang, X.; Ouyang, Q.; Hao, N.; Luo, C. Protein expression patterns of the yeast mating response. *Integr. Biol. (UK)* **2016**, *8*, 712–719. [CrossRef]
13. Shao, B.; Yuan, H.; Zhang, R.; Wang, X.; Zhang, S.; Ouyang, Q.; Hao, N.; Luo, C. Reconstructing the regulatory circuit of cell fate determination in yeast mating response. *PLoS Comput. Biol.* **2017**, *13*, e1005671. [CrossRef] [PubMed]
14. Yarmush, M.L.; King, K.R. Living-Cell Microarrays. *Annu. Rev. Biomed. Eng.* **2009**, *11*, 235–257. [CrossRef] [PubMed]
15. Park, M.C.; Hur, J.Y.; Cho, H.S.; Park, S.H.; Suh, K.Y. High-throughput single-cell quantification using simple microwell-based cell docking and programmable time-course live-cell imaging. *Lab Chip* **2011**, *11*, 79–86. [CrossRef] [PubMed]
16. Minc, N.; Boudaoud, A.; Chang, F. Mechanical Forces of Fission Yeast Growth. *Curr. Biol.* **2009**, *19*, 1096–1101. [CrossRef] [PubMed]

17. Falconnet, D.; Niemistö, A.; Taylor, R.J.; Ricicova, M.; Galitski, T.; Shmulevich, I.; Hansen, C.L. High-throughput tracking of single yeast cells in a microfluidic imaging matrix. *Lab Chip* **2011**, *11*, 466–473. [CrossRef]
18. Groisman, A.; Lobo, C.; Cho, H.; Campbell, J.K.; Dufour, Y.S.; Stevens, A.M.; Levchenko, A. A microfluidic chemostat for experiments with bacterial and yeast cells. *Nat. Methods* **2005**, *2*, 685–689. [CrossRef]
19. Ryley, J.; Pereira-Smith, O.M. Microfluidics device for single cell gene expression analysis inSaccharomyces cerevisiae. *Yeast* **2006**, *23*, 1065–1073. [CrossRef]
20. Bell, L.; Seshia, A.; Lando, D.; Laue, E.; Palayret, M.; Lee, S.F.; Klenerman, D. A microfluidic device for the hydrodynamic immobilisation of living fission yeast cells for super-resolution imaging. *Sens. Actuators B Chem.* **2014**, *192*, 36–41. [CrossRef]
21. Carlo, D.D.; Wu, L.Y.; Lee, L.P. Dynamic single cell culture array. *Lab Chip* **2006**, *6*, 1445–1449. [CrossRef]
22. Barbulovic-Nad, I.; Lucente, M.; Sun, Y.; Zhang, M.; Wheeler, A.R.; Bussmann, M. Bio-microarray fabrication techniques—A review. *Crit. Rev. Biotechnol.* **2006**, *26*, 237–259. [CrossRef] [PubMed]
23. Guillemot, F.; Souquet, A.; Catros, S.; Guillotin, B.; Lopez, J.; Faucon, M.; Pippenger, B.; Bareille, R.; Rémy, M.; Bellance, S.; et al. High-throughput laser printing of cells and biomaterials for tissue engineering. *Acta Biomater.* **2010**, *6*, 2494–2500. [CrossRef] [PubMed]
24. Bean, G.J.; Jaeger, P.A.; Bahr, S.; Ideker, T. Development of ultra-high-density screening tools for microbial "omics". *PLoS ONE* **2014**, *9*, e85177. [CrossRef] [PubMed]
25. Schaack, B.; Reboud, J.; Combe, S.; Fouqué, B.; Berger, F.; Boccard, S.; Odile Filhol-Cochet, F.C. A "DropChip" Cell Array for DNA and siRNA Transfection Combined with Drug Screening. *Nanobiotechnology* **2005**, *1*, 183–189. [CrossRef]
26. Ringeisen, B.R.; Othon, C.M.; Barron, J.A.; Young, D.; Spargo, B.J. Jet-based methods to print living cells. *Biotechnol. J.* **2006**, *1*, 930–948. [CrossRef]
27. Roth, E.A.; Xu, T.; Das, M.; Gregory, C.; Hickman, J.J.; Boland, T. Inkjet printing for high-throughput cell patterning. *Biomaterials* **2004**, *25*, 3707–3715. [CrossRef]
28. Ferris, C.J.; Gilmore, K.G.; Wallace, G.G.; In Het Panhuis, M. Biofabrication: An overview of the approaches used for printing of living cells. *Appl. Microbiol. Biotechnol.* **2013**, *97*, 4243–4258. [CrossRef]
29. Demirci, U.; Montesano, G. Cell encapsulating droplet vitrification. *Lab Chip* **2007**, *7*, 1428–1433. [CrossRef]
30. Lee, W.; Debasitis, J.C.; Lee, V.K.; Lee, J.H.; Fischer, K.; Edminster, K.; Park, J.K.; Yoo, S.S. Multi-layered culture of human skin fibroblasts and keratinocytes through three-dimensional freeform fabrication. *Biomaterials* **2009**, *30*, 1587–1595. [CrossRef]
31. Gonzalez-Macia, L.; Morrin, A.; Smyth, M.R.; Killard, A.J. Advanced printing and deposition methodologies for the fabrication of biosensors and biodevices. *Analyst* **2010**, *135*, 845–867. [CrossRef]
32. Li, J.; Rossignol, F.; Macdonald, J. Inkjet printing for biosensor fabrication: Combining chemistry and technology for advanced manufacturing. *Lab Chip* **2015**, *15*, 2538–2558. [CrossRef] [PubMed]
33. Huh, W.K.; Falvo, J.V.; Gerke, L.C.; Carroll, A.S.; Howson, R.W.; Weissman, J.S.; O'Shea, E.K. Global analysis of protein localization in budding yeast. *Nature* **2003**, *425*, 686–691. [CrossRef] [PubMed]
34. Schindelin, J.; Arganda-Carreras, I.; Frise, E.; Kaynig, V.; Longair, M.; Pietzsch, T.; Preibisch, S.; Rueden, C.; Saalfeld, S.; Schmid, B.; et al. Fiji: An open-source platform for biological-image analysis. *Nat. Methods* **2012**, *9*, 676–682. [CrossRef] [PubMed]
35. Preibisch, S.; Saalfeld, S.; Tomancak, P. Globally optimal stitching of tiled 3D microscopic image acquisitions. *Bioinformatics* **2009**, *25*, 1463–1465. [CrossRef]
36. Pemberton, L.F. Preparation of yeast cells for live-cell imaging and indirect immunofluorescence. *Methods Mol. Biol.* **2014**, *1205*, 79–90.
37. Zhang, R.; Yuan, H.; Wang, S.; Ouyang, Q.; Chen, Y.; Hao, N.; Luo, C. High-throughput single-cell analysis for the proteomic dynamics study of the yeast osmotic stress response. *Sci. Rep.* **2017**, *7*, 42200. [CrossRef]
38. Mirzaei, M.; Pla-Roca, M.; Safavieh, R.; Nazarova, E.; Safavieh, M.; Li, H.; Vogel, J.; Juncker, D. Microfluidic perfusion system for culturing and imaging yeast cell microarrays and rapidly exchanging media. *Lab Chip* **2010**, *10*, 2449–2457. [CrossRef]
39. Kang, D.H.; Han, W.B.; Choi, N.; Kim, Y.J.; Kim, T.S. Tightly Sealed 3D Lipid Structure Monolithically Generated on Transparent SU-8 Microwell Arrays for Biosensor Applications. *ACS Appl. Mater. Interfaces* **2018** *10*, 40401–40410. [CrossRef]

40. Wu, Z.Z.; Zhao, Y.; Kisaalita, W.S. Interfacing SH-SY5Y human neuroblastoma cells with SU-8 microstructures. *Colloids Surf. B Biointerfaces* **2006**, *52*, 14–21. [CrossRef]
41. Calvert, P. Inkjet printing for materials and devices. *Chem. Mater.* **2001**, *13*, 3299–3305. [CrossRef]
42. Derby, B. Bioprinting: Inkjet printing proteins and hybrid cell-containing materials and structures. *J. Mater. Chem.* **2008**, *18*, 5717–5721. [CrossRef]
43. Newman, J.D.; Turner, A.P.F.; Marrazza, G. Ink-jet printing for the fabrication of amperometric glucose biosensors. *Anal. Chim. Acta* **1992**, *262*, 13–17. [CrossRef]
44. Montenegro-Nicolini, M.; Miranda, V.; Morales, J.O. Inkjet Printing of Proteins: An Experimental Approach. *AAPS J.* **2017**, *19*, 234–243. [CrossRef] [PubMed]
45. Matsugaki, A.; Yamazaki, D.; Nakano, T. Selective patterning of netrin-1 as a novel guiding cue for anisotropic dendrogenesis in osteocytes. *Mater. Sci. Eng. C* **2020**, *108*, 110391. [CrossRef]
46. Xu, T.; Petridou, S.; Lee, E.H.; Roth, E.A.; Vyavahare, N.R.; Hickman, J.J.; Boland, T. Construction of high-density bacterial colony arrays and patterns by the ink-jet method. *Biotechnol. Bioeng.* **2004**, *85*, 29–33. [CrossRef] [PubMed]
47. Zheng, Q.; Lu, J.; Chen, H.; Huang, L.; Cai, J.; Xu, Z. Application of inkjet printing technique for biological material delivery and antimicrobial assays. *Anal. Biochem.* **2011**, *410*, 171–176. [CrossRef] [PubMed]
48. Lukacs, G.; Maloney, N.; Hegner, M. Ink-jet printing: Perfect tool for cantilever array sensor preparation for microbial growth detection. *J. Sensors* **2012**, *2012*. [CrossRef]
49. Zhang, J.; Chen, F.; He, Z.; Ma, Y.; Uchiyama, K.; Lin, J.M. A novel approach for precisely controlled multiple cell patterning in microfluidic chips by inkjet printing and the detection of drug metabolism and diffusion. *Analyst* **2016**, *141*, 2940–2947. [CrossRef]
50. Collart, M.A.; Struhl, K. NOT1(CDC39), NOT2(CDC36), NOT3, and NOT4 encode a global-negative regulator of transcription that differentially affects TATA-element utilization. *Genes Dev.* **1994**, *8*, 525–537. [CrossRef]
51. Kruk, J.A.; Dutta, A.; Fu, J.; Gilmour, D.S.; Reese, J.C. The multifunctional Ccr4-Not complex directly promotes transcription elongation. *Genes Dev.* **2011**, *25*, 581–593. [CrossRef]
52. Ghaemmaghami, S.; Huh, W.K.; Bower, K.; Howson, R.W.; Belle, A.; Dephoure, N.; O'Shea, E.K.; Weissman, J.S. Global analysis of protein expression in yeast. *Nature* **2003**, *425*, 737–741. [CrossRef] [PubMed]
53. Benton, B.K.; Tinkelenberg, A.; Gonzalez, I.; Cross, F.R. Cla4p, a Saccharomyces cerevisiae Cdc42p-activated kinase involved in cytokinesis, is activated at mitosis. *Mol. Cell. Biol.* **1997**, *17*, 5067–5076. [CrossRef] [PubMed]
54. Madden, K.; Snyder, M. Cell Polarity and Morphogenesis in Budding Yeast. *Annu. Rev. Microbiol.* **1998**, *52*, 687–744. [CrossRef]
55. Mino, A.; Tanaka, K.; Kamei, T.; Umikawa, M.; Fujiwara, T.; Takai, Y. Shs1p: A novel member of septin that interacts with Spa2p, involved in polarized growth in Saccharomyces cerevisiae. *Biochem. Biophys. Res. Commun.* **1998**, *251*, 732–736. [CrossRef] [PubMed]
56. Visintin, R.; Craig, K.; Hwang, E.S.; Prinz, S.; Tyers, M.; Amon, A. The phosphatase Cdc14 triggers mitotic exit by reversal of Cdk-dependent phosphorylation. *Mol. Cell* **1998**, *2*, 709–718. [CrossRef]
57. Mendenhall, M.D.; Hodge, A.E. Regulation of Cdc28 cyclin-dependent protein kinase activity during the cell cycle of the yeast Saccharomyces cerevisiae. *Microbiol. Mol. Biol. Rev.* **1998**, *62*, 1191–1243. [CrossRef] [PubMed]
58. Chymkowitch, P.; Eldholm, V.; Lorenz, S.; Zimmermann, C.; Lindvall, J.M.; Bjørås, M.; Meza-Zepeda, L.A.; Enserink, J.M. Cdc28 kinase activity regulates the basal transcription machinery at a subset of genes. *Proc. Natl. Acad. Sci. USA* **2012**, *109*, 10450–10455. [CrossRef]
59. Weig, M.; Brown, A.J.P. Genomics and the development of new diagnostics and anti-Candida drugs. *Trends Microbiol.* **2007**, *15*, 310–317. [CrossRef]
60. Howson, R.; Huh, W.K.; Ghaemmaghami, S.; Falvo, J.V.; Bower, K.; Belle, A.; Dephoure, N.; Wykoff, D.D.; Weissman, J.S.; O'Shea, E.K. Construction, verification and experimental use of two epitope-tagged collections of budding yeast strains. *Comp. Funct. Genom.* **2005**, *6*, 2–16. [CrossRef]
61. Cohen, A.A.; Geva-Zatorsky, N.; Eden, E.; Frenkel-Morgenstern, M.; Issaeva, I.; Sigal, A.; Milo, R.; Cohen-Saidon, C.; Liron, Y.; Kam, Z.; et al. Dynamic proteomics of individual cancer cells in response to a drug. *Science* **2008** *322*, 1511–1516. [CrossRef]

62. Michael, C.A.; Dominey-Howes, D.; Labbate, M. The Antimicrobial Resistance Crisis: Causes, Consequences, and Management. *Front. Public Health* **2014**, *2*, 145. [CrossRef] [PubMed]
63. Willaert, R. Micro- and Nanoscale Approaches in Antifungal Drug Discovery. *Fermentation* **2018**, *4*, 43. [CrossRef]

9

Adding Flavor to Beverages with Non-Conventional Yeasts

Davide Ravasio [1,†], Silvia Carlin [2], Teun Boekhout [3,4], Marizeth Groenewald [3], Urska Vrhovsek [2], Andrea Walther [1,‡] and Jürgen Wendland [1,5,*]

1 Carlsberg Research Laboratory, Yeast & Fermentation, J. C. Jacobsens Gade 4, DK-1799 Copenhagen V, Denmark; davide.ravasio@gmail.com (D.R.); awa@novozymes.com (A.W.)
2 Fondazione Edmund Mach, Research and Innovation Centre, Food Quality and Nutrition Department, Via E. Mach 1, I-38010 S. Michele all'Adige, Italy; Silvia.carlin@fmach.it (S.C.); urska.vrhovsek@fmach.it (U.V.)
3 Westerdijk Fungal Biodiversity Institute, 3584 CT Utrecht, The Netherlands; t.boekhout@cbs.knaw.nl (T.B.); m.groenewald@westerdijkinstitute.nl (M.G.)
4 Institute of Biodiversity and Ecosystem Dynamics (IBED), University of Amsterdam, 1000 GG Amsterdam, The Netherlands
5 Vrije Universiteit Brussel, Research Group of Microbiology (MICR)—Functional Yeast Genomics, BE-1050 Brussels, Belgium
* Correspondence: jurgen.wendland@vub.be
† Current address: Evolva SA, Duggingerstrasse 23, CH-4153 Reinach, Switzerland.
‡ Current address: Novozymes A/S, Krogshoejvej 36, DK-2880 Bagsvaerd, Denmark.

Abstract: Fungi produce a variety of volatile organic compounds (VOCs) during their primary and secondary metabolism. In the beverage industry, these volatiles contribute to the the flavor and aroma profile of the final products. We evaluated the fermentation ability and aroma profiles of non-conventional yeasts that have been associated with various food sources. A total of 60 strains were analyzed with regard to their fermentation and flavor profile. Species belonging to the genera *Candida*, *Pichia* and *Wickerhamomyces* separated best from lager yeast strains according to a principal component analysis taking alcohol and ester production into account. The speed of fermentation and sugar utilization were analysed for these strains. Volatile aroma-compound formation was assayed via gas chromatography. Several strains produced substantially higher amounts of aroma alcohols and esters compared to the lager yeast strain *Weihenstephan* 34/70. Consequently, co-fermentation of this lager yeast strain with a *Wickerhamomyces anomalus* strain generated an increased fruity-flavour profile. This demonstrates that mixed fermentations utilizing non-*Saccharomyces cerevisiae* biodiversity can enhance the flavour profiles of fermented beverages.

Keywords: aroma profiling; solid-phase microextraction–gas chromatography/mass spectrometry (SPME–GC/MS); yeast; *Saccharomycetes*; fermentation; volatile organic compound (VOC); aroma

---

1. Introduction

Beer is one of the most widely consumed alcoholic beverages in the world. In 2003, worldwide beer production reached around 1.82 billion hectoliters and increased to a volume of 1.93 billion hectoliters in 2013 according to the Kirin Beer University Report of 2014. Production is divided into several beer styles of which ale and lager beers are most prominent. Generally, ale is produced by top-fermenting yeasts at temperatures between 15–30 °C. Ales are known for their fruity aromas which are regarded as a distinctive characteristic of top-fermenting beers. Lager beers, however, are produced by a distinct group of bottom-fermenting yeasts at fermentation temperatures between 10–15 °C. The aroma of lager beers is more neutral compared to ale-type beers as they contain lower amounts of fruity flavors.

Top-fermenting yeasts generally are *S. cerevisiae* strains. At the end of fermentation, these yeasts rise to the surface of the fermenter, creating a thick cell layer. Bottom-fermenting yeasts belong to two distinct groups of lager yeasts [1]. Lager yeasts are hybrids between *S. cerevisiae* and *S. eubayanus* [2]. Lager yeasts can be divided into two groups, group I or *Saaz* type, and group II or *Frohberg* type, that are distinguished at the molecular level by ploidy differences, characteristic chromosomal rearrangements and chromosome losses [3,4].

In contrast to lager beers, lambic beers with sometimes exceptional flavor compositions are based on a larger biodiversity, including acetic and lactic acid bacteria and various yeasts, e.g., *S. cerevisiae*, *S. pastorianus* and *Brettanomyces bruxellensis* [5].

Beer is a complex product consisting of volatile and non-volatile components that form the final aroma. The contribution of ale yeasts to the final flavor bouquet is generally higher than that of lager yeasts. This has been attributed to the greater diversity of ale yeasts compared to the limited diversity of the two groups of lager yeast [6]. Other process parameters also influence volatile-compound formation. Specifically, sugar concentration of the wort and different aeration regimes influence the production of flavor-active esters. These esters mainly contribute to the fruitiness of the product. Dominant esters are acetate esters such as ethyl acetate (fruity), isoamyl acetate (banana) and 2-phenyl acetate (rose) and ethyl or medium-chain fatty acid esters such as ethyl hexanoate, ethyl octanoate and ethyl decanoate, which provide a fruity apple- or wine-like flavor to the beer [7]. Among the higher alcohols, n-propanol, isobutanol, 2-phenylethanol and isoamyl alcohol are most abundant. Higher alcohols such as isobutanol can contribute a rum-like aroma which gives a warm mouth feeling, while 2-phenylethanol and isoamyl alcohol are prevalent for their sweet/rose and fruity/banana-like aromas, respectively [8].

Not all flavors are desirable. Strecker aldehydes (aged flavour), aldehydes of the *Maillard* reaction, e.g., furfural, and aldehydes of fatty acid oxidation, e.g., trans-2-nonenal, are regarded as off-flavors in beer [9]. The ketone diacetyl (2,3-butanedione) is monitored during the lager beer brewing process, in particular, as it imparts an undesirable buttery flavor with a low flavour threshold [10]. Given the low concentration and volatile nature of these aroma compounds, gas chromatography coupled with mass spectrometry (GC/MS) offers an optimal technique to analyze the flavor profile of beer.

Recently, non-conventional yeasts or non-*cerevisiae* yeasts have gained importance for fermented alcoholic beverages [11,12]. They produce various mixtures of volatile compounds and so contribute to the aroma profile of beverages [6,13]. Mixed fermentations using *S. cerevisiae* in combination with a non-conventional yeast strain e.g., belonging to the genera *Lachancea*, *Pichia* or *Hanseniaspora*, could provide novel beverages with improved ester profiles [14,15]. On the other hand, the synergistic effects on aromatic compound production were observed in co-cultures with *Metschnikowia pulcherrima* and *S. cerevisiae* [16]. Non-conventional yeasts could contribute to satisfying the demand of novel and distinctive, yet natural, flavors in fermented beverages [17].

A vast number of non-conventional yeast strains has been isolated from various food sources and deposited in culture collection such as the CBS yeast collection of the *Westerdijk* Fungal Biodiversity Institute (formerly, CBS, Centraalbureau voor Schimmelcultures, Utrecht, Netherlands). In this study, we aimed at covering a broad spectrum of species isolated from different substrates like berries, fruits, cheese, fruit flies or even soil and spanning a broad evolutionary distance within the *Saccharomycotina* in order to identify species that could contribute with their particular flavour to lager beer fermentations.

2. Materials and Methods

2.1. Strains and Media

Yeast strains used in this study are shown in Table 1, including their CBS reference number. Each strain was coded based on its coordinates in a 96-well plate. Yeast strains were subcultured in YPD (1% yeast extract, 2% peptone, 2% glucose) at room temperature overnight.

Table 1. Strains used in this study. Each selected strain was assigned to a code based on its coordinates on a 96-well plate.

Position	Strain Number	Taxon Name	Substrate of Isolation	Origin
B2	CBS 10151	*Candida alimentaria*	Cured ham	Norway
B3	CBS 12367	*Candida alimentaria*	Brie Régalou cheese	
B4	CBS 4074	*Candida diversa*	Grape must	Japan
B5	CBS 8058	*Candida kofuensis*	Berries of *Vitis coignetiae*	Japan
B6	CBS 1760	*Candida versatilis*	Pickling vat with 22% brine	USA
B7	CBS 2649	*Candida stellate*	Grape juice	France
B8	CBS 6936	*Clavispora lusitaniae*	Citrus essence	Israel
B9	CBS 4373	*Debaryomyces fabryi*	Dry white wine	South Africa
B10	CBS 767	*Debaryomyces hansenii*		
B11	CBS 2659	*Debaryomyces subglobosus*	Apple	Italy
C2	CBS 8139	*Dekkera anomala*		Netherlands
C3	CBS 615.84	*Geotrichum candidum*	Brie cheese	France
C4	CBS 95	*Hanseniaspora guilliermondii*	Fermenting bottled tomatoes	Netherlands
C5	CBS 6783	*Hanseniaspora occidentalis var. citrica*	Orange juice	Italy
C6	CBS 2585	*Hanseniaspora uvarum*	Sour dough	Portugal
C7	CBS 2568	*Hanseniaspora vineae*	*Drosophila persimilis* (fruit fly)	
C8	CBS 2352	*Hyphopichia burtonii*	Pollen, carried by wild bees	
C9	CBS 4311	*Kazachstania servazii*	Soil	Finland
C10	CBS 3019	*Kazachstania spencerorum*	Soil	South Africa
C11	CBS 2186	*Kazachstania transvaalensis*	Soil	South Africa
D2	CBS 398	*Kazachstania unispora*		
D3	CBS 7775	*Kluyveromyces aestuarii*	Neotredo reynei (shipworm)	Brazil
D4	CBS 8530	*Kluyveromyces dobzhanskii*	*Drosophila* sp.	Canada
D5	CBS 1557	*Kluyveromyces marxianus*	Stracchino cheese	Italy
D6	CBS 7005	*Lachancea fermentati*	Alpechín	Spain
D7	CBS 3082	*Lachancea kluyveri*	*Drosophila pinicola* (fruit fly)	
D8	CBS 7703	*Lachancea waltii*	Either fruit or leaf of fruit tree	
D9	CBS 5833	*Metschnikowia pulcherrima*	Berries of *Vitis labrusca* (Concord grapes)	USA
D10	CBS 2030	*Meyerozyma guilliermondii*	Insect frass on *Ulmus americana* (elm tree)	USA
D11	CBS 8417	*Meyerozyma guilliermondii*	Brine bath in cheese factory	Netherlands
E2	CBS 7720	*Nakaseomyces bacillisporus*	Exudate of *Quercus emoryi* (Emory oak)	USA
E3	CBS 2170	*Nakaseomyces delphensis*	Sugary deposit on dried figs	South Africa
E4	CBS 8255	*Pichia*	Kefyr	
E5	CBS 2020	*Pichia farinosa*	Fermenting cacao	Trinidad and Tobago
E6	CBS 2057	*Pichia fermentans*	Brewers yeast	
E7	CBS 188	*Pichia kluyveri*	Olives	
E8	CBS 5147	*Pichia kudriavzevii*	Fruit juice	
E9	CBS 191	*Pichia membranifaciens*	Wine	Italy
E10	CBS 429	*Saccharomyces cerevisiae*	Fermenting must of champagne grapes	
E11	CBS 1250	*Saccharomyces cerevisiae*	Sherry	Spain
F2	CBS 1782	*Saccharomyces cerevisiae*	Super-attenuated beer	
F3	CBS 820	*Saccharomycodes ludwigii*	Grape must	Germany
F4	CBS 2863	*Schwanniomyces occidentalis*	Soil of vineyard	Spain
F5	CBS 6741	*Schwanniomyces polymorphus var. africanus*	Soil	South Africa
F6	CBS 133	*Torulaspora delbrueckii*	Ragi	Indonesia
F7	CBS 427	*Torulaspora microellipsoides*	Apple juice	Germany
F8	CBS 248	*Wickerhamomyces anomalus*	Red currants	Netherlands
F9	CBS 249	*Wickerhamomyces anomalus*	Berries	
F10	CBS 261	*Wickerhamomyces anomalus*	Ragi	Indonesia
F11	CBS 262	*Wickerhamomyces anomalus*	Beer	
G2	CBS 4689	*Zygosaccharomyces bailii var. bailii*	Grape must	Italy
G3	CBS 1082	*Zygosaccharomyces bisporus*	Tea-beer fungus	Indonesia
G4	CBS 726	*Zygosaccharomyces mellis*	Wine grapes	Germany
G5	C1030	*Saccharomyces pastorianus*	Brewers' yeast	
G6	C1039	*Saccharomyces cerevisiae*	Wine yeast	
G7	C746	*Saccharomyces cerevisiae*	Brewers' yeast	
G8	CBS 1513	*Saccharomyces carlsbergensis*	Brewers' yeast	
G9	C482	*Saccharomyces cerevisiae*	Brewers' yeast	
G10	WS34/70	*Saccharomyces pastorianus*	Brewers' yeast	
G11	C598	*Saccharomyces cerevisiae*	Laboratory strain	

To evaluate growth at different temperatures or the utilization of maltose, cells were spotted on solid media plates in 10-fold dilution series and incubated for 5 days before evaluation. Growth at the different dilutions was 10-color coded and is presented as a heat map.

2.2. Fermentation Conditions

Lab-scale fermentations were carried out in 50 mL tubes filled with 40 mL nutrient-rich YPD with the glucose concentration adjusted to 16 °Plato at 20 °C. Each fermentation (300 rpm, using a triangular magnetic stirrer) was started with a cell density corresponding to OD600 = 0.2. For co-fermentations, the amounts of cells used equaled OD600 = 0.1. The fermentation progress was monitored for up to

14 days by daily measurement of CO_2 release. Sugar content was measured by a DMA 35 Anton Paar densitometer (medium gravity in °P). Plato measurements were taken at the beginning and at the end of each fermentation process. The fermentation was defined as finished when the CO_2 loss did not increase any further and the residual sugar concentration remained constant for 2 days. Ethanol was measured after the end of fermentation using an Anton Paar DMA 4500 M Alcolyzer. Aroma profiles were analysed by headspace gas chromatography.

2.3. Sample Preparation for Solid-Phase MicroExtraction–Gas Chromatography/Mass Spectrometry (SPME-GC/MS) Analysis

2.5 mL of the samples were put in 20 mL vials with the addition of sodium chloride (final concentration 40 mg/ml), 50 µL NaN_3 0.1%, 25 µL of the internal standard (2-octanol final concentration of 200 µg/L) and ascorbic acid (final concentration of 20 mg/mL).

All samples were incubated for 10 min at 40 °C, then the volatile compounds were collected on a divinylbenzene/carboxen/polydimethylsiloxane fiber (DVB-CAR-PDMS) coating 50/30 µm, and 2-cm length SPME fibre purchased from Supelco (Sigma-Aldrich, Milan, Italy) for 40 min.

2.4. Analytical Methods for GC/MS

GC analysis was performed on a Trace GC Ultra gas chromatograph coupled with a TSQ Quantum Tandem mass spectrometer (Thermo Electron Corporation, Waltham, MA USA), with adaptations as described in Ravasio et al. [18]. GC separation was performed on a 30 m Solgelwax PEG capillary column with an internal diameter of 0.25 mm and a film thickness of 0.25 m (SGE Analytical Science, Melbourne, Australia). The GC oven was kept at 40 °C for 4 min and then increased by 6 °C/min to 250 °C and kept at the final temperature for 5 min. The injector and transfer-line temperatures were kept at 250 °C as well. Helium was used as the carrier gas with a flow rate of 1.2 mL/min. The time for thermal desorption of analytes was 4 min. The MS detector was operated in full scan mode at 70 eV with a scan range from 35 to 350 m/z. Data analysis was performed using the software ThermoXcalibur (Version 2.2 SP1.48, Thermo scientific, Waltham, MA, USA). Identification of compounds was based on comparison with a mass spectral database (NIST version 2.0, Gaithersburg, MD, USA) and with reference standards when available. The relative amount of each volatile was expressed as µg/L of 2-octanol.

2.5. Multivariate Data Analysis

Multivariate data analysis was performed using StatSoft, Inc. STATISTICA version 8.0 (data analysis software system, 2007, StatSoft (Europe) GmbH, Berikon, Switzerland). Principal component analysis was employed to simplify data interpretation. The matrix initially contained the 60 strains considered in this study and the average of the relative 62 VOCs detected and was later reduced to the sub-selection of 19 strains studied further.

3. Results

3.1. Strain Selection and Identification of Representative Isolates

In order to cover a wide range of the biodiversity of non-conventional yeasts, we selected 60 strains from 48 different species which were obtained from the strain collection of the Westerdijk Institute. The focus was on strains that were previously isolated from various fermentations, e.g., from fermented liquids, fruits, vegetables, or meat. These strains, therefore, may have evolved superior or desirable features, e.g., the production of high levels of ethanol or may have been recognized based on their contribution to flavors. All strains were run in lab-scale fermentation trials in nutrient-rich broth (YPD with adjusted glucose content at 16 °Plato) to assess their aroma production. Flavor profiles were analyzed using GC/MS and compared to a set of brewing and wine yeast strains (Supplementary Table S1).

3.2. Aroma Profiles of Fermentations

In total, we identified 62 different volatiles in the samples of all strains (Table 2). Major volatile-aroma compounds that were detected included esters, alcohols, aldehydes, ketones and acids. Most species harbor the ability to produce a large variety of flavors. Thus, we did not identify single species that produced only very few compounds or species with a very high diversity, as shown in Figure 1 for a selection of species. However, there was an enormous difference in the amounts of specific volatiles produced. Esters were the most prominent group of volatiles. In total, 22 different esters could be identified and within this group ethyl-esters dominated, such as ethyl hexanoate and ethyl acetate, associated with fruity wine/apple-like and sweet pear drop flavors, respectively. Alcohols comprised the second major group of compounds. Besides ethanol, we identified 14 different alcohols. Yet, only two compounds, 2-phenylethanol, perceived as rose flavor, and isoamyl alcohol, a banana-like flavor, were produced by all strains analyzed. During anaerobic fermentations the formation of aroma alcohols is favored over the production of aroma acids [19]. In line with this, we identified only six acids, of which acetic acid and butyric acid were prominent.

Table 2. List of 62 volatiles that were detected by GC/MS measurement.

Number	Alcohols
1	Benzyl alcohol
2	Butanol
3	Dodecanol
4	Fenchyl alcohol
5	Furaneol
6	Isoamyl alcohol
7	Propanol
8	2-Ethyl-1-hexanol
9	2-Furanmethanol
10	2-Methyl propanol
11	2-Nonanol
12	2-Phenyl ethanol
13	3 Ethoxy-1-Propanol
14	3-(Methylthio)-1-propanol
	Esters
15	Butyl acetate
16	Ethyl (4E)-4-decenoate
17	Ethyl 2-methylbutyrate
18	Ethyl acetate
19	Ethyl butanoate
20	Ethyl decanoate
21	Ethyl dodecanoate
22	Ethyl heptanoate
23	Ethyl hexadecanoate
24	Ethyl hexanoate
25	Ethyl isobutyrate
26	Ethyl isovalerate
27	Ethyl octanoate
28	Ethyl propanoate
29	Ethyl tetradecanoate
30	Isoamyl acetate
31	Isoamyl butyrate
32	Isobutyl acetate
33	Isobutyl butanoate
34	Phenethyl acetate
35	*S*-methyl thioacetate
36	2-Methyl propanoate

Table 2. *Cont.*

Number	Alcohols
	Acids
37	Acetic acid
38	Butyric acid
39	Decanoic acid
40	Hexanoic acid
41	Isovaleric acid
42	Octanoic acid
	Aldehydes
43	Acetaldehyde
44	Benzaldehyde
45	Furfural
46	Phenyl acetaldehyde
47	1-Decanal
48	1-Nonanal
49	3-Methyl butanal
50	4-Methyl benzaldehyde
51	5 Methyl furfural
52	5-Hydroxymethylfurfural
	Ketons
53	Acetoin
54	Diacetyl
55	Pyranone
56	2-Cyclopentene-1,4-dione
57	2-Dodecanone
58	2-Methyltetrahydrothiophen-3-one
59	2-Nonanone
60	2-Undecanone
	Pyrazines
61	2,5-Dimethyl-3-ethylpyrazine
62	2,6-Dimethylpyrazine

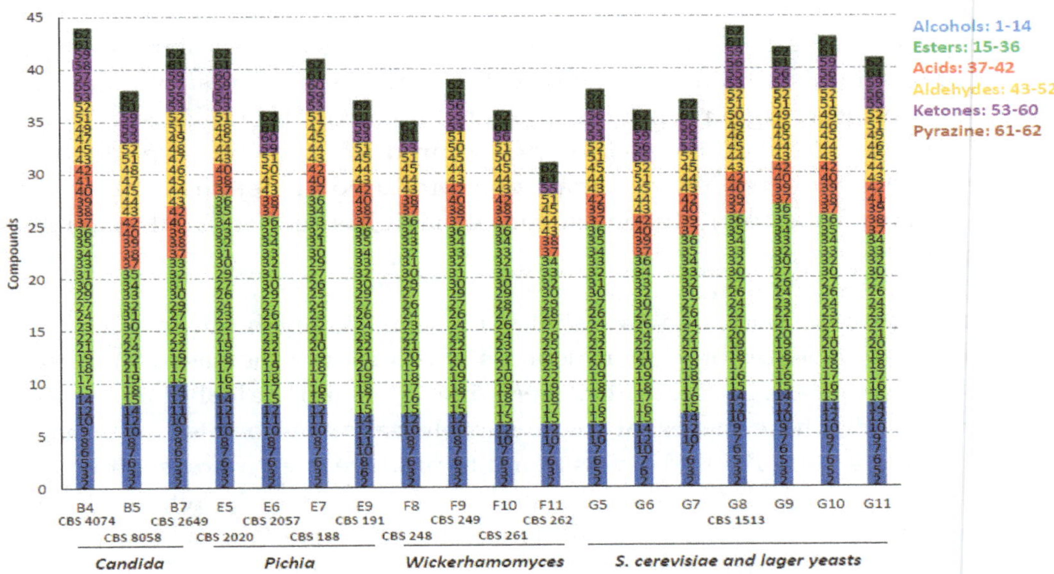

Figure 1. Number of volatile organic compounds (VOCs) produced during fermentation of selected strains. The compounds were grouped, numbered and colored according to their chemical class. The complete list of compounds is shown in Table S1 (see Table 1 for nomenclature of strains).

We were most interested in species that separated well from *S. cerevisiae* or lager yeast strains. Therefore, we narrowed down the selection of strains to those that showed clear separation in their flavour profiles to the set of *S. cerevisiae* and lager yeast strains based on principal component analysis (Figure 2). This identified species belonging to the genera *Candida*, *Pichia* and *Wickerhamomyces*.

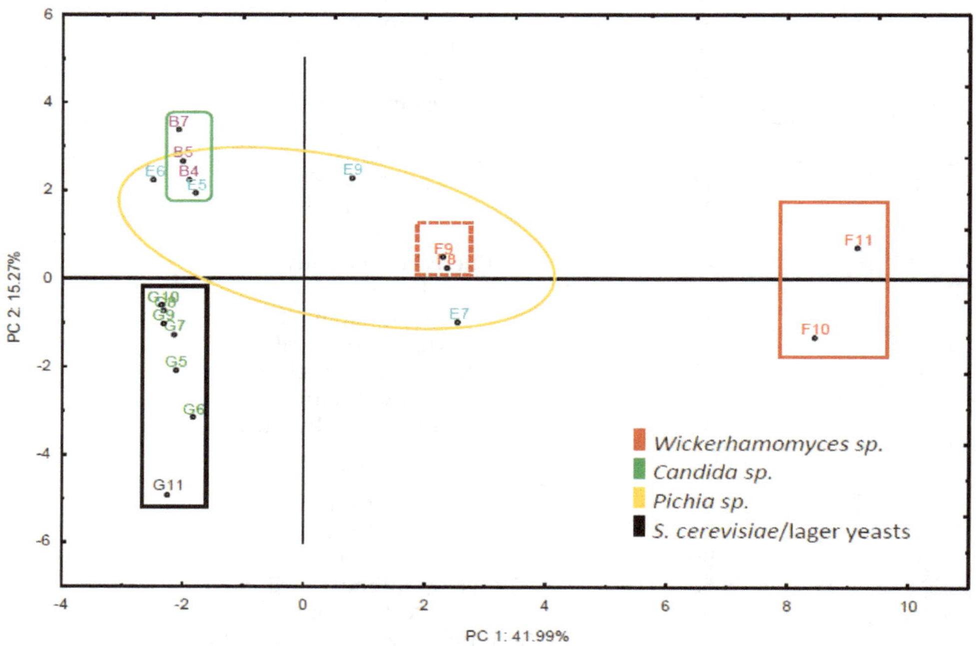

Figure 2. Principal component analysis of the set of selected strains and their chemical compounds produced. The matrix is based on the set of 19 strains and the VOCs detected. Strains are presented with their assigned coordinates (see Table 1). Species belonging to the same genus are represented by the same color.

Based on the GC data, we generated relative quantifications of the volatile aroma compounds formed. The set of strains was then compared to the lager yeast *S. pastorianus/Weihenstephan* to identify strains that produced higher amounts of esters and alcohols (Tables 3 and 4). One strain of *Saccharomyces cerevisiae* (CBS 1250; E11) was a very good producer of alcohols, particularly of isoamyl alcohol, by contrast with lager yeast. Yet, this strain only produced moderate amounts of esters, mostly ethyl acetate. Its origin from sherry production suggests that it has been selected as the preferred strain for these fermentations.

The *Wickerhamomyces anomalus* isolates showed a remarkably high amount of both alcohol and ester production. Production of higher alcohols was strongly increased in these *W. anomalus* strains compared to lager yeast. Interestingly, here mainly isoamyl alcohol and 2-phenyl ethanol production was increased. These represent two desirable flavors associated with banana and rose flavours. The sherry isolate (CBS 1250, E11) was found to produce even more isoamyl alcohol than the *W. anomalus* strains, but far less 2-phenylethanol than these strains (Table 3). Yet, the *W. anomalus* strains produced abundant amounts of esters, particularly ethyl acetate, isoamyl acetate and 2-phenylethyl acetate. Overall, these strains produced up to 10-fold more esters than the *Weihenstephan* lager yeast strain. (Table 4). Another highly aromatic strain identified in this collection of strains was a *Pichia kluyveri* strain, CBS 188. This strain produced almost fourfold more esters than lager yeast, while its aroma alcohols profile was similar to lager yeast with the exception of fourfold higher 2-phenylethanol production.

Table 3. Strains with higher aroma alcohol production than the *Weihenstephan* lager yeast strain.

Strain Identifier	CBS1250	CBS262	CBS1082	CBS726	CBS261	CBS191	CBS188	CBS2568	CBS2649
Position	E11	F11	G3	G4	F10	E9	E7	C7	B7
Species	*S. cerevisiae*	*W. anomalus*	*Z. bisporus*	*Z. mellis*	*W. anomalus*	*P. membrani-faciens*	*P. kluyveri*	*H. vineae*	*C. stellata*
Butanol	4.15	3.23	4.57	2.98	1.89	1.80	1.75	1.27	1.49
Furaneol	1.81	0.00	3.49	2.07	0.00	0.00	0.00	0.00	1.82
Isoamyl alcohol	3.83	2.45	1.86	2.04	1.12	1.91	0.90	0.79	0.76
Propanol	4.28	1.18	2.19	2.37	1.68	0.00	2.04	1.87	0.00
2-Furanmethanol	0.00	0.00	3.11	2.10	0.00	0.00	0.00	0.00	9.13
2-Methyl propanol	9.10	2.92	2.78	3.62	2.31	3.61	1.24	0.59	2.08
2-Phenyl ethanol	1.42	4.26	1.98	1.94	4.81	2.19	2.97	1.53	1.45
3-(Methylthio)-1-propanol	2.38	0.00	0.00	2.98	0.00	1.83	0.00	12.58	1.98

The values reported represent the relative abundances of the specific subset of metabolites compared to the lager yeast strain (G10).

Table 4. Strains with higher aroma ester production than the *Weihenstephan* lager yeast strain.

Strain Identifier	CBS261	CBS262	CBS188	CBS249	CBS248	CBS1082	CBS133
Position	F10	F11	E7	F9	F8	G3	F6
Species	*W. anomalus*	*W. anomalus*	*P. kluyveri*	*W. anomalus*	*W. anomalus*	*Z. bisporus*	*T. delbrueckii*
Butyl acetate	23.64	16.84	9.18	8.76	11.73	3.84	3.56
Ethyl (4E)-4-decenoate	0.00	0.00	0.17	0.00	0.04	0.56	0.07
Ethyl 2-methylbutyrate	50.89	368.00	36.00	62.34	39.06	2.94	10.63
Ethyl acetate	31.81	29.17	15.88	13.09	10.02	8.72	5.70
Ethyl butanoate	16.59	26.92	8.00	9.26	10.75	2.58	2.71
Ethyl decanoate	0.14	0.00	2.07	0.46	0.45	2.67	0.72
Ethyl dodecanoate	0.91	0.00	2.60	0.88	1.44	2.69	0.98
Ethyl heptanoate	1.17	2.19	6.96	3.14	1.77	1.58	2.82
Ethyl hexadecanoate	12.19	13.62	16.05	9.10	8.22	11.00	2.36
Ethyl hexanoate	7.23	12.89	5.07	6.90	7.47	4.33	2.72
Ethyl isovalerate	11.84	120.99	33.97	14.43	15.00	0.00	10.79
Ethyl octanoate	0.56	0.26	1.59	0.59	0.32	0.64	0.55
Isoamyl acetate	9.36	5.84	3.21	2.92	3.10	1.27	0.99
Isobutyl acetate	152.96	59.59	24.70	40.69	26.57	6.75	9.93
Isobutyl butanoate	209.13	167.38	29.44	88.87	52.26	10.77	13.62
2-Phenylethyl acetate	5.88	4.07	2.70	2.68	2.90	0.44	0.91
methyl thioacetate	0.00	0.00	0.00	0.00	0.00	0.21	0.00
2-Methyl propanoate	0,89	0,00	1,16	1,13	0,46	0,00	0,44

The values reported represent the relative abundances of the specific subset of metabolites compared to the lager yeast strain (G10).

3.3. Fermentation Performance

We compared the fermenting capacity of the strains with very strong volatile compound formation to the lager yeast strain *S. pastorianus WS34/70* and other *S. cerevisiae* strains. All strains were adjusted to the same optical density prior to the start of fermentation to allow for comparison of the speed of fermentation between strains. Fermentation rates were followed by measuring the CO_2 release daily (Figure 3). All strains were able to ferment the liquid within one week except for the *W. anomalus* strains CBS 248; CBS 249 and CBS 261. However, in contrast to the other strains, the *W. anomalus* strains showed an almost linear fermentation curve with equal amounts of CO_2 released per day. During fermentation the *W. anomalus* strains produced oily top layers and biofilms of cells. This oily phase may have reduced the CO_2 release as we observed the lowest pH in the *W. anomalus* fermentations measured among all strains: for strain CBS 261 the pH reached 4.79 and for strain CBS 262 the pH was 4.59, while the liquids fermented by other yeast strains showed a pH of >5. This suggests that the dissolved CO_2 contributed to the acidification of the liquid in the *W. anomalus* fermentations. The most rapid fermentations as represented by fast CO_2 loss were observed with the *S. cerevisiae* and lager beer production strains. Yet, in addition to the *Wickerhamomyces* strains, several other strains showed prolonged CO_2 loss, including *P. kluyveri* CBS 188 (Figure 3).

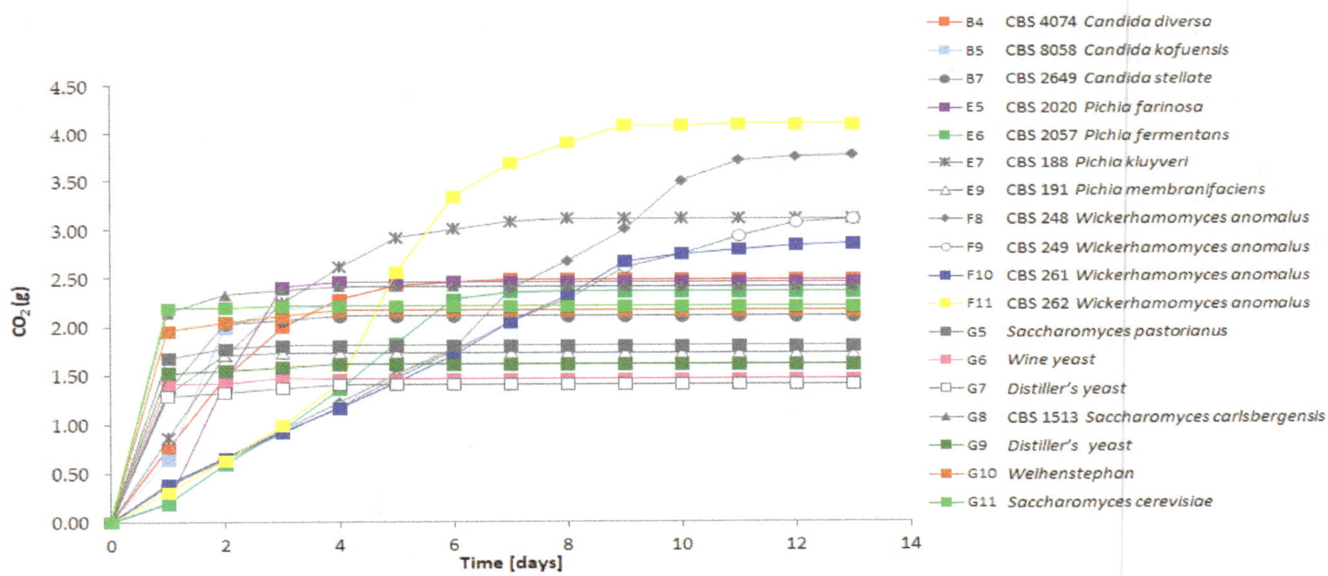

Figure 3. Fermentation profiles of the selected strains representing the detected flavor diversity. The CO_2 release was measured daily for each individual strain (see Table 1 for the nomenclature of strains). Fermentations were followed for 13 days.

For most of the strains we could observe a correlation between fermentation speed, residual sugar concentration and ethanol production. The *S. cerevisiae* and lager yeast strains left 2–2.5% of sugar in the medium, while the residual sugar in *Candida* and *Pichia* strains was between 3–3.5%. *Wickerhamomyces anomalus* strains left 4.5–5% of sugars unfermented (Figure 4). Concomitantly, *S. cerevisiae* and lager yeast strains produced the highest amount of alcohol (up to 8%) and the *W. anomalus* strains only 4%.

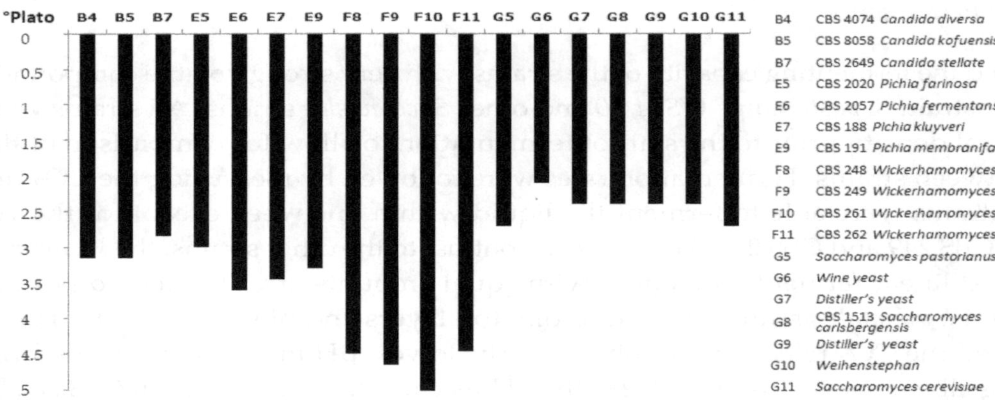

Figure 4. Final sugar content of the selected strains at the end of fermentation. Sugar concentration was measured in °P after 13 days of fermentation.

3.4. Growth on Other Carbon Sources and Temperatures

The selected species were analysed for growth at low (10 °C) and high (37 °C) temperatures as well as for their ability to utilize maltose using serial-dilution plate assays. Growth at elevated temperatures was mainly restricted to *Saccharomyces* and *Zygosaccharomyces* strains, while growth at lower temperatures was often better in non-*cerevisiae* strains. Typically, all strains grew well at intermediate temperatures. Lager beer fermentation relies on maltose utilization. However, in this screening, good maltose utilization was restricted to lager yeast strains, with the exception of *Candida diversa* CBS 4074 (Figure 5), which also showed very good growth on maltose plates and *Clavispora lusitaniae* CBS 6936.

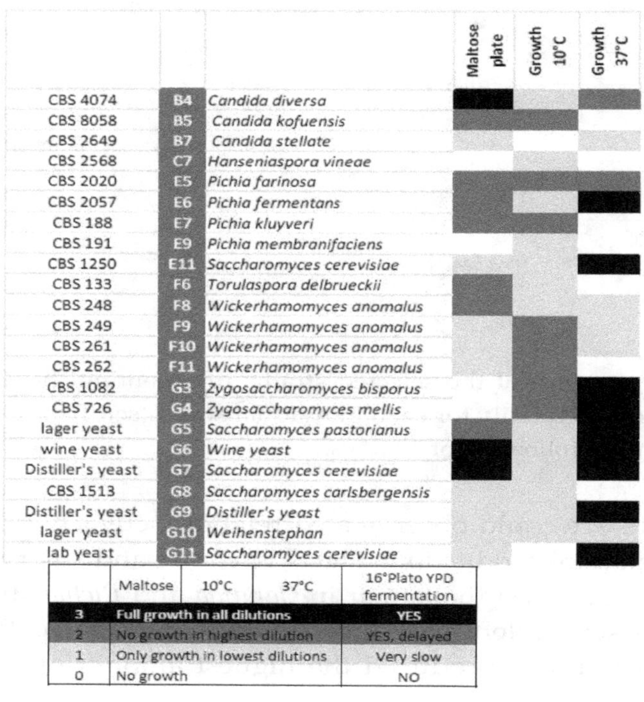

Figure 5. Heat map displaying maltose utilization and growth at the indicated temperatures of the selected strains.

3.5. Utilization of Mixed Fermentations for Flavor Improvement

Co-fermentations are one way to improve the flavour composition of fermented beverages. We plotted volatile compound formation of several candidate strains against lager yeast (Figure 6a). This demonstrated the superior flavour-generation capacity of several strains, most notably of *W. anomalus*. Therefore, we used the same nutrient-rich high Plato fermentation broth and inoculated the *Weihenstephan* lager yeast WS34/70 in a 1:1 ratio with *W. anomalus* CBS 261. Volatiles of this mixed fermentation were determined at the end of fermentation. This clearly demonstrated an enhancement of ethyl hexadecanoate, isoamyl alcohol and 2-phenyl ethanol, improving the fruity flavour perception of the fermented liquid (Figure 6b). *W. anomalus* CBS 261 is a very strong producer of ethyl acetate, which was also pronounced in the co-fermentation.

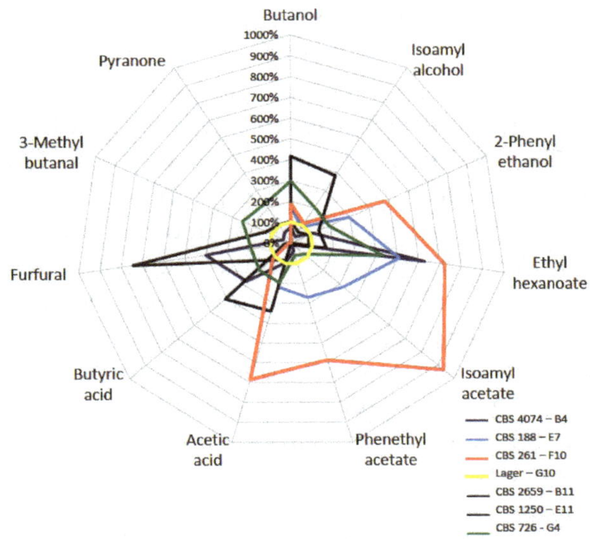

(A)

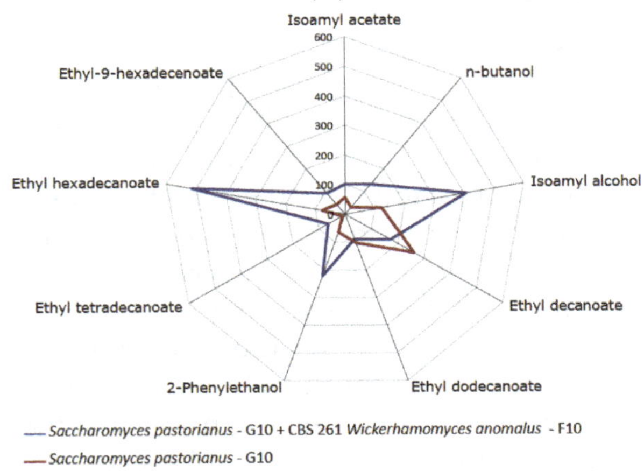

(B)

Figure 6. (**A**) Comparisons of selected volatiles of selected strains that produced exceptionally high concentrations of organic volatiles with the lager yeast reference. Strains used were: *Candida diversa* (CBS 4074, B4); *Debaryomyces subglobosus* (CBS 2659, B11); *Pichia kluyveri* (CBS 188, E7); *Saccharomyces cerevisiae* (CBS 1250; E11); *Wickerhamomyces anomalus* (CBS 261, F10); *Zygosaccharomyces mellis* (CBS 726, G4); *Saccharomyces pastorianus* (G10, reference). (**B**) Concentration of volatiles at the end of a co-fermentation using *Wickerhamomyces anomalus* (CBS 261, F10) and the *Weihenstephan* lager yeast (WS, *Saccharomyces pastorianus* G10, reference).

4. Discussion

In our study we have screened non-*Saccharomyces* cerevisiae yeast biodiversity in order to identify strains with more pronounced volatile compound formation than present in lager yeast strains. Pronounced differences in aroma alcohol production were identified between *S. cerevisiae* and lager yeast strains, as expected from a clean pilsner beer produced by lager yeasts versus the more complex flavors produced by ale and wine yeasts. In our screening, we identified *W. anomalus* strains as the most dominant flavor producers, which also included production of substantial amounts of acetic acid. The formation of floating cell layers such as seen for the *W. anomalus* strains has been reported as a typical phenomenon in stored wines [20]. Yeast biodiversity holds a plethora of strains that show useful characteristics such as ethanol production and flavor formation. This requires a detailed evaluation of the initially identified favorable strains under different conditions, particularly in co-fermentation regimes [21]. It will be challenging to identify the most suitable co-fermentation setups, as flavor profiles will certainly be influenced by different ratios of non-conventional yeasts versus standard brewing strains.

In earlier studies, *W. anomalus* strains have been isolated from a range of cereal-based sources. It has been reported from sourdoughs and was found as the dominating yeast in sourdough microbial ecosystems next to *S. cerevisiae*. The prevalence of the fungus was associated with its osmotolerance and increased acid tolerance in comparison to *S. cerevisiae* [22]. Furthermore, it was shown that *W. anomalus* provides some antimicrobial activity, e.g., mycocin production, that can be used to prolong the shelf life of bread [23,24]. Other *Wickerhamomyces* species have been put to use in baking using microbread baking platforms. Specifically, bread obtained with *Kazachstania gamospora* and *Wickerhamomyces subpelliculosus* provided added broader aroma profiles compared to control baker's yeast [25].

P. kluyveri CBS 188 produced a total of 41 volatiles during our fermentations. Yet, it was outstanding in its ester profile, e.g., ethyl acetate, isoamyl acetate or phenethyl acetate that was several folds higher than in the reference lager yeast strain *WS34/70*. *P. kluyveri* strains are found in "wild ferments" of wine but it is also commercially available to boost flavour production through sequential fermentations. *P. kluyveri* strains together with strains of *K. marxianus* were also presented as potential starter yeasts for controlled cocoa fermentation [26].

Based on increased demand for natural flavors produced from yeasts, the use of non-conventional yeasts as platform strains for the production of aroma volatile has gained attention. Hence, *K. marxianus* was suggested as a cell factory for flavor and fragrance production based on several advantageous traits, e.g., thermotolerance and the wide array of volatile molecules it produces [27]. We had one strain of *K. marxianus* (CBS 1557) in our collection. However, this strain was as low in aroma-compound production as our lager yeast reference. This indicates the need to obtain a larger collection of strains of one species and also to analyse volatile compound formation under different nutritional regimes and fermentation conditions [28]. It further requires the implementation of high-throughput screening tools and assays to identify suitable strains, which was actually successfully shown for *K. marxianus* strains producing ethyl acetate [29]. Additionally, it requires more effort to acquire genomic, transcriptomic and metabolomic datasets of non-conventional yeasts in order to bridge the knowledge gap about *S. cerevisiae* [12,30]. That said, our approach to screening non-conventional yeasts in non-industrial fermentation broths only provides an initial glance of the volatile production capabilities of the tested strains. These capabilities will certainly vary with fermentation and process conditions, particularly at larger scales.

The strong interest in volatile-compound formation by non-conventional yeasts may also open a new perspective on yeast ecology. Yeasts occur in diverse niches and interact with other microbes, insects and plants. These interactions may present selective forces for the production of specific volatile compounds. Interestingly, strains of *Debaryomyces hansenii* were shown to produce methyl salicylate (MeS) as a major compound under very specific conditions using pine weevil (*Hylobius abietis*) frass broth [31]. This compound essentially acts as a deterrent for pine weevil. In a follow-up case, 2-phenylethanol was identified as a strong anti-feedant compound against the pine weevil [32].

Several of the strains we analyzed produced large amounts of 2-phenylethanol, including strains from *S. cerevisiae* but most prominently the *W. anomalus* strains. *W. anomalus* strains are often isolated from tree habitats, insects and insect frass [33]. Elucidating this fascinating interplay of yeasts, insects and plants will provide substantial insight into yeast biology and ecology in the future.

The results obtained in this study indicate that yeast biodiversity harbors a large variety of strains that could enter diverse beverage production pipelines and provide additional all-natural flavor variants to improve the taste and sensory perception of lager beers and, beyond that, other fermented beverages. We used a non-industrial platform to assay strains. In future research, more detailed fermentations using specific industrial fermentation broths, e.g., wort, grape must, and other juices, should be explored with non-conventional yeasts.

Acknowledgments: This research was supported in part by the European Union Marie Curie Initial Training Networks Cornucopia 264717 (http://www.yeast-cornucopia.se/) and Aromagenesis 764364. Yeast strains were obtained from the Westerdijk Fungal Biodiversity Institute, Utrecht, The Netherlands.

Author Contributions: Davide Ravasio, Andrea Walther, Marizeth Groenewald, Teun Boekhout and Jürgen Wendland conceived and designed the experiments; Davide Ravasio, Silvia Carlin, Urska Vrhovsek and Andrea Walther performed the experiments; Davide Ravasio, Silvia Carlin, Urska Vrhovsek, Andrea Walther and Jürgen Wendland analyzed the data; all authors contributed to writing the paper.

References

1. Dunn, B.; Sherlock, G. Reconstruction of the genome origins and evolution of the hybrid lager yeast *Saccharomyces pastorianus*. *Genome Res.* **2008**, *18*, 1610–1623. [CrossRef] [PubMed]
2. Libkind, D.; Hittinger, C.T.; Valerio, E.; Goncalves, C.; Dover, J.; Johnston, M.; Goncalves, P.; Sampaio, J.P. Microbe domestication and the identification of the wild genetic stock of lager-brewing yeast. *Proc. Natl. Acad. Sci. USA* **2011**, *108*, 14539–14544. [CrossRef] [PubMed]
3. Nakao, Y.; Kanamori, T.; Itoh, T.; Kodama, Y.; Rainieri, S.; Nakamura, N.; Shimonaga, T.; Hattori, M.; Ashikari, T. Genome sequence of the lager brewing yeast, an interspecies hybrid. *DNA Res.* **2009**, *16*, 115–129. [CrossRef] [PubMed]
4. Walther, A.; Hesselbart, A.; Wendland, J. Genome sequence of *Saccharomyces carlsbergensis*, the world's first pure culture lager yeast. *G3 (Bethesda)* **2014**, *4*, 783–793. [CrossRef] [PubMed]
5. Spitaels, F.; Wieme, A.D.; Janssens, M.; Aerts, M.; Daniel, H.M.; Van Landschoot, A.; De Vuyst, L.; Vandamme, P. The microbial diversity of traditional spontaneously fermented lambic beer. *PLoS ONE* **2014**, *9*, e95384. [CrossRef] [PubMed]
6. Mertens, S.; Steensels, J.; Saels, V.; De Rouck, G.; Aerts, G.; Verstrepen, K.J. A large set of newly created interspecific *Saccharomyces* hybrids increases aromatic diversity in lager beers. *Appl. Environ. Microbiol.* **2015**, *81*, 8202–8214. [CrossRef] [PubMed]
7. Verstrepen, K.J.; Derdelinckx, G.; Dufour, J.P.; Winderickx, J.; Thevelein, J.M.; Pretorius, I.S.; Delvaux, F.R. Flavor-active esters: Adding fruitiness to beer. *J. Biosci. Bioeng.* **2003**, *96*, 110–118. [CrossRef]
8. Pires, E.J.; Teixeira, J.A.; Branyik, T.; Vicente, A.A. Yeast: The soul of beer's aroma—A review of flavour-active esters and higher alcohols produced by the brewing yeast. *Appl. Microbiol. Biotechnol.* **2014**, *98*, 1937–1949. [CrossRef] [PubMed]
9. Vanderhaegen, B.; Neven, H.; Coghe, S.; Verstrepen, K.J.; Verachtert, H.; Derdelinckx, G. Evolution of chemical and sensory properties during aging of top-fermented beer. *J. Agric. Food Chem.* **2003**, *51*, 6782–6790. [CrossRef] [PubMed]
10. Duong, C.T.; Strack, L.; Futschik, M.; Katou, Y.; Nakao, Y.; Fujimura, T.; Shirahige, K.; Kodama, Y.; Nevoigt, E. Identification of sc-type *ilv6* as a target to reduce diacetyl formation in lager brewers' yeast. *Metab. Eng.* **2011**, *13*, 638–647. [CrossRef] [PubMed]
11. Gamero, A.; Quintilla, R.; Groenewald, M.; Alkema, W.; Boekhout, T.; Hazelwood, L. High-throughput screening of a large collection of non-conventional yeasts reveals their potential for aroma formation in food fermentation. *Food Microbiol.* **2016**, *60*, 147–159. [CrossRef] [PubMed]

12. Varela, C. The impact of non-*saccharomyces* yeasts in the production of alcoholic beverages. *Appl. Microbiol. Biotechnol.* **2016**, *100*, 9861–9874. [CrossRef] [PubMed]
13. Steensels, J.; Verstrepen, K.J. Taming wild yeast: Potential of conventional and nonconventional yeasts in industrial fermentations. *Annu. Rev. Microbiol.* **2014**, *68*, 61–80. [CrossRef] [PubMed]
14. Viana, F.; Gil, J.V.; Genoves, S.; Valles, S.; Manzanares, P. Rational selection of non-*saccharomyces* wine yeasts for mixed starters based on ester formation and enological traits. *Food Microbiol.* **2008**, *25*, 778–785. [CrossRef] [PubMed]
15. Comitini, F.; Gobbi, M.; Domizio, P.; Romani, C.; Lencioni, L.; Mannazzu, I.; Ciani, M. Selected non-*saccharomyces* wine yeasts in controlled multistarter fermentations with *Saccharomyces cerevisiae*. *Food Microbiol.* **2011**, *28*, 873–882. [CrossRef] [PubMed]
16. Sadoudi, M.; Tourdot-Marechal, R.; Rousseaux, S.; Steyer, D.; Gallardo-Chacon, J.J.; Ballester, J.; Vichi, S.; Guerin-Schneider, R.; Caixach, J.; Alexandre, H. Yeast-yeast interactions revealed by aromatic profile analysis of sauvignon blanc wine fermented by single or co-culture of non-*saccharomyces* and *Saccharomyces* yeasts. *Food Microbiol.* **2012**, *32*, 243–253. [CrossRef] [PubMed]
17. Canonico, L.; Comitini, F.; Ciani, M. Torulaspora delbrueckii contribution in mixed brewing fermentations with different saccharomyces cerevisiae strains. *Int. J. Food Microbiol.* **2017**, *259*, 7–13. [CrossRef] [PubMed]
18. Ravasio, D.; Wendland, J.; Walther, A. Major contribution of the Ehrlich pathway for 2-phenylethanol/rose flavor production in *Ashbya gossypii*. *FEMS Yeast Res.* **2014**, *14*, 833–844. [CrossRef] [PubMed]
19. Hazelwood, L.A.; Daran, J.M.; van Maris, A.J.; Pronk, J.T.; Dickinson, J.R. The Ehrlich pathway for fusel alcohol production: A century of research on *Saccharomyces cerevisiae* metabolism. *Appl. Environ. Microbiol.* **2008**, *74*, 2259–2266. [CrossRef] [PubMed]
20. Alexandre, H. Flor yeasts of *Saccharomyces cerevisiae*—Their ecology, genetics and metabolism. *Int. J. Food Microbiol.* **2013**, *167*, 269–275. [CrossRef] [PubMed]
21. Viana, F.; Belloch, C.; Valles, S.; Manzanares, P. Monitoring a mixed starter of *Hanseniaspora vineae-Saccharomyces cerevisiae* in natural must: Impact on 2-phenylethyl acetate production. *Int. J. Food Microbiol.* **2011**, *151*, 235–240. [CrossRef] [PubMed]
22. Daniel, H.M.; Moons, M.C.; Huret, S.; Vrancken, G.; De Vuyst, L. *Wickerhamomyces anomalus* in the sourdough microbial ecosystem. *Antonie Van Leeuwenhoek* **2011**, *99*, 63–73. [CrossRef] [PubMed]
23. Coda, R.; Di Cagno, R.; Rizzello, C.G.; Nionelli, L.; Edema, M.O.; Gobbetti, M. Utilization of African grains for sourdough bread making. *J. Food Sci.* **2011**, *76*, M329–M335. [CrossRef] [PubMed]
24. Tay, S.T.; Lim, S.L.; Tan, H.W. Growth inhibition of *Candida* species by *Wickerhamomyces anomalus* mycocin and a lactone compound of *Aureobasidium pullulans*. *BMC Complement. Altern. Med.* **2014**, *14*, 439. [CrossRef] [PubMed]
25. Zhou, N.; Schifferdecker, A.J.; Gamero, A.; Compagno, C.; Boekhout, T.; Piskur, J.; Knecht, W. *Kazachstania gamospora* and *Wickerhamomyces subpelliculosus*: Two alternative baker's yeasts in the modern bakery. *Int. J. Food Microbiol.* **2017**, *250*, 45–58. [CrossRef] [PubMed]
26. Crafack, M.; Mikkelsen, M.B.; Saerens, S.; Knudsen, M.; Blennow, A.; Lowor, S.; Takrama, J.; Swiegers, J.H.; Petersen, G.B.; Heimdal, H.; et al. Influencing cocoa flavour using *Pichia kluyveri* and *Kluyveromyces marxianus* in a defined mixed starter culture for cocoa fermentation. *Int. J. Food Microbiol.* **2013**, *167*, 103–116. [CrossRef] [PubMed]
27. Morrissey, J.P.; Etschmann, M.M.; Schrader, J.; de Billerbeck, G.M. Cell factory applications of the yeast *Kluyveromyces marxianus* for the biotechnological production of natural flavour and fragrance molecules. *Yeast* **2015**, *32*, 3–16. [PubMed]
28. Urit, T.; Li, M.; Bley, T.; Loser, C. Growth of *Kluyveromyces marxianus* and formation of ethyl acetate depending on temperature. *Appl. Microbiol. Biotechnol.* **2013**, *97*, 10359–10371. [CrossRef] [PubMed]
29. Lobs, A.K.; Lin, J.L.; Cook, M.; Wheeldon, I. High throughput, colorimetric screening of microbial ester biosynthesis reveals high ethyl acetate production from *Kluyveromyces marxianus* on c5, c6, and c12 carbon sources. *Biotechnol. J.* **2016**, *11*, 1274–1281. [CrossRef] [PubMed]
30. Masneuf-Pomarede, I.; Bely, M.; Marullo, P.; Albertin, W. The genetics of non-conventional wine yeasts: Current knowledge and future challenges. *Front. Microbiol.* **2015**, *6*, 1563. [CrossRef] [PubMed]

31. Azeem, M.; Terenius, O.; Rajarao, G.K.; Nagahama, K.; Nordenhem, H.; Nordlander, G.; Borg-Karlson, A.K. Chemodiversity and biodiversity of fungi associated with the pine weevil *Hylobius abietis*. *Fungal Biol.* **2015**, *119*, 738–746. [CrossRef] [PubMed]
32. Axelsson, K.; Konstanzer, V.; Rajarao, G.K.; Terenius, O.; Seriot, L.; Nordenhem, H.; Nordlander, G.; Borg-Karlson, A.K. Antifeedants produced by bacteria associated with the gut of the pine weevil *Hylobius abietis*. *Microb. Ecol.* **2017**, *74*, 177–184. [CrossRef] [PubMed]
33. James, S.A.; Barriga, E.J.; Barahona, P.P.; Harrington, T.C.; Lee, C.F.; Bond, C.J.; Roberts, I.N. Wickerhamomyces arborarius f.A.; sp. Nov., an ascomycetous yeast species found in arboreal habitats on three different continents. *Int. J. Syst. Evol. Microbiol.* **2014**, *64*, 1057–1061. [CrossRef] [PubMed]

10

The Interaction of Two *Saccharomyces cerevisiae* Strains Affects Fermentation-Derived Compounds in Wine

Frida S. Gustafsson [1], Vladimir Jiranek [2], Marissa Neuner [1], Chrystal M. Scholl [1], Sydney C. Morgan [1] and Daniel M. Durall [1,*]

[1] The University of British Columbia (UBC), Okanagan, Biology Department, 1177 Research Rd., Kelowna, BC V1V 1V7, Canada; fridasofiegustafsson@gmail.com (F.S.G.); marissa.neuner@ubc.ca (M.N.); chrystalmarie.scholl@gmail.com (C.M.S.); sydney-morgan@hotmail.com (S.C.M.)

[2] Department of Wine and Food Science, The University of Adelaide, PMB1, Glen Osmond SA 5064, Australia; vladimir.jiranek@adelaide.edu.au

* Correspondence: daniel.durall@ubc.ca

Academic Editor: Ronnie G. Willaert

Abstract: Previous winery-based studies showed the strains Lalvin® RC212 (RC212) and Lalvin® ICV-D254 (D254), when present together during fermentation, contributed to >80% relative abundance of the *Saccharomyces cerevisiae* population in inoculated and spontaneous fermentations. In these studies, D254 appeared to out-compete RC212, even when RC212 was used as the inoculant. In the present study, under controlled conditions, we tested the hypotheses that D254 would out-compete RC212 during fermentation and have a greater impact on key fermentation-derived chemicals. The experiment consisted of four fermentation treatments, each conducted in triplicate: a pure culture control of RC212; a pure culture control of D254; a 1:1 co-inoculation ratio of RC212:D254; and a 4:1 co-inoculation ratio of RC212:D254. Strain abundance was monitored at four stages. Inoculation ratios remained the same throughout fermentation, indicating an absence of competitive exclusion by either strain. The chemical profile of the 1:1 treatment closely resembled pure D254 fermentations, suggesting D254, under laboratory conditions, had a greater influence on the selected sensory compounds than did RC212. Nevertheless, the chemical profile of the 4:1 treatment, in which RC212 dominated, resembled that of pure RC212 fermentations. Our results support the idea that co-inoculation of strains creates a new chemical profile not seen in the pure cultures. These findings may have implications for winemakers looking to control wine aroma and flavor profiles through strain selection.

Keywords: *Saccharomyces cerevisiae*; strain interaction; fermentation-derived compounds

---

1. Introduction

In spontaneous fermentations conducted at commercial wineries, it is common to find more than one *Saccharomyces cerevisiae* strain fermenting the wine must [1]; however, multiple strains have also been detected even in inoculated fermentations [2,3]. It is well documented that different wine strains of *S. cerevisiae* affect flavor and aroma properties differently [1]. Although the sensory influence of co-inoculation between non-*Saccharomyces* and a single *S. cerevisiae* strain has been widely studied [4–8], fewer studies have reported on the co-inoculation of multiple *S. cerevisiae* strains [9–13]. The commercial active dry yeast (ADY) strains, Lalvin® Bourgorouge RC212 (RC212) and Lalvin® ICV-D254 (D254), are frequently used to ferment Pinot Noir and Chardonnay musts, respectively. Together, they have been found to dominate operational fermentations, with an overall relative abundance of >80% in both inoculated (where RC212 was used as the sole inoculum and D254 entered as a contaminant) and spontaneous Pinot Noir fermentations [14,15]. Furthermore, D254 was the dominant strain at the end of these fermentations, even when tanks were inoculated with RC212 [3]. These findings suggest, when observing their dynamics during operationally conducted fermentations, that D254 out-competes RC212. Originally, the strain RC212 was selected by the Burgundy Wine Board (BIVB) to extract and protect the polyphenols of Pinot Noir. In the information supplied by the manufacturer, it is claimed that wines fermented by RC212 have good structure with fruity and spicy characteristics (Lallemand Inc., Montreal, QC, Canada). The strain D254 is commonly used in both red and white wines. Red wines fermented with D254 contribute to high fore-mouth volume, smooth tannins, intense fruits and a slightly spicy finish (Lallemand Inc., Montreal, QC, Canada). Nevertheless, there is a lack of information on the sensorial attributes when these two strains co-exist during fermentation. Given that there are many factors that can affect the interactions of these two strains under operational conditions, it is important to determine how these two strains interact and affect key fermentation-derived chemicals under controlled conditions.

The formation of aroma and flavor compounds is dependent on the nutrient availability, the physicochemical properties of the fermentation, and the yeast strains present, especially *S. cerevisiae* strains. Higher alcohols and esters are usually yeast-derived and can greatly contribute to the aroma and flavor profile of the wine [16]. Many of these flavor compounds are derivatives of amino acids, and it has been shown that amino acid uptake by yeasts is strain-dependent [11,17]. Other wine aroma and flavor compounds include pyrazines, terpenes, lactones, sulfur-containing compounds, phenols, organic acids, and aldehydes, which are usually not strain-dependent. The concentration of these other compounds is strongly influenced by varietal, grape ripeness, non-*Saccharomyces* organisms, aging, and winemaking practices [16,18]. Several studies have concluded that different strains of *S. cerevisiae* produce strain-specific metabolites [19,20]. For example, higher alcohols and esters can differ with varying dominance of two or more strains [11,19,20]. At low concentrations, higher alcohols contribute to increased aroma complexity, but at high concentrations (>300 mg/L), their presence can be undesirable [21,22]. At low concentrations (<100 mg/L), ethyl esters, such as ethyl acetate, often contribute fruity aromas, but at high concentrations they can produce undesirable solvent-like aromas and flavors [16,23]. In the present study, we targeted only compounds that are known to be fermentation-derived and are integral to aroma and flavor development.

Knowledge of the competitive interaction between different *S. cerevisiae* strains and its effect on aroma and flavor compounds will guide winemakers in choosing commercial yeasts, because final wine composition may be enhanced with the use of the most suitable combination of yeast strains [11]. In addition, we are not aware of any competition or metabolomic studies that have conducted co-fermentations with RC212 and D254 strains in grape must. For our study, competition between two strains, which ultimately results in competitive exclusion, is defined at the end of a co-inoculated fermentation, where one strain has a greater relative abundance than it did when it was inoculated.

The aim of this study was to generate and test hypotheses that were based on observations from operational settings and from the literature. We tested, under controlled conditions, the hypotheses that: (1) D254 will out-compete RC212 when inoculated as a 1:1 or as a 4:1 RC212:D254 ratio; (2) D254 will have a greater impact than RC212 on key fermentation-derived chemicals when the inoculation abundances of the two strains are equal; and (3) D254 will have a greater impact than RC212 on key fermentation-derived chemicals when the inoculation is administered in a 4:1 RC212:D254 ratio. Our results indicate that no competitive exclusion occurred in the co-inoculated treatments, but rather the inoculated ratios remained constant throughout fermentation. Furthermore, we found that the chemical profile of the 1:1 RC212:D254 treatment closely resembled the chemical profile of the pure D254 fermentations, but the 4:1 RC212:D254 treatment more closely resembled the chemical profile of the pure RC212 fermentations. We conclude that although D254 does not appear to competitively exclude RC212 under controlled conditions, it has a relatively larger impact on the sensory profile of the resulting wines than RC212.

2. Materials and Methods

2.1. Experimental Design

The experiment consisted of four fermentation treatments: a pure culture control of RC212; a pure culture control of D254; a 1:1 co-inoculation ratio of RC212:D254; and a 4:1 co-inoculation ratio of RC212:D254. Each treatment was replicated using three separate fermentation flasks for a total of 12 flasks, with each flask containing 100 mL Pinot Noir juice. Each flask was sampled for strain abundance at the start (180 g/L sugar, 0 h), early (83–102 g/L sugar, 24 h), mid (64–73 g/L, 32 h), and end stages (<2 g/L sugar, 97 h) of the 100 h fermentation. Samples for chemical analysis were taken only at the end stage of fermentation. The co-inoculation treatments represented one situation where the two strains were inoculated in equal abundance (1:1 ratio) and another where RC212 was inoculated at a higher proportion than D254 (4:1 ratio); these two co-inoculation treatments, along with their pure-culture controls, allowed us to adequately test all of our hypotheses.

2.2. Juice Preperation

Pinot Noir juice was obtained from WineExpert™ (Port Coquitlam, BC, Canada). The juice was prepared by centrifugation for 45 min at 3500× g and was subsequently filtered through a series of filters, which had a decreasing pore size: 2.7 μm glass fiber filter (GF), 1 μm GF, 0.45 μm mixed cellulose ester membrane filter (MCE), and 0.22 μm MCE and polyvinylidene difluoride filter (PVDF). The filtered juice was adjusted to 180 g/L sugar with sterile Milli-Q water and stored at −20 °C until it was needed for the experiment. We selected this concentration because it was within the typical range (180–220 g/L) at which grape juice fermentation commences [24]. The filtered juice, following the adjustment to 180 g/L sugar, had a pH of 3.8 and its sterility was confirmed by plating 0.1 mL onto yeast extract peptone dextrose (YEPD) media and observing an absence of colonies after 4 days of incubation at 28 °C. The adjusted filtered juice (>2 L) was used as the source to make the RC212 and D254 inoculated solutions, described in the section below.

2.3. Inoculation and Fermentations

For both strains, (~10 mg) ADY inoculum was rehydrated in 25 mL liquid YEPD media and was shaken for eight hours (120 rpm) at 28 °C. Yeast abundance (cells/mL) in the rehydrated suspension was counted using a hemocytometer. Rehydrated yeasts were added in a quantity of 1×10^6 cells/mL to 100 mL diluted Pinot Noir grape juice (1:1 juice:sterile Milli-Q H_2O) for each strain. Once yeast cell count was determined in each solution, the RC212 and D254 solutions were added separately to 1.2 L and 700 mL of the filtered Pinot Noir juice, respectively, to produce a concentration of 5×10^6 cells/mL. The resulting master mixes of each strain were combined in the appropriate ratios to obtain 300 mL of each co-inoculation treatment. Subsequently, for each co-inoculation treatment, the resulting solution was divided into three independent flasks (each containing 100 mL juice). The pure-culture controls were treated the same way; however, the RC212 and D254 solutions were not mixed. For all treatments, the final inoculation concentration was 5×10^6 cells/mL.

Fermentations (100 mL per flask) were conducted in 250 mL fermentation flasks, which contained sampling ports and air-locks. The flasks were shaken (120 rpm) at 28 °C until the end of the fermentation. To monitor the progression of fermentation and to identify strains, 0.5 mL samples were collected aseptically at the start, early, mid, and end stages of fermentation. At the end of fermentation, all wines contained <2 g/L residual sugar, as indicated with a D-Glucose/D-Fructose sugar assay kit (Megazyme, Bray, Ireland). The wine was clarified by centrifugation (1200× g; 2 min) and filtered (0.45 μm) at the end of fermentation. At the end stage, 40 mL were transferred to glass vials and stored at −80 °C until chemical analysis was performed.

2.4. Yeast Strain Identification

Wine must samples from each stage were plated on YEPD agar and incubated at 28 °C for 48 h. Twenty colonies from each plate (960 colonies total) were randomly chosen for DNA analysis. Extraction and amplification of the DNA followed the methods of Lange *et al.* [15], except that amplification of the isolates was performed with primer sets for the microsatellite loci C11 and SCYOR267c [25]. These two loci were chosen because RC212 is heterozygous and D254 is homozygous at both of these loci, resulting in two fragments for RC212 and one fragment for D254 [14]. Additionally, the size of the two loci was separated by 78 base pairs, which allowed for simultaneous analysis.

2.5. Chemical Analysis

A total of 11 fermentation-derived compounds were selected based on reports of their importance to Pinot Noir, their importance to flavor and aroma, and whether they were yeast strain-dependent. Four of these compounds (ethyl butyrate, isoamyl acetate, 1-hexanol, and phenethyl alcohol) were quantified with gas chromatography mass spectrometry (GC-MS) at UBC Okanagan. A Varian/Agilent CP-3800 GC equipped with a VF-5MS 30 m × 0.25 mm FactorFour capillary column and with a CP-8400 auto sampler was used for a splitless analysis. The injector was ramped from 40 to 100 °C at 10 °C/min. The oven was ramped from 40 to 240 °C at 10 °C/min and a solvent delay of 2.5 min was used. Samples were extracted with liquid-liquid extraction using a 1:1 ratio of the solvents pentane and diethyl ether. A combination of 5 mL sample, 5 μL of 1.615 mg/L methyl isobutyl carbinol (MIBC), and 5 mL solvent were shaken vigorously in large test tubes. The solution settled for 1 h and the extract was transferred from the top layer to GC–MS vials. The other seven compounds (ethyl acetate, acetaldehyde, methanol, 1-propanol, isobutanol, amyl alcohol, and isoamyl alcohol) were quantified by ETS laboratories (St. Helena, CA, USA), using a gas chromatography flame ionization detector (GC-FID), as per the methods of the American Association for Laboratory Accreditation.

2.6. Data Analysis

Strain ratios at the start of the fermentation were compared with expected ratios and with pooled data from subsequent stages by performing a Chi-square goodness of fit test. Relative abundance of strains was compared between treatments and controls by performing a one-way analysis of variance (ANOVA) on data that had fermentation stages pooled, as well as a one-way ANOVA on the end-stage of each treatment. Furthermore, the relative abundance of RC212 in the co-inoculated fermentations was compared between fermentation stages of the same treatment by performing one-way ANOVAs. When significance was indicated, a Tukey–Kramer honest significant difference (HSD) *post-hoc* test was performed. The relationship between the abundance of RC212 and the concentrations of fermentation-derived compounds was determined using regression analysis. Hierarchical cluster analysis, based on Ward's method with euclidean distance, was used to group treatments [9,26]; these results were visualized using a Principal Component Analysis (PCA). The statistical analyses mentioned above were conducted using JMP® 11.0.1. The hierarchical cluster analysis and PCA employed an R 2.0 platform add-in. The concentrations of fermentation-derived chemical compounds were compared between inoculation treatments by performing one-way ANOVAs. When significant differences were detected, Tukey–Kramer HSD *post-hoc* tests were performed to determine differences between treatments. Statistical analysis of chemical compounds was performed using the Rcmdr package in RStudio version 3.1.1. All results were considered significant at $p < 0.05$.

3. Results

The starting proportions of RC212 to D254, sampled immediately after co-inoculation, were not different from their expected ratios (1:1 treatment: $X^2 = 0.563$, $p = 0.453$; 4:1 treatment: $X^2 = 0.039$, $p = 0.844$) (Table 1). This indicated that our inoculation treatments were accurate, which was important in order to make conclusions about the competition between these two strains and about the specificity of chemical compounds to one strain or the other. The co-fermentation treatments differed significantly in their proportion of yeast strains from both control treatments and from each other, when all fermentation stages were pooled (F = 436.1, $p < 0.0001$). Furthermore, the yeast ratios at the end stage of fermentation differed significantly between the two co-fermentation treatments (F = 171.4, $p < 0.0001$), but the yeast proportions of each co-inoculation treatment were constant throughout fermentation (1:1 treatment: F = 0.50, $p = 0.70$; 4:1 treatment: F = 1.6, $p = 0.27$). These results confirm that the proportions of RC212 and D254 differed between all co-inoculation and control treatments at both the beginning and throughout fermentation, and that the inoculated yeast ratios remained constant over the course of fermentation for both co-inoculated treatments.

Table 1. Percent relative abundance of RC212. Chi-square tests were performed to compare pooled data from the early, mid, and end stage ratios with the start ratio of a given treatment. Statistics were only run on the two co-inoculated treatments and not on the pure culture treatments. Any bolded results indicate significance at $p < 0.05$.

RC212:D254	**Fermentation Stage**				**Chi-Square Results**	
Treatments	**Start**	**Early**	**Mid**	**End**	X^2	***p*-Value**
1:1	46 ± 0.07	43 ± 0.11	35 ±0.07	50 ± 0.09	3.131	0.077
4:1	77 ± 0.02	88 ± 0.04	73 ± 0.04	76 ± 0.04	0.620	0.431
1:0	100 ± 0	100 ± 0	100 ± 0	100 ± 0		
0:1	0 ± 0	0 ± 0	0 ± 0	0 ± 0		

There was a positive linear relationship between the abundance of RC212 and the quantity of four compounds present during fermentations. These compounds were acetaldehyde, 1-propanol, isobutanol, and isoamyl alcohol (Table 2). Alternatively, there was a negative linear relationship between the abundance of RC212 and the quantity of ethyl acetate, amyl alcohol, and isoamyl acetate. We considered that a positive relationship indicated specificity towards RC212 and a negative relationship indicated specificity towards D254. The compounds, ethyl butyrate and phenethyl alcohol, while detected, were not significantly correlated with the relative abundance of RC212 (Table 2). The compounds 1-hexanol and methanol were not detected in any treatment (Table 3).

Table 2. Regression analysis between chemical concentrations and abundances of RC212. Chemicals having a positive linear relationship with RC212 abundance indicate RC212 strain specificity. Chemicals having a negative linear relationship with RC212 abundance indicate D254 specificity. Any bolded results indicate significance at $p < 0.05$.

Chemical	**Correlation with RC212**	***p*-Value**	**R^2 Value**
Acetaldehyde	+	**0.0003**	0.740
1-Propanol	+	**<0.0001**	0.878
Isobutanol	+	**<0.0001**	0.938
Isoamyl alcohol	+	**<0.0001**	0.791
Ethyl acetate	−	**0.0166**	0.452
Amyl alcohol	−	**0.0011**	0.674
Isoamyl acetate	−	**0.0132**	0.475
Ethyl butyrate	None	0.2706	0.120
Phenethyl alcohol	None	0.7611	0.010

In our study, RC212 produced significantly higher levels of isobutanol than did D254. Production of this compound by RC212 was also evident in the two co-inoculated treatments, which both contained higher levels of this compound than the pure D254 treatment, but lower levels than the pure RC212 treatment (Table 3). In the pure RC212 cultures, the concentration of this compound approached the sensory threshold of 300 mg/L [18]. For all treatments, isoamyl alcohol was detected at concentrations approaching its bitter sensory threshold of 300 mg/L, although its production was significantly higher in the pure RC212 treatment than the pure D254 treatment (Table 3). Acetaldehyde and 1-propanol concentrations were well below their aroma thresholds of 100–125 mg/L for all treatments [27,28] (Table 3). Ethyl acetate was detected in all treatments at levels above its detection threshold but well below its solvent-like threshold of 100 mg/L, and above the sensory threshold for fruitiness [23]. Unlike previous studies [10,29,30], no strain specificity in the production of ethyl butyrate or phenethyl alcohol was detected (Table 3).

Table 3. Summary of fermentation-derived compounds in concentration (mg/L) for all controlled fermentation treatments. Values are means ± S.E. (n = 3). Different superscript letters indicate significant differences between treatments at $p < 0.05$. Each compound was analyzed separately.

Compounds	Treatments (RC212:D254)			
	1:0	4:1	1:1	0:1
Acetaldehyde	22.0 ± 4.6 ^{a}	17.7 ± 2.5 ^{a,b}	13.7 ± 1.5 ^{b,c}	10.0 ± 1.0 ^{c}
Higher alcohols				
1-Hexanol	ND	ND	ND	ND
Isoamyl alcohol	280.7 ± 6.2 ^{a}	254.7 ± 10.5 ^{a,b}	261.3 ± 1.8 ^{a,b}	249.7 ± 1.8 ^{b}
Isobutanol	270.7 ± 8.3 ^{a}	173.0 ± 14.2 ^{b}	138.3 ± 3.2 ^{c}	89.0 ± 1.7 ^{d}
Methanol	ND	ND	ND	ND
Amyl alcohol	47.0 ± 3.5 ^{a,b}	45.3 ± 4.5 ^{b}	51.3 ± 1.5 ^{a,b}	53.7 ± 0.6 ^{a}
Phenethyl alcohol	23.3 ± 5.5 ^{a}	27.3 ± 6.1 ^{a}	18.4 ± 4.1 ^{a}	22.1 ± 3.5 ^{a}
1-Propanol	47.3 ± 3.2 ^{a}	41.0 ± 2.0 ^{b}	39.0 ± 1.0 ^{b,c}	34.0 ± 1.0 ^{c}
Esters				
Ethyl acetate	17.0 ± 3.6 ^{b}	22.3 ± 5.1 ^{a,b}	27.0 ± 2.7 ^{a}	25.3 ± 1.2 ^{a,b}
Ethyl butyrate	0.49 ± 0.006 ^{b}	0.53 ± 0.006 ^{a}	0.49 ± 0.005 ^{b}	0.50 ± 0.007 ^{b}
Isoamyl acetate	1.4 ± 0.0 ^{b}	1.6 ± 0.0 ^{a}	1.5 ± 0.0 ^{a,b}	1.5 ± 0.1 ^{a,b}

ND: not detected.

A Principal Components Analysis (PCA) showed the chemical profile of the 4:1 RC212:D254 co-inoculation treatment clustering with the chemical profile of the pure RC212 fermentations, while the profiles of the 1:1 ratio co-inoculation treatment clustered with the D254 pure culture (Figure 1). The pure RC212 culture fermentations, as well as the 4:1 RC212:D254 co-inoculated fermentations, were correlated with the presence of 1-propanol, acetaldehyde, isobutanol, and isoamyl alcohol. The D254 pure culture fermentations, as well as the 1:1 RC212:D254 co-inoculated fermentations, were correlated with the presence of isoamyl acetate, amyl alcohol, and ethyl acetate.

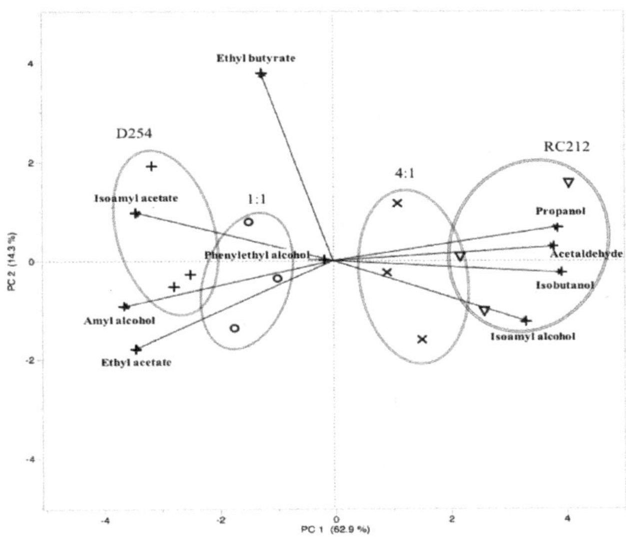

Figure 1. Principal Component Analysis of fermentation-derived compounds detected in each fermentation treatment. The variation (62.9%) among chemical profiles for all treatments can be attributed to a primary principal component (PC1) that differentiates the treatments into two unique chemical groups: (1) D254 pure culture and 1:1 ratio fermentations; and (2) RC212 pure culture and 4:1 ratio fermentations.

4. Discussion

The finding that the proportion of RC212 to D254 remained constant throughout fermentation in both co-inoculation treatments suggests that there was a lack of competitive exclusion under controlled conditions between RC212 and D254, which does not support our first hypothesis that D254 would out-compete RC212 even when RC212 was inoculated in a 4:1 RC212:D254 ratio. Our original hypothesis was based on winery-based studies [3,14], where physical, chemical, and microbial conditions likely differ from in-lab fermentations. We are not aware of any other in-lab studies that have followed the interaction of these two strains during co-fermentation. Nevertheless, one study has followed mixtures of different *S. cerevisiae* strains throughout fermentation and showed both strain exclusion as well as situations where inoculated ratios remained the same throughout fermentation [12]. A second co-inoculation study, using three different commercial strains, observed one strain (Anchor® Vin7) competively excluding Anchor® Vin13 and Lalvin® QA23 [13].

Production of isobutanol was highest in the pure RC212 treatment and lowest in the pure D254 treatment. Thus, the presence of D254 in the co-inoculated treatments appeared to have an inhibiting effect on the production of isobutanol by RC212, as evidenced by the decrease in isobutanol concentration with increasing relative abundance of D254 in the co-fermentations. Although we are not aware of any study that has worked with these two strains in grape must, one other study has shown levels of both n-butanol and isobutanol to differ between some *S. cerevisiae* pure cultures and their mixtures, indicating a significant production trend due to strain interactions [10]. In our RC212 pure cultures, the concentration of isobutanol approached the sensory threshold of 300 mg/L [18], where it could produce bitter flavors; however, solvent-like aromas and flavors probably would not be produced until it neared concentrations of 400 mg/L [18,24]. Although isoamyl alcohol showed the same trend as isobutanol, with respect to pure cultures, the co-inoculations did not result in a significant trend. We are not aware of any studies that have observed the effects of *S. cerevisiae* strain interactions on isoamyl alcohol. As with isobutanol, isoamyl alcohol could produce bitter flavors at the concentrations we found, but not solvent-like aromas and flavors. Both isobutanol and isoamyl alcohol are derivatives of amino acids, so the high concentrations of these compounds were likely, in part, a reflection of the amino acid content in the initial must [9,17]. We did not find a significant interaction trend for isoamyl acetate in both co-inoculation treatments. Our results were similar to another study where ethyl ester concentrations of strain mixtures were similar or slightly higher than those of pure cultures [10]. Supporting our results, this previously conducted study found ethyl esters, including ethyl acetate, above their sensory thresholds for fruity aromas, but not for solvent-like characteristics [10]. Acetaldehyde and 1-propanol concentrations were well below their aroma thresholds of 100–125 mg/L for all treatments, and thus they did not likely contribute directly to the sensorial characteristics of these wines. Many of the compounds we evaluated were below their detection limits, but it is important to note that our study reports on only a small portion of chemicals that are important in contributing to the sensory profile of wine. A full metabolomics study may reveal other chemicals that are important in the interaction of these two strains.

The results of PCA cluster analysis revealed the RC212 pure culture and the 4:1 co-inoculation treatment shared similar chemical profiles, separate from the D254 pure culture and the 1:1 co-inoculation treatment, which also shared similar chemical profiles. This suggests that when the two strains were equally abundant, D254 had a greater effect on the chemical profile than did RC212. This also suggests that the presence of D254 reduced the chemical profile that was contributed by RC212, as evidenced by the reduction in production of a number of chemicals positively correlated with RC212, including acetaldehyde, isobutanol, and 1-propanol. These results support our second hypothesis that D254 would have a greater impact on the chemical profile than RC212 when the cell numbers of the two strains were equal. Nevertheless, our results did not support our third hypothesis that D254 would have a greater impact than RC212 on key fermentation-derived chemicals when the inoculation was administered in a 4:1 RC212:D254 ratio. In our study, both co-inoculated abundance ratios remained constant throughout the fermentations, and when a 4:1 RC212:D254 ratio was in place, the chemical profile resembled the RC212 pure culture more than the D254 pure culture. Nevertheless, the chemical profiles of the co-inoculations shared some of the characteristics of both pure culture fermentations, which supports the results of other studies showing that chemical profiles differ between co-inoculation and pure culture fermentations [9–12]. This indicates that the interaction between two or more strains creates a new chemical profile not seen in the pure cultures. The interactions of multiple strains during fermentation can have synergistic or antagonistic effects on the final sensory attributes of wine [9,12,31], which makes strain selection an important consideration for commercial winemakers. Our results, along with those of Saberi *et al.* [10], suggest that by increasing the number of different strains in a fermentation, a more complex wine, in terms of chemical profile, can be achieved and managed due to multiple interactions between different strains of yeasts. Further research is necessary to determine whether increasing the number of strains in fermentation has an additive effect on the complexity of the wine's chemical profile.

5. Conclusions

In contrast to our original prediction, RC212 and D254 maintained their original inoculation ratios throughout the bench-top fermentations, suggesting that neither RC212 nor D254 competitively excluded the other strain under controlled conditions. The chemical profiles of both co-inoculated fermentations shared some characteristics of each pure culture fermentation. Nevertheless, when the two strains were equally abundant, D254 had a greater impact on the chemical profile than did RC212; this is in support of our hypothesis that D254 would have a relatively greater impact than RC212 on the chemical profile of wine. This is the first report to show that the co-fermentations of these two commercial strains can result in chemical profiles that are different than what is found when each strain is fermenting in pure culture.

Acknowledgments: This work was supported by Quails' Gate Estate Winery and the Natural Sciences and Research Engineering Council (NSERC) through an NSERC Collaborative Research Development (CRD) grant CRDPJ 406796-10, as well as by the UBC internal wine grant. We thank Grant Stanley and David Ledderhof for providing valuable assistance throughout the study. We also graciously thank Lallemand Inc. for the donation of Lalvin® yeasts.

Author Contributions: Frida S. Gustafsson, Vladimir Jiranek, and Daniel M. Durall conceived and designed the experiment. Frida S. Gustafsson conducted the experiment. Marissa Neuner, Chrystal M. Scholl, Sydney C. Morgan, and Daniel M. Durall contributed to the analysis, interpretation, and visualization of data, and the writing of this manuscript.

Abbreviations

The following abbreviations are used in this manuscript:

RC21	Lalvin® Bourgorouge RC212
D254	Lalvin® ICV-D254

References

1. Bisson, L.F.; Joseph, C.M.; Yeasts, L. *Biology of Microorganisms on Grapes, in Must and in Wine*; Konig, H., Unden, G., Frohlich, J., Eds.; Springer Berlin Heidelberg: Berlin, Germany, 2009; pp. 47–60.
2. Clavijo, A.; Calderón, I.L.; Paneque, P. Effect of the Use of Commercial *Saccharomyces* Strains in a Newly Established Winery in Ronda (Málaga, Spain). *Int. J. Gen. Mol. Microbiol.* **2011**, *99*, 727–731. [CrossRef] [PubMed]
3. Lange, J.N.; Faasse, E.; Tantikachornkiat, M.; Gustafsson, F.S.; Halvorsen, L.C.; Kluftinger, A.; Ledderhof, D.; Durall, D.M. Implantation and Persistence of Yeast Inoculum in Pinot Noir Fermentations at Three Canadian Wineries. *Int. J. Food Microbiol.* **2014**, *180*, 56–61. [CrossRef] [PubMed]
4. Lee, P.-R.; Saputra, A.; Yu, B.; Curran, P.; Liu, S.-Q. Effects of Pure and Mixed-Cultures of *Saccharomyces cerevisiae* and *Williopsis saturnus* on the Volatile Profiles of Grape Wine. *Food Biotechnol.* **2012**, *26*, 307–325. [CrossRef]

5. Sadoudi, M.; Tourdot-Maréchal, R.; Rousseaux, S.; Steyer, D.; Gallardo-Chacón, J.J.; Ballester, J.; Vichi, S.; Guérin-Schneider, R.; Caixach, J.; Alexandre, H. Yeast-Yeast Interactions Revealed by Aromatic Profile Analysis of Sauvignon Blanc Wine Fermented by Single or Co-Culture of Non-*Saccharomyces* and *Saccharomyces* Yeasts. *Food Microbiol.* **2012**, *32*, 243–253. [CrossRef] [PubMed]
6. Comitini, F.; Gobbi, M.; Domizio, P.; Romani, C.; Lencioni, L.; Mannazzu, I.; Ciani, M. Selected Non-*Saccharomyces* Wine Yeasts in Controlled Multistarter Fermentations with *Saccharomyces cerevisiae*. *Food Microbiol.* **2011**, *28*, 873–882. [CrossRef] [PubMed]
7. Domizio, P.; Romani, C.; Lencioni, L.; Comitini, F.; Gobbi, M.; Mannazzu, I.; Ciani, M. Outlining a Future for Non-*Saccharomyces* Yeasts: Selection of Putative Spoilage Wine Strains to Be Used in Association with *Saccharomyces cerevisiae* for Grape Juice Fermentation. *Int. J. Food Microbiol.* **2011**, *147*, 170–180. [CrossRef] [PubMed]
8. Viana, F.; Gil, J.V.; Genovés, S.; Vallés, S.; Manzanares, P. Rational Selection of Non-*Saccharomyces* Wine Yeasts for Mixed Starters Based on Ester Formation and Enological Traits. *Food Microbiol.* **2008**, *25*, 778–785. [CrossRef] [PubMed]
9. King, E.S.; Kievit, R.L.; Curtin, C.; Swiegers, J.H.; Pretorius, I.S.; Bastian, S.E.P.; Francis, I.L. The Effect of Multiple Yeasts Co-Inoculations on Sauvignon Blanc Wine Aroma Composition, Sensory Properties and Consumer Preference. *Food Chem.* **2010**, *122*, 618–626. [CrossRef]
10. Saberi, S.; Cliff, M.A.; van Vuuren, H.J.J. Impact of Mixed *S. cerevisiae* Strains on the Production of Volatiles and Estimated Sensory Profiles of Chardonnay Wines. *Food Res. Int.* **2012**, *48*, 725–735. [CrossRef]
11. Barrajón, N.; Arévalo-Villena, M.; Úbeda, J.; Briones, A. Enological Properties in Wild and Commercial *Saccharomyces cerevisiae* Yeasts: Relationship with Competition during Alcoholic Fermentation. *World J. Microbiol. Biotechnol.* **2011**, *27*, 2703–2710. [CrossRef]
12. Howell, K.S.; Cozzolino, D.; Bartowsky, E.J.; Fleet, G.H.; Henschke, P.A. Metabolic Profiling as a Tool for Revealing *Saccharomyces* Interactions during Wine Fermentation. *FEMS Yeast Res.* **2006**, *6*, 91–101. [CrossRef] [PubMed]
13. King, E.S.; Swiegers, J.H.; Travis, B.; Francis, I.L.; Bastian, S.E.P.; Pretorius, I.S. Coinoculated Fermentations Using *Saccharomyces* Yeasts Affect the Volatile Composition and Sensory Properties of *Vitis vinifera* L. cv. Sauvignon Blanc Wines. *J. Agric. Food Chem.* **2008**, *56*, 10829–10837. [CrossRef] [PubMed]
14. Hall, B.; Durall, D.M.; Stanley, G. Population Dynamics of *Saccharomyces cerevisiae* during Spontaneous Fermentation at a British Columbia Winery. *Am. J. Enol. Vitic.* **2011**, *62*, 66–72. [CrossRef]
15. Lange, J.N. Yeast Population Dynamics During Inoculated and Spontaneous Fermentations at Three Local British Columbia Wineries. Master Thesis, University of British Columbia, Kelowna, Canada, December 2012.
16. Bisson, L.F.; Karpel, J.E. Genetics of Yeast Impacting Wine Quality. *Annu. Rev. Food Sci. Technol.* **2010**, *1*, 139–162. [CrossRef] [PubMed]
17. Jiranek, V.; Langridge, P.; Henschke, P.A. Amino Acid and Ammonium Utilization by *Saccharomyces cerevisiae* Wine Yeasts from a Chemically Defined Medium. *Am. J. Enol. Vitic.* **1995**, *46*, 75–83.
18. Swiegers, J.H.; Bartowsky, E.J.; Henschke, P.A.; Pretorius, I.S. Yeast and Bacterial Modulation of Wine Aroma and Flavour. *Aust. J. Grape Wine Res.* **2005**, *11*, 139–173. [CrossRef]
19. Romano, P.; Fiore, C.; Paraggio, M.; Caruso, M.; Capece, A. Function of Yeast Species and Strains in Wine Flavour. *Int. J. Food Microbiol.* **2003**, *86*, 169–180. [CrossRef]
20. Styger, G.; Prior, B.; Bauer, F.F. Wine Flavor and Aroma. *J. Ind. Microbiol. Biotechnol.* **2011**, *38*, 1145–1159. [CrossRef] [PubMed]
21. Suárez-Lepe, J.A.; Morata, A. New Trends in Yeast Selection for Winemaking. *Trends Food Sci. Technol.* **2012**, *23*, 39–50. [CrossRef]
22. Ugliano, M.; Henschke, P.A. Yeasts and Wine Flavour. In *Wine Chemistry and Biochemistry*; Moreno-Arribas, M.V., Polo, M.C., Eds.; Springer: New York, NY, USA, 2009; pp. 313–374.
23. Sumby, K.M.; Grbin, P.R.; Jiranek, V. Microbial Modulation of Aromatic Esters in Wine: Current Knowledge and Future Prospects. *Food Chem.* **2010**, *121*, 1–16. [CrossRef]
24. Ribereau-Gayon, P.; Maujean, A.; Dubourdieu, D. *Handbook of Enology*, 2nd ed.; John Wiley and Sons Ltd.: Chichester, UK, 2006.
25. Legras, J.L.; Ruh, O.; Merdinoglu, D.; Karst, F. Selection of Hypervariable Microsatellite Loci for the Characterization of *Saccharomyces cerevisiae* Strains. *Int. J. Food Microbiol.* **2005**, *102*, 73–83. [CrossRef] [PubMed]

26. Capece, A.; Romaniello, R.; Poeta, C.; Siesto, G.; Massari, C.; Pietrafesa, R.; Romano, P. Control of Inoculated Fermentations in Wine Cellars by Mitochondrial DNA Analysis of Starter Yeast. *Ann. Microbiol.* **2011**, *61*, 49–56. [CrossRef]
27. Grosch, W. Evaluation of the Key Odorants of Foods by Dilution Experiments, Aroma Models and Omission. *Chem. Senses* **2001**, *26*, 533–545. [CrossRef] [PubMed]
28. Liu, S.-Q.; Pilone, G.J. An Overview of Formation and Roles of Acetaldehyde in Winemaking with Emphasis on Microbiological Implications. *Int. J. Food Sci. Technol.* **2000**, *35*, 49–61. [CrossRef]
29. Lilly, M.; Bauer, F.F.; Styger, G.; Lambrechts, M.G.; Pretorius, I.S. The Effect of Increased Branched-Chain Amino Acid Transaminase Activity in Yeast on the Production of Higher Alcohols and on the Flavour Profiles of Wine and Distillates. *FEMS Yeast Res.* **2006**, *6*, 726–743. [CrossRef] [PubMed]
30. Saerens, S.M.G.; Delvaux, F.; Verstrepen, K.J.; Van Dijck, P.; Thevelein, J.M.; Delvaux, F.R. Parameters Affecting Ethyl Ester Production by *Saccharomyces cerevisiae* during Fermentation. *Appl. Environ. Microbiol.* **2008**, *74*, 454–461. [CrossRef] [PubMed]
31. Favale, S.; Pietromarchi, P.; Ciolfi, G. Metabolic Activity and Interactions between Two Strains, *Saccharomyces cerevisiae* r.f. *bayanus* (SBC2) and *Saccharomyces cerevisiae* r.f. *uvarum* (S6u), in Pure and Mixed Culture Fermentations. *Vitis J. Grapevine Res.* **2007**, *46*, 39–43.

Evolution of Aromatic Profile of *Torulaspora delbrueckii* Mixed Fermentation at Microbrewery Plant

Laura Canonico [1,*], Enrico Ciani [2], Edoardo Galli [1], Francesca Comitini [1] and Maurizio Ciani [1,*]

1 Department of Life and Environmental Sciences, Marche Polytechnic University, Via Brecce Bianche, 60131 Ancona, Italy; edogah@hotmail.it (E.G.); f.comitini@univpm.it (F.C.)
2 Birra dell'Eremo, Microbrewery, Via Monte Peglia, 5, 06081 Assisi, Italy; enrico@birradelleremo.it
* Correspondence: l.canonico@univpm.it (L.C.); m.ciani@univpm.it (M.C.)

Abstract: Nowadays, consumers require quality beer with peculiar organoleptic characteristics and fermentation management has a fundamental role in the production of aromatic compounds and in the overall beer quality. A strategy to achieve this goal is the use of non-conventional yeasts. In this context, the use of *Torulaspora delbrueckii* was proposed in the brewing process as a suitable strain to obtain a product with a distinctive aromatic taste. In the present work, *Saccharomyces cerevisiae/T. delbrueckii* mixed fermentation was investigated at a microbrewery plant monitoring the evolution of the main aromatic compounds. The results indicated a suitable behavior of this non-conventional yeast in a production plant. Indeed, the duration of the process was very closed to that exhibited by *S. cerevisiae* pure fermentation. Moreover, mixed fermentation showed an increase of some aromatic compounds as ethyl hexanoate, α-terpineol, and β-phenyl ethanol. The enhancement of aromatic compounds was confirmed by the sensory evaluation carried out by trained testers. Indeed, the beers produced by mixed fermentation showed an emphasized note of fruity/citric and fruity/esters notes and did not show aroma defects.

Keywords: *Torulaspora delbrueckii*; craft beer; microbrewery plant; mixed fermentation; aroma profile

1. Introduction

In the last years, there has been a worldwide growth in microbreweries, which leads to competition in the beer market to find new beers and also those that are characterized by peculiar aroma taste. To achieve this, the brewers paid attention to the ingredients which are water, malts, hops, and yeast [1–4]. In particular, the brewers focused their attention on the yeast strains to use in brewing fermentation which are selected not only for their good fermentation efficiency but also for their characteristic aroma and flavors.

In this regard, several recent investigations were focused on the selection of non-conventional yeasts [5–8]. Non-*Saccharomyces* yeasts represent a large source of biodiversity to produce new beer styles. In the last years, different non-*Saccharomyces* yeasts were proposed in brewing, such as *Brettanomyces bruxellensis*, *Torulaspora delbrueckii*, *Candida shehatae*, *Candida tropicalis*, *Zygosaccharomyces rouxii*, *Lachancea thermotolerans*, *Saccharomycodes ludwigii*, and *Pichia kluyveri* [9–13]. *T. delbrueckii* is one of the most well-known non-*Saccharomyces* yeasts and it can be found in wild environments such as plants and soils as well as in wine or in fermented food processes. In the brewing process, *T. delbrueckii* received particular attention due to its ability to ferment maltose, produce ester compounds, and biotransform the monoterpenoid flavor compounds of hops [12,14–16]. In particular, *T. delbrueckii* can improve the amount of different fruity aromas, such as β-phenyl ethanol ("rose" flavors), n-propanol, iso-butanol, amyl alcohol ("solvent brandy" aroma), and ethyl acetate [17–19].

Canonico et al. [6,12] evaluated the use of *T delbrueckii* for beer production, both pure and in mixed cultures with different *S. cerevisiae* starter strains. *T. delbrueckii* in mixed fermentation with different *S. cerevisiae* starter strains showed different behavior and resulting in beers with distinctive flavors. Generally, the main aromatic compounds that were affected by *T. delbrueckii* are some fruity esters. Furthermore, in mixed fermentation, *T. delbrueckii* provided higher levels of higher alcohols, in contrast to data obtained in winemaking, where higher alcohols had lower levels. Moreover, beers obtained with *T. delbrueckii* pure cultures were characterized by a distinctive analytical, aromatic profile, and a low alcohol content (2.66% *v/v*) [12].

Michel et al. [16] investigated different *T. delbrueckii* strains coming from different habitats. One strain was able to produce a fruity and floral aroma (β-phenyl ethanol) and amyl alcohols. Furthermore, two strains were found to be suitable for producing low-alcohol beer owing to their inability to ferment maltose and maltotriose but still produced good flavor. However, investigation into the use of non-conventional yeasts in the brewing process has been performed at a laboratory scale or at a pilot scale while validation trials are lacking at the industrial level, which would give a more accurate assessment of their brewing ability. For this reason and based on the results of previous investigations [6,12] in this study, the contribution of *T. delbreuckii* in mixed fermentation with *S. cerevisiae* starter strain at inoculum ratio 1:20 was assessed at the microbrewery plant. The effect of this non-conventional yeast in mixed fermentation on the evolution of biomass and aroma profile as well as on the final beer composition was evaluated. The sensorial profile of the final beers was also tested.

2. Materials and Methods

2.1. Yeast Strains

T. delbrueckii DiSVA 254 comes from the Yeast Collection of the Department of Life and Environmental Sciences (DiSVA) of the Polytechnic University of Marche (Italy). *T. delbrueckii* strain DiSVA 254 and *S. cerevisiae* commercial strain US-05 (Fermentis, Lesaffre, Marcq En Baroeul, France) were used in mixed fermentation at inoculum ratio 20:1 as reported in a previous study [12]. The US-05 was rehydrated following the manufacturer's instructions and was plated on YPD agar medium at 25 °C, by spreading 0.1 mL yeast suspension onto the surface of the medium.

The yeast strains were maintained on yeast extract (10 g/L), peptone (20 g/L), dextrose (20 g/L), (YPD) agar (18 g/L) at 4 °C, for short-term storage, and in YPD liquid with 80% (*w/v*) glycerol at −80 °C for long-term storage.

2.2. Wort Production and Fermentation Condition

The wort used for the trials was produced at Birra dell'Eremo Microbrewery (Assisi, Italy) from a batch of 1500 L in duplicate fermentations. The wort was made with pilsner malt (100%), the Cascade hop variety, and produced according to the scheme reported by Canonico et al. [6]. The main analytical characters of this wort were pH 5.5, specific gravity 12.3° GPlato, and 20 IBU. The fermentation process was carried out in 2 different batches of 1500 L at 20 °C.

2.3. Growth Kinetics

The biomass evolution was monitored during the fermentation process using viable cell counts on WL Nutrient Agar (Oxoid, Hampshire, UK) and Lysine Agar (Oxoid, Hampshire, UK). Lysine Agar is a medium unable to support the growth of *S. cerevisiae* [20] for the differentiation of *T. delbrueckii* yeast from *S. cerevisiae* US-05 starter strain.

2.4. Bottle Conditioning

At the end of the fermentation process, the beers obtained were transferred into 500-mL bottles, adding 5.5 g/L of sucrose. The secondary fermentation in the bottle was carried out at 18–20 °C for 7–10 days.

2.5. Analytical Procedures

The contents of acetaldehyde, ethyl acetate, higher alcohols (n-propanol, isobutanol, amyl alcohol, isoamyl alcohol) were determined by direct injection into a gas–liquid chromatography system. The volatile compounds were determined by the solid-phase microextraction (HS-SPME) method. Five ml of each sample was placed in a vial containing 1g NaCl closed with a septum-type cap. HS-SPME was carried out under magnetic stirring for 10 min at 25 °C. After this period, an amount of 3-octanol as the internal standard (1.6 mg/L) was added and the solution was heated to 40 °C and extracted with a fiber Divinylbenzene/Carboxen/Polydimethylsiloxane (DVB/CAR/PDMS) for 30 min by insertion into the vial headspace. The compounds were desorbed by inserting the fiber into the Shimadzu gas chromatograph GC injector for 5 min. A glass capillary column was used: 0.25 μm Supelcowax 10 (length, 60 m; internal diameter, 0.32 mm). The fiber was inserted in split–splitless mode: 60 s splitless; the temperature of injection, 220 °C; the temperature of detector, 250 °C; carrier gas, with nitrogen; flow rate, 2.5 mL/min. The temperature program was 50 °C for 5 min, 3 °C/min to 220 °C, and then 220 °C for 20 min. The compounds were identified and quantified by comparisons with external calibration curves for each compound.

2.6. Sensorial Analysis

At the end of the fermentation process, the beers obtained were transferred into 330-mL bottles, adding 5.5 g/L sucrose. The secondary fermentation in the bottle was carried out at 18–20 °C for 7–10 days. After this period, the beers were stored at 4 °C underwent sensory analysis using a scale from 1 to 10 (Analytica EBC, 1997). This was carried out by a group of 14 trained testers, that evaluated the main aromatic notes regarding the olfactory and gustatory perception and structural features. The data were elaborated with statistical analyses to obtained information about the contribution of each descriptor on the organoleptic quality of beer.

2.7. Statistical Analysis

Analysis of variance (ANOVA) was applied to the main characteristics of the beers. The means were analyzed using the STATISTICA 7 software. The significant differences were determined by the means of Duncan tests, and the results were considered significant if the associated *p*-Values were < 0.05. The results of the sensory analysis were also subjected to Fisher ANOVA, to determine the significant differences with a *p*-Value < 0.05.

3. Results

3.1. Yeast Species Evolution

The growth kinetics of *T. delbrueckii* in mixed fermentations and *S. cerevisiae* pure culture were reported in Figure 1.

The growth kinetics of the *S. cerevisiae* US-05 pure cultures achieved ca. 10^7 CFU/mL at 3 days of fermentation and decreased at 10^6 CFU/mL until the end of fermentation. Regarding the mixed fermentation, *S. cerevisiae* reached cell concentrations <106 CFU/mL at 3 days of fermentation and decreased at 10^5 CFU/mL, while *T. delbrueckii*, started at a concentration $>10^6$ CFU/mL, achieved the maximum cell concentration at 3 days of fermentation (107 CFU/mL), and decreased at the end of fermentation (10^6 CFU/mL). The results for mixed fermentation indicated that *T. delbrueckii* at

20-folds higher than *S. cerevisiae* dominated the fermentation process and highlighted a high level of competitiveness of *T. delbrueckii* towards *S. cerevisiae* commercial strain.

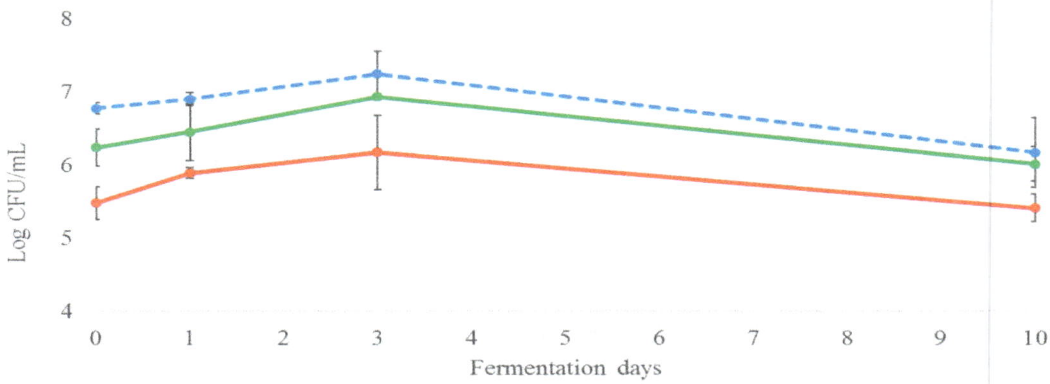

Figure 1. Growth kinetics of pure and mixed fermentation. Pure culture of *S. cerevisiae* ( ), *S. cerevisiae* ( ), and *T. delbrueckii* ( ) individually for the mixed fermentation.

3.2. Main Analytical Profile

The analytical compositions of the beers are reported in Table 1.

Table 1. The main analytical characteristics of the beer produced by pure and mixed fermentations.

Fermentation	Wort Gravity Attenuation (°P)	Ethanol % *v/v*	Residual Sucrose g/L	Residual Glucose g/L	Residual Maltose g/L
S. cerevisiae pure culture	2.85 ± 0.00	4.75 ± 0.03	0.00 ± 0.00	0.00 ± 0.00	0.06 ± 0.02
S. cerevisiae/T. delbrueckii	2.79 ± 0.17	4.68 ± 0.05	0.24 ± 0.02	0.02 ± 0.01	1.36 ± 0.14

Data are means ± standard deviation. The initial composition of the sugars in the wort was sucrose 5.9 g/L; glucose 8.2 g/L; maltose 61.76 g/L. The wort gravity at the start was 12.3 °P.

Both trials finished the process on the 10th day of fermentation highlighting that *T. delbrueckii* in the condition used at the microbrewery plant did not influence the time of the fermentation process.

S. cerevisiae/T. delbrueckii mixed fermentation and *S. cerevisiae* pure fermentation, produce beer with a comparable amount of ethanol content and final values of °P. Regarding the residual sugar, both fermentation trials consumed all sucrose and glucose content, while beer brewed by *S. cerevisiae/T. delbrueckii* mixed fermentation exhibited a slightly higher amount of maltose.

3.3. By-Products and Volatile Compounds

The main volatile compound by-products are reported in Table 2.

For the main volatile compound by-products, *S. cerevisiae* US-05/*T. delbrueckii* mixed fermentations showed different profiles to those produced by *S. cerevisiae* US-05 pure fermentation. In particular, the evolution of the main aroma compounds during the fermentation process showed that β-phenyl ethanol significantly increases in mixed fermentation in all steps of the fermentation process if compared with *S. cerevisiae* starter strain pure culture. Differently, there were no significant differences between the trials for amyl and isoamyl alcohol content with the exception of *S. cerevisiae* pure culture trials, which exhibited a lower amount of these two alcohols at the first step of fermentation (after one day).

Table 2. Evolution of the main volatile compounds (mg/L) in the beer produced by pure (*S. cerevisiae*) and mixed fermentation (*S. cerevisiae/T. delbrueckii*) during the fermentation process. Data are means ± standard deviation. Data with different superscript letters (^{a,b,c,d,}) within each column are significantly different (Duncan tests; $p < 0.05$). ND: not detected.

Fermentation	Alcohols					Carbonyl Compound	Esters						Terpene
	n-Propanol	Isobutanol	Amyl Alcohol	Isoamyl Alcohol	β-Phenyl Ethanol	Acetaldehyde	Ethyl Acetate	Ethyl Butyrate	Phenylethyl Acetate	Ethyl Hexanoate	Ethyl Octanoate	Isoamyl Acetate	α-Terpineol
S. cerevisiae 1 day fermentation	ND	4.51 ± 0.30^{c}	2.56 ± 0.25^{b}	17.85 ± 0.06^{b}	3.5 ± 0.02^{e}	15.72 ± 0.43^{a}	ND	0.144 ± 0.012^{b}	ND	ND	0.002 ± 0.001^{b}	1.70 ± 0.75^{ab}	ND
S. cerevisiae 3 days of fermentation	20.86 ± 0.29^{c}	13.93 ± 0.31^{b}	6.42 ± 0.60^{a}	45.82 ± 0.52^{a}	15.77 ± 1.08^{c}	11.44 ± 1.03^{b}	3.83 ± 0.45^{b}	0.738 ± 0.123^{a}	ND	ND	0.043 ± 0.001^{a}	2.79 ± 0.32^{ab}	ND
S. cerevisiae 10 days fermentation	28.09 ± 0.16^{a}	27.54 ± 0.24^{a}	7.28 ± 0.13^{a}	46.72 ± 0.48^{a}	17.39 ± 2.1^{c}	4.44 ± 0.98^{d}	7.42 ± 0.35^{a}	0.754 ± 0.048^{a}	ND	0.05 ± 0.03^{b}	0.044 ± 0.003^{a}	4.00 ± 0.24^{a}	ND
S. cerevisiae/T. delbrueckii 1 day of fermentation	24.47 ± 0.40^{b}	3.25 ± 0.15^{d}	6.85 ± 0.17^{a}	46.27 ± 0.19^{a}	2.15 ± 0.29^{d}	7.94 ± 1.36^{c}	5.62 ± 0.97^{a}	0.15 ± 0.03^{b}	0.060 ± 0.001^{c}	0.054 ± 0.006^{b}	0.001 ± 0.001^{b}	0.32 ± 0.0^{b}	0.082 ± 0.01^{b}
S. cerevisiae/T. delbrueckii 3 days fermentation	26.28 ± 0.22^{ab}	27.44 ± 0.33^{a}	7.06 ± 0.15^{a}	46.49 ± 0.25^{a}	36.37 ± 1.02^{b}	6.19 ± 0.88^{c}	6.52 ± 0.84^{a}	0.367 ± 0.12^{ab}	0.405 ± 0.023^{b}	0.143 ± 0.08^{a}	0.002 ± 0.002^{b}	3.85 ± 0.14^{a}	0.123 ± 0.03^{b}
S. cerevisiae/T. delbrueckii 10 days of fermentation	25.38 ± 0.17^{b}	27.52 ± 0.17^{a}	6.96 ± 0.12^{a}	46.38 ± 0.89^{a}	42.71 ± 0.98^{a}	7.07 ± 0.85^{c}	6.07 ± 0.73^{a}	0.770 ± 0.03^{a}	0.987 ± 0.124^{a}	0.157 ± 0.09^{a}	0.02 ± 0.002^{b}	3.56 ± 0.25^{a}	0.163 ± 0.03^{a}

Acetaldehyde content showed a different trend: pure culture trials showed a progressive reduction of this carbonyl compound during the fermentation, while the *S. cerevisiae/T. delbrueckii* mixed fermentation exhibited the same acetaldehyde content during the process. Ethyl acetate and ethyl hexanoate were detected only in mixed fermentation until the beginning of fermentation. The same trend was also exhibited by α-terpineol. Moreover, regarding ethyl hexanoate and α-terpineol, there was a significant increase at the end of fermentation. For isoamyl acetate content, the results did not show a significant difference between the two fermentations.

3.4. Sensory Analysis

The beers obtained by pure and mixed fermentations underwent sensory analysis, and the results were illustrated in Figure 2.

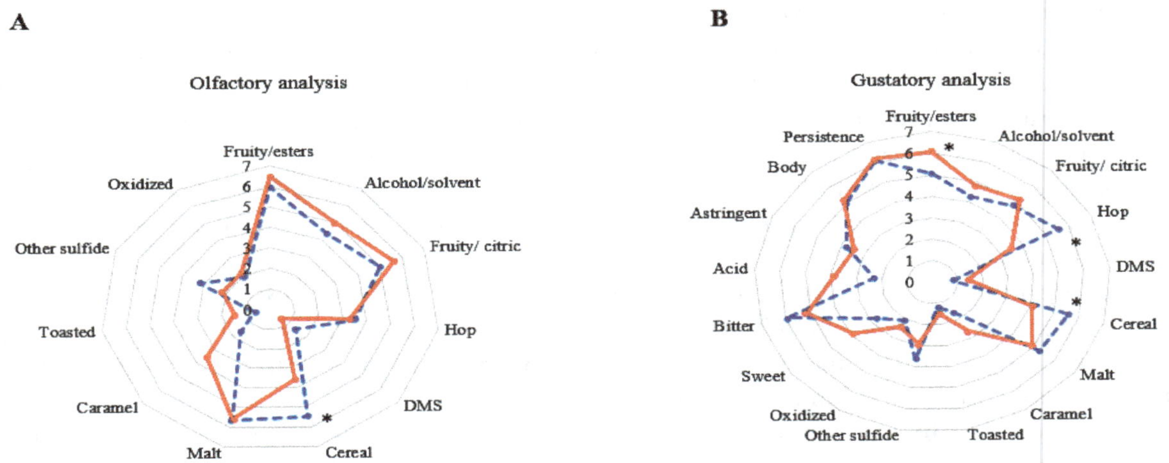

Figure 2. Sensory analysis of beer produced in the microbrewery plant by the *T. delbrueckii* mixed fermentation. From pure cultures of *S. cerevisiae* (blue dashed line) and mixed cultures of *S. cerevisiae/T. delbrueckii* (orange line). (**A**) Olfactory analysis; (**B**) gustatory analysis. DMS, dimethyl sulfide. * = Significantly different (Fisher ANOVA).

All of the beers analyzed showed significant differences for their main aromatic notes regarding the olfactory and gustatory analysis. In particular, for the main sensorial descriptors, the data showed that the beer obtained with the mixed fermentation was significantly different from that of the *S. cerevisiae* US-05 starter strain for a variety of the sensorial characteristics. Regarding olfactory analysis (Figure 2A), beers brewed with *S. cerevisiae/T. delbrueckii* mixed fermentation showed a bouquet with notes that emphasized the fruity/esters, fruity/citric, and caramel. Moreover, the perception of DMS (dimethyl sulfide) and other sulfide compounds shows they are less well perceived than beers obtained by *S. cerevisiae* pure culture. However, the only significant difference between the two beers was exhibited by the cereal note, which resulted in the emphasis of the product brewed with the *S. cerevisiae* starter strain.

Regarding gustatory analysis (Figure 2B), the beers obtained by mixed fermentation are characterized by the significant perception of fruity/esters notes. The beers obtained with *S. cerevisiae* pure culture were significantly characterized by hop and cereal notes.

In addition, the beers produced by *T. delbrueckii* mixed fermentation were characterized by a pale yellow color, clarity, and persistent and compact foam, which are very important features in the assessment of the quality of a beer (data not shown).

4. Discussion

The use of non-conventional yeasts in the brewing process was recently proposed with the aim to produce beers with distinctive aromatics note or to develop a new technology to increase the typicity of specialty beers such as low-calorie beer, low alcohol beer, novel flavored beer, and gluten-free beer [8,12,13,16,21–24]. In previous studies, the use of *T. delbrueckii* (strain DiSVA 254) in mixed fermentation with *S. cerevisiae* starter strains was investigated at a laboratory scale [6,12]. The results indicated a promising behavior of this yeast for use at microbrewery plants. Indeed, the interactions between *S. cerevisiae* and *T. delbrueckii* produced beers characterized by a distinctive aromatic profile (fruity/citric notes, fruity/esters notes, and full-bodied attributes) [12]. For these reasons, the application of *S. cerevisiae*/*T. delbrueckii* was assessed at the microbrewery plant evaluating the evolution of the volatile compounds during the fermentation process. The first relevant aspect for its application at the industrial level was the duration of fermentation. Similarly, to the laboratory-scale trials, the brewing process carried out with *T. delbrueckii* mixed fermentation showed a comparable fermentation time to that exhibited by the *S. cerevisiae* starter strain showing good competitiveness with *S. cerevisiae* in the co-culture. This aspect is crucial for its application in a microbrewery where for economic reasons the fermentation process should not exceed 10–15 days.

Regarding the evolution of aroma compounds, generally higher alcohols did not show a significant difference between mixed and pure fermentations. Regarding β-phenyl ethanol content, known for the rose and floral aroma with an odor threshold of 10 mg/L [18], the results showed an increase in mixed fermentation exhibiting a different trend by a previous study [6,12]. An increase of β-phenyl ethanol was observed by Toh et al. [25] and Drosou et al. [26] highlighted that the production of this compound was determined by *T. delbrueckii* and *S. cerevisiae* strains used in fermentation but also by the fermentation condition.

Esters compounds produced by esterification between alcohol and short- or long-chain fatty acids are important compounds that can affect the aroma of the beer [27]. Phenyl ethyl acetate is known for the floral, sweet, honey, and fruity aroma with a threshold of 3.8 mg/L [16] was significantly affected by the presence of *T. delbrueckii* as previously reported [6,12,26].

This study, confirming a previous study [12], showed the significant increase of ethyl hexanoate, fruity esters associated with apple flavor [27], when *T. delbrueckii* was used in mixed fermentation while a different trend was observed with different *T. delbrueckii* and *S. cerevisiae* strains [25,26]. α-terpineol, the terpene responsible for balsamic/fruit notes, was detected only in mixed fermentation and highlighted that these aroma compounds were related to *T. delbrueckii*.

Regarding the evolution of the main aroma compounds during fermentation, the *S. cerevisiae* pure culture and mixed fermentation exhibited a different trend. In particular, the evolution of acetaldehyde content is related to a different metabolic pathway of *S. cerevisiae* and *T. delbrueckii*. Indeed, in *S. cerevisiae* fermentation the acetaldehyde content decreased during the fermentation process, while in mixed fermentation the content of this carbonyl compound remains similar from beginning to end. The same trend was also exhibited for the main alcohol compounds.

Few works are present in the literature regarding the application of *T. delbrueckii* in the brewing process and there are no data about its use at the industrial level. These results confirming the fermentation behavior of *T. delbrueckii* in mixed fermentation, emphasize and reinforce its possible use at the industrial level allowing one to obtain beers with characteristics different from those obtained with *S. cerevisiae* starter strains and with a sensory profile appreciated by tasters.

Author Contributions: L.C., E.C., E.G., F.C., and M.C. contributed equally to this manuscript. All authors participated in the design and discussion of the research. L.C. carried out the experimental part of the work. E.C. carried out the fermentation in the microbrewery. L.C., E.C., E.G., F.C., and M.C. carried out the analysis of the data and wrote the manuscript. All authors have read and agreed to the published version of the manuscript.

Acknowledgments: The authors wish to thank the Birra dell'Eremo Microbrewery (Assisi, Italy) for making the microbrewery available and for supporting the technical experimental design. Moreover, thanks go to the UNIONBIRRAI association (Milano, Italy) and the trained testers belonging to UNINOBIRRAI BEER TASTERS (UBT Marche region) for helping the authors to complete the study for an industrial application.

References

1. Bernstein, J.M. *Imbibe Magazine*; Portland, OR, USA, 2010; Volume 2017. Available online: https://www.bjcp.org/ (accessed on 20 November 2019).
2. Brungard, M. Calcium and magnesium in brewing water. *New Brew.* **2014**, *31*, 80–88.
3. So, A. Developing new barely varities: A work in progress. *New Brew.* **2014**, *31*, 60–68.
4. Osburn, K.; Ahmad, N.N.; Bochman, M.L. Bio-prospecting, selection, and analysis of wild yeasts for ethanol fermentation. *Zymurgy* **2016**, *39*, 81–88.
5. van Rijswijck, I.M.; Wolkers-Rooijackers, J.C.; Abee, T.; Smid, E.J. Performance of non-conventional yeasts in co-culture with brewers' yeast for steering ethanol and aroma production. *Microb. Biotechnol.* **2017**, *10*, 1591–1602. [CrossRef] [PubMed]
6. Canonico, L.; Comitini, F.; Ciani, M. *Torulaspora delbrueckii* contribution in mixed brewing fermentations with different *Saccharomyces cerevisiae* strains. *Int. J. Food Microbiol.* **2017**, *259*, 7–13. [CrossRef] [PubMed]
7. Osburn, K.; Amaral, J.; Metcalf, S.R.; Nickens, D.M.; Rogers, C.M.; Sausen, C.; Caputo, R.; Miller, J.; Li, H.; Tennessen, J.M.; et al. Primary souring: A novel bacteria-free method for sour beer production. *Food Microbial.* **2018**, *70*, 76–84. [CrossRef]
8. Canonico, L.; Galli, E.; Ciani, E.; Comitini, F.; Ciani, M. Exploitation of Three Non-Conventional Yeast Species in the Brewing Process. *Microorganisms* **2019**, *7*, 11. [CrossRef]
9. Sarens, S.; Swiegers, J.H. Enhancement of Beer Flavor by a Combination of *Pichia* Yeast and Different Hop Varieties. U.S. Patent Application No. 14/241,761, 21 August 2014.
10. De Francesco, G.; Turchetti, B.; Sileoni, V.; Marconi, O.; Perretti, G. Screening of new strains of *Saccharomycodes ludwigii* and *Zygosaccharomyces rouxii* to produce low-alcohol beer. *J. Inst. Brew.* **2015**, *121*, 113–121. [CrossRef]
11. Domizio, P.; House, J.F.; Joseph, C.M.L.; Bisson, L.F.; Bamforth, C.W. *Lachancea thermotolerans* as an alternative yeast for the production of beer. *J. Inst. Brew.* **2016**, *122*, 599–604. [CrossRef]
12. Canonico, L.; Agarbati, A.; Comitini, F.; Ciani, M. *Torulaspora delbrueckii* in the brewing process: A new approach to enhance bioflavour and to reduce ethanol content. *Food Microbial.* **2016**, *56*, 45–51. [CrossRef]
13. Holt, S.; Mukherjee, V.; Lievens, B.; Verstrepen, K.J.; Thevelein, J.M. Bioflavoring by non-conventional yeasts in sequential beer fermentations. *Food Microbial.* **2018**, *72*, 55–66. [CrossRef] [PubMed]
14. King, A.; Dickinson, J.R. Biotransformation of monoterpene alcohols by *Saccharomyces cerevisiae*, *Torulaspora delbrueckii* and *Kluyveromyces lactis*. *Yeast* **2000**, *16*, 499–506. [CrossRef]
15. Tataridis, P.; Kanelis, A.; Logotetis, S.; Nerancis, E. Use of non-*Saccharomyces Torulaspora delbrueckii* yeast strains in winemaking and brewing. *Zb. Matitse Srp. Prir. Nauke* **2013**, *124*, 415–426. [CrossRef]
16. Michel, M.; Kopecká, J.; Meier-Dörnberg, T.; Zarnkow, M.; Jacob, F.; Hutzler, M. Screening for new brewing yeasts in the non-*Saccharomyces* sector with *Torulaspora delbrueckii* as model. *Yeast* **2016**, *33*, 129–144. [CrossRef]
17. Pires, E.J.; Teixeira, J.A.; Brányik, T.; Vincente, A.A. Yeast: The soul of beer's aroma and a review of flavour-active esters and higher alcohols produced by the brewing yeast. *Appl. Microbiol. Biotechnol.* **2014**, *98*, 1937–1949. [CrossRef]
18. Etschmann, M.; Huth, I.; Walisko, R.; Schuster, J.; Krull, R.; Holtmann, D.; Wittmann, C.; Schrader, J. Improving 2-phenylethanol and 6-pentyl-α-pyrone production with fungi by microparticle-enhanced cultivation (MPEC). *Yeast* **2015**, *32*, 145–157. [CrossRef]
19. Basso, R.F.; Alcarde, A.R.; Portugal, C.B. Could non-*Saccharomyces* yeasts contribute on innovative brewing fermentations? *Food Res. Int.* **2016**, *86*, 112–120. [CrossRef]
20. Lin, Y. Detection of wild yeasts in brewery. Efficiency of differential media. *J. Inst. Brew.* **1975**, *81*, 410–417. [CrossRef]
21. Petruzzi, L.; Carbo, M.; Sinigaglia, M.; Bevilacqua, A. Brewers' yeast in controlled and uncontrolled fermentation, with a focus on novel, non-conventional and superior strains. *Food Rev. Int.* **2016**, *32*, 341–363. [CrossRef]
22. Michel, M.; Meier-Dörnberg, T.; Jacob, F.; Schneiderbanger, H.; Haselbeck, K.; Zarnkow, M.; Hutzler, M. Optimization of beer fermentation with a novel brewing strain *Torulaspora delbrueckii* using response surface methodology. *MBAA TQ* **2017**. [CrossRef]
23. Senkarcinova, B.; Dias, I.A.G.; Nespor, J.; Branyik, T. Probiotic alcohol-free beer made with *Saccharomyces cerevisiae* var. boulardii. *LWT* **2019**, *100*, 362–367. [CrossRef]

24. Sannino, C.; Mezzasoma, A.; Buzzini, P.; Turchetti, B. Non-conventional Yeasts for Producing Alternative Beers. In *Non-Conventional Yeasts: From Basic Research to Application*; Springer: Cham, Germany, 2019; pp. 361–388.
25. Toh, D.W.K.; Chua, J.Y.; Liu, S.Q. Impact of simultaneous fermentation with *Saccharomyces cerevisiae* and *Torulaspora delbrueckii* on volatile and non-volatile constituents in beer. *LWT* **2018**, *91*, 26–33. [CrossRef]
26. Drosou, F.; Tataridis, P.; Oreopoulou, V.; Dourtoglou, V. Use of wine Non-*Saccharomyces* yeasts in brewing. In Proceedings of the 36th European Brwery Convention, Antwerpen, Belgium, 2–6 June 2019.
27. Nykänen, L.; Suomalainen, H. *Aroma of Beer, Wine and Distilled Alcoholic Beverages*; Springer Science & Business Media, Reidel: Dordrecht, The Netherlands, 1983.

12

Combinatorial Engineering of *Yarrowia lipolytica* as a Promising Cell Biorefinery Platform for the *de novo* Production of Multi-Purpose Long Chain Dicarboxylic Acids

Ali Abghari [1,*], Catherine Madzak [2] and Shulin Chen [1,*]

[1] Department of Biological Systems Engineering, Bioprocessing and Bioproducts Engineering Laboratory, Washington State University, Pullman, WA 99163, USA

[2] UMR 782 GMPA, CBAI, INRA—AgroParisTech-Université Paris-Saclay, F-78850 Thiverval Grignon, France; Catherine.Madzak@grignon.inra.fr

[*] Correspondence: ali.abghari@wsu.edu (A.A.); chens@wsu.edu (S.C.)

Abstract: This proof-of-concept study establishes *Yarrowia lipolytica* (*Y. lipolytica*) as a whole cell factory for the *de novo* production of long chain dicarboxylic acid (LCDCA-16 and 18) using glycerol as the sole source of carbon. Modification of the fatty acid metabolism pathway enabled creating a pool of fatty acids in a β-oxidation deficient strain. We then selectively upregulated the native fatty acid ω-oxidation pathway for the enhanced terminal oxidation of the endogenous fatty acid precursors. Nitrogen-limiting conditions and leucine supplementation were employed to induce fatty acid biosynthesis in an engineered Leu$^-$ modified strain. Our genetic engineering strategy allowed a minimum production of 330 mg/L LCDCAs in shake flask. Scale up to a 1-L bioreactor increased the titer to 3.49 g/L. Our engineered yeast also produced citric acid as a major by-product at a titer of 39.2 g/L. These results provide basis for developing *Y. lipolytica* as a safe biorefinery platform for the sustainable production of high-value LCDCAs from non-oily feedstock.

Keywords: *Yarrowia lipolytica*; long chain dicarboxylic acid; building blocks; citric acid; glycerol; genetic and metabolic engineering; fermentation; bioconversion

---

1. Introduction

Growing demand for petroleum dependent chemicals, in addition to surging environmental concerns, has inspired increased use of renewable resources. The sustainable production of monomers and polymers is of particular interest for reducing petroleum feedstock dependency and CO_2 emissions. The environmental cost of using petrochemical polymers can be lowered by replacing them with bio-based polymers, produced through fermentation using biomass or byproducts as feedstock [1]. Glycerol is the main byproduct of biodiesel production process, and its efficient valorization would help offset biodiesel production costs. Such bioconversion also decreases the environmental impact of waste streams. Despite various technological successes, it is still challenging to establish an integrated bioprocess, at commercial scale, for the valorization of such hydrophilic substrates into desirable compounds [2]. Nonetheless, the development of high productive strains and high value co-products advance the commercialization of such bioprocesses.

Bulk monomers of high value are commonly targeted co-products for industrial biotechnology. LCDCA monomers have a much higher value than regular free fatty acids. These building block monomers are used to make various polyamides, polyester and, polyurethanes [3] in the bio-plastic and coating industries. Their use enables the synthesis of various novel polymers with enhanced hydrolytic stability, solvent resistance, optical clarity, and flexibility properties [4]. They can also be used in the automotive, food, medical, and chemical industries as adhesives, corrosion inhibitors, lubricants, fragrances, surfactants, antiseptics, and personal care ingredients [5]. Non-renewable petroleum-derived alkanes or plant-derived fatty acids that compete with food production are currently the main hydrophobic feedstock for the production of these multi-purpose building blocks [6–8].

A chemical-based approach for the production of these monomers from hydrophobic substrates has already been established. For instance, the chemical degradation of plant oil fatty acids to LCDCAs via self-metathesis reaction is one of the major chemical-based production routes [9]. The chemical-based production of LCDCA is an energy-intensive and environmentally unfriendly process, which requires expensive catalysts and cost intensive purification steps due to the generation of various by-products. Contrary to chemical approaches, the bio-based approach offers higher selectivity and sustainability [10]. The microbial production of short to medium chain DCAs has been reviewed and partly industrialized [10–13]. In terms of LCDCAs, the major producing strain is a pathogenic yeast, *Candida tropicalis* [14].

The biosynthetic pathway to LCDCA monomers is mediated by the P450 system. In general, cytochrome P450 monooxygenases (P450 or CYP) systems are involved in the oxygenation of various hydrophobic compounds. For instance, the CYP52 family performs terminal ω-oxidation of alkanes, alcohols, and fatty acids of various chain lengths [15]. This family belongs to yeast species, including *Candida* and *Yarrowia*, and is comprised of members whose products have different specificities for alkanes versus fatty acids [16,17]. The yeast CYP52 has higher productivity than the bacterial ω-hydroxylases [18]. Terminal oxidation by P450 is carried out at high regioselectivity and stereoselectivity. This can result in the synthesis of high value compounds whose production is nearly impossible through chemical approaches due to the instability of double bonds or the generation of unwanted byproducts [19]. Several examples of cytochrome P450 monooxygenases with terminal ω-oxidation activity toward fatty acids have been reported [20–27]. Some are also capable of fatty acid over-oxidation to their corresponding dicarboxylic acids [23,26,28]. The mechanism of a P450 mediated ω-oxidation process has been discussed [29], and is briefly presented here.

This nicotinamide adenine dinucleotide phosphate (NADPH) dependent reaction takes place under mild conditions and requires molecular oxygen. In summary, an active intermediate is formed when the heme group of P450 monoxygenase is coupled to molecular oxygen. As a result, oxygen is added to the attached substrate, the C-H bond of fatty acid. Subsequently, another atom of oxygen is released in the form of a water molecule, and a hydrogen atom is transferred from the substrate to the heme group, leading to the generation of hydroxylated substrates. During this process, the P450-diflavin reductase partner (CPR) is bound to the NADPH cofactor to derive electrons from that while positioning the redox active nicotinamide ring and flavin adenine dinucleotide (FAD) at a close distance. The FAD accepts hydride ions from the NADPH and transfers electrons to the flavin mononucleotide (FMN) to reduce the heme group of the P450 monooxygenase [30]. This complex enzymatic system has some limitations, such as low activity and stability, dependency on cofactor and redox partner proteins, and low product yields for broad industrial usage [19,31]. Moreover, eukaryotic CYPs are often associated with ER membranes [32] and have lower solubility and stability compared to bacterial ones. High-density cultivation of P450-carrying cells or modification of the P450 system for higher expression and stability can help overcome these limitations in a bioreactor application [15].

Yarrowia lipolytica is an advantageous biorefinery platform for the production of fatty acid-based bioproducts [33] and for P450 catalytic bioconversion [34]. The P450-dependent hydroxylation of alkane, fatty alcohol, and fatty acid precursors is predominant in this yeast [35,36]. This assists with detoxification and degradation of these potentially toxic molecules. The overexpression of homologous and heterologous P450 has been employed in *Y. lipolytica* [37] resulting in a remarkable proliferation of ER [38]. There are at least 17 P450 genes and some electron transfer proteins in this oleaginous yeast [38]. Among them, twelve P450 Alk genes (*YlALK*) and a single NADPH P450 reductase gene,

YlCPR, have been isolated and characterized [32,36,38,39]. They have various substrate requirements with respect to length, degree of unsaturation, and functional groups [40]. For instance, *YlALK 3*, *YlALK 5*, and *YlALK 7* can catalyze the ω-oxidation of fatty acids [41]. The resulting ω-hydroxy fatty acids undergo further steps of oxidation that could also constitute rate limiting steps in this yeast. These steps can be carried out by fatty alcohol oxidase (*FAO1*; YALI0B14014g) and by the alcohol dehydrogenase genes (*FADHs*) in the ER and peroxisome [42,43]. The fatty acid precursors and their corresponding oxygenated fatty acid monomers are susceptible to the effective peroxisomal β-oxidation degradation in *Y. lipolytica*, resulting in the generation of energy, water, and CO_2. The *POX* genes are involved in the degradation of these monomers [44,45]. The disruption of the β-oxidation pathway via deletion of *POX* genes is commonly employed to prevent fatty acid and LCDCA degradation and unwanted chain modification [14,46,47]. In this yeast, the overproduction of fatty acids can be efficiently achieved in the form of free fatty acids rather than fatty acyl-CoAs, since the production of the latters are highly regulated by feedback inhibition [48,49]. In fact, the ω-oxidation of free fatty acids can favor their degradation and alleviate their toxicity. The ω-oxidation pathway acts as a rescue route in the peroxisome-deficient cell by generating oxygenated fatty acids of higher solubility in water [50,51].

Engineering *Y. lipolytica* for the *de novo* production of LCDCAs enables utilization of various unrelated feedstock, including sugars, glycerol, and short chain volatile fatty acids, for this process. *Y. lipolytica* has a preference for using glycerol as a carbon source [32]. Glycerol and glucose have a transcription-repressive effect, with various degrees of strength, on background P450 activities in this yeast [52]. Glycerol has a strong repressor effect, so the overexpression of selected P450 members in the presence of this substrate can enable higher selectivity through the minimization of background ω-oxidation activity. *Y. lipolytica* accumulates a larger proportion of endogenous unsaturated long chain fatty acids when grown on glucose [53]. The preservation of double bonds through the terminal ω-oxidation results in the generation of multi-functional unsaturated LCDCAs. These double bonds enable the attachment of additional functional groups [54] and the generation of cross-linked degradable polymer networks with adjustable properties [55].

In this study, we engineered *Y. lipolytica*, a "safe-to-use" yeast [56], for the *de novo* production of high value LCDCA monomers. This involved a one-step bioprocess, using glycerol as the sole source of carbon. This was done by inactivating the endogenous cytosolic fatty acyl-CoA synthetase, *YlFAA1* YALI0D17864g, and redirecting the resulting pool of free fatty acids toward an engineered ω-oxidation pathway, in a deficient β-oxidation background. The ω-oxidation pathway was engineered by overexpressing P450 reductase (YALI0D04422g), fatty alcohol oxidase *YlFAO1*, and a selected member of P450 monoxygenases (*YlALK5*, YALI0B13838g). After preliminary screening and shake flask cultivations, we used a 1-L bioreactor for the P450-based biocatalytic system and achieved a productivity of 0.04 g/L·h, which is more than the process productivity of 0.001 g/L·h acceptable for P450-based biocatalytic production of pharmaceuticals [19]. Our results demonstrated the promise of *Y. lipolytica* for use as an oleaginous yeast cell factory for LCDCA production. The safe status of this yeast enhances the suitability of the resulting LCDCA monomers for food and pharmaceutical uses, where high security standards should be met. To the best of our knowledge, this is the first reported study on the biosynthesis of LCDCAs from glycerol by engineered *Y. lipolytica*. The findings of this work present a unique biosynthetic route that is expected to advance *Yarrowia* platform for the sustainable production of these multi-purpose long chain building blocks.

2. Materials and Methods

2.1. Strains and Culture Conditions

Escherichia coli top 10 was used for plasmid DNA construction and propagation. According to the standard protocols, Luria-Bertani (LB) broth or agar medium was made and supplemented with ampicillin at the concentration of 100 $\mu g \cdot mL^{-1}$ [57]. The recombinant *Y. lipolytica* strains of this study were all derived from *Y. lipolytica* H222 (a potential citric acid-producer wild-type German strain) [42] and are presented in Table 1

Table 1. *Y. lipolytica* strains used in this study.

Y. lipolytica Strains Names	Strain Genotypes	Gene Configurations	Reference
H222 (W) [wild type]	*MatA*		[42]
H222ΔP (HP-U) [*leu*$^+$, *ura*$^-$]	*MatA ura3-302::SUC2* Δ*pox1* Δ*pox2* Δ*pox3* Δ*pox4* Δ*pox5* Δ*pox6*		[42]
H222ΔPΔLΔSΔF (F) [*leu*$^-$, *ura*$^+$]	Same as HP-U, +Δ*leu2* Δ*snf1* Δ*faa1* Δ*snf1::URA3*	*loxR-URA3-loxP* flanked by *SNF1* upstream and downstream	This study
H222ΔPΔLΔSΔF ΔL +ALK5 YlCPR YlFAO1 (P) [*leu*$^-$, *ura*$^+$]	Same as F, +Δ*leu2::URA3* *YlALK5 YlCPR YFAO1*	*loxR-URA3-loxP* flanked by *SNF1* upstream and downstream	This study
H222ΔPΔLΔSΔF ΔL +ALK5 YlCPR YlFAO1 ++ALK5 (M) [*leu*$^-$, *ura*$^+$]	Same as P, +*YlALK5*	Multiple-copy integration of *YlALK5* using zeta based integrative vector pINA1291(this strain was selected after screening of 10–20 transformants for their growth and production capacities)	This study

The yeast nitrogen base (YNB) medium with following composition was used to select auxotrophic Ura$^-$ and Leu$^-$ transformants and to prepare the seed cultures: 6.7 g/L yeast nitrogen base (without amino acids w/ammonium sulfate) (Becton, Dickinson and Company, Sparks, MD, USA), 20 g/L glucose, and 1.92 g/L synthetic mix minus uracil (YNB-Ura) or 1.62 g/L synthetic mix minus leucine (YNB-Leu) (US Biological, Salem, MA, USA). YNB-Leu was also used to grow the strain carrying the plasmid with the *LEU2* marker and *CRE* gene to restore the uracil auxotrophic derivatives of the F and P strains. Rich medium (YPD) containing 20 g/L glucose, 20 g/L bacto peptone (BD), and 10 g/L bacto yeast extract (BD) was used for non-selective propagation of strains. Agar (US Biological, Salem, MA, USA) was added at a concentration of 20 g/L for preparing plates.

Production media in shake flasks and bioreactor were performed in minimal media with the following defined composition: 1.7 g/L Yeast Nitrogen Base (without amino acids and without ammonium sulphate) (BD), 1.5 g/L $MgSO_4$. $7H_2O$, and1.92 g/L drop-out synthetic mix minus uracil. The $(NH4)_2SO_4$ was used as the major source of nitrogen at a concentration of 1.9 g/L. The production media formulations contained 52 g/L glycerol, to reach the target carbon to nitrogen (C/N) ratio of 50. The $CaCO_3$ was added to the shake flasks at a concentration of 6 g/L to provide buffering conditions. This carbonate plays a role in maintaining a stable pH [58]. The shake flask cultivations were carried out in 250 mL Erlenmeyer flasks containing 50 mL of the medium at an agitation rate of 180 $\pm$ 5 rpm and a temperature of 28 $\pm$ 1 °C. Fresh colonies from the YNB-Ura plates were precultured in YNB-Ura broth. Exponentially growing cells were harvested by centrifugation, washed with water, and then resuspended in water. The production media were inoculated with the resuspended cells to an initial optical density (OD_{600}) of 0.1.

2.2. Batch Fermentation

Batch cultivation was carried out in a 1-L benchtop fermenter BioFlo 110 (New Brunswick Scientific, Enfield, CT, USA). Cells from a 24 h-shake flask YNB-Ura culture were harvested, washed using water, and inoculated into 700 mL of the fermentation medium (with C/N of 50) to reach a minimum OD_{600} of 0.1. The temperature was kept at 28 °C. The pH was controlled and maintained at 6 during the biomass propagation (typically for 1 day) and then was increased to 8 during the production phase using 5 M NaOH. This pH adjustment was also beneficial to lessen the rate of citric acid production. A minimum dissolved oxygen level of 50% was maintained by cascading with an agitation ranging from 250 to 800 rpm, and by supplying sterile air at a flow rate of 2 vvm. Samples with a volume of 15–25 mL were taken daily during 4 days of fermentation. An antifoam Y-30 emulsion (Sigma, St. Louis, MO, USA) solution of 10% was made and added at the beginning of each batch cultivation, as well as during the fermentation, when required.

2.3. Genetic Techniques

Standard molecular manipulation techniques were used to develop the vectors [57]. All plasmids and their functions are presented in Table 2.

Table 2. Vectors used in this study.

Vector Names	Map	Features
Cre (CR)	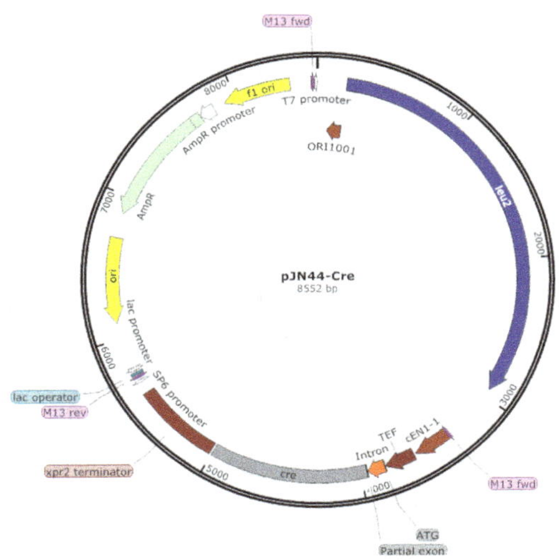	Cre recombinase flanked by TEFin promoter and Xpr2 terminator
pGR12 *YlALK5* (5)	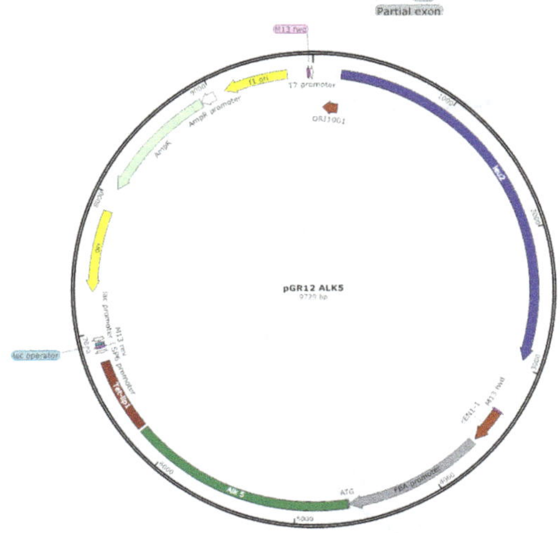	Centromeric (CEN) replicative vector, low-copy CEN plasmids (1–2 copies/cell~1.6 plasmid copies/cell), *YlALK5* controlled by P_{FBA}-Tlip1, leucine selection marker

Table 2. *Cont.*

Vector Names	Map	Features
pGR51 *YlCPR* (C)	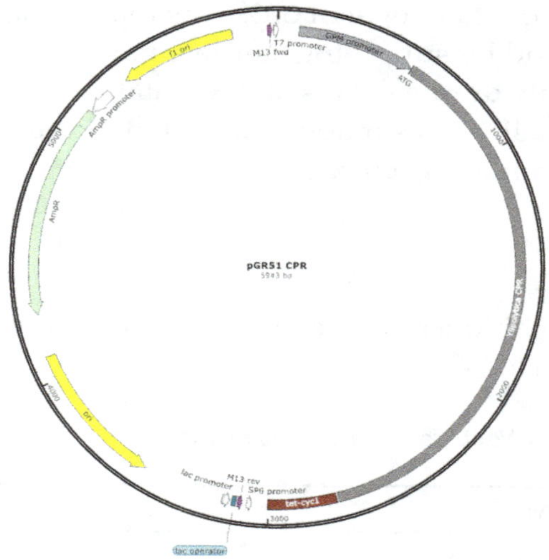	Centromeric (CEN) replicative vector, *YlCPR* controlled by P_{GPM}-Tcyc1
pJN44 *YlFAO* (F)		Centromeric (CEN) replicative vector, *YlFAO1* controlled by P_{TEFin}-Txpr2
ALK5 CPR FAO (5CF)		CEN replicative vector, P_{FBA}-*YlALK5* Tlip1 P_{GPM}-*YlCPR* Tcyc1 P_{TEFin}-*YlFAO1* Txpr2

Table 2. *Cont.*

Vector Names	Map	Features
Leu 5cf (L5CF)	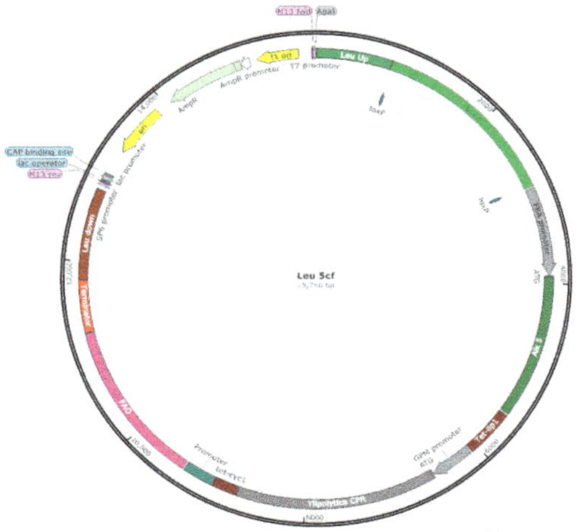	Uracil selection marker flanked by *LEU2* upstream and P_{FBA}-*YlALK5* Tlip1 P_{GPM}-*YlCPR* Tcyc1 P_{TEFin}-*YlFAO1* Txpr2 *LEU2* downstream sequences
pINA1291 ALK5 (Z5)	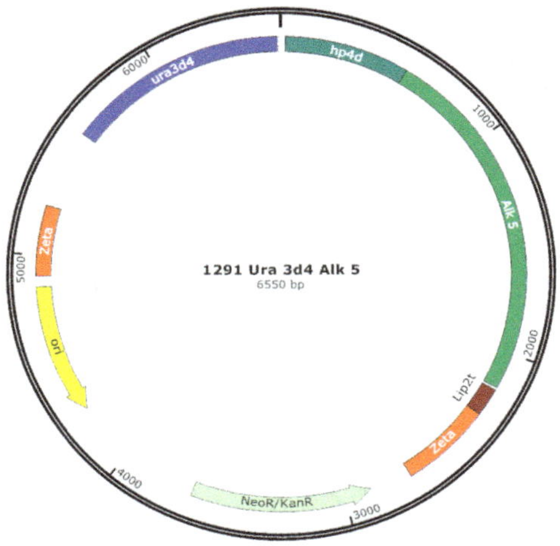	Integrative vector, hp4d- *YlALK5* Tlip2, ura3d4 defective marker, zeta region for multi-copy integration
LEU (LU)	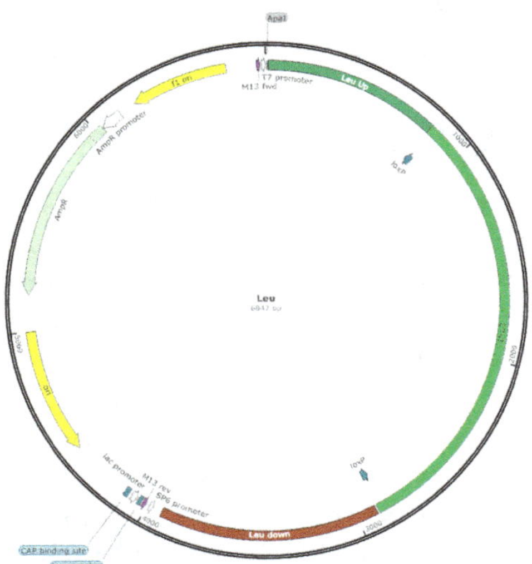	Uracil selection marker flanked by *LEU2* upstream and downstream sequences

Table 2. *Cont.*

Vector Names	Map	Features
SNF (SU)	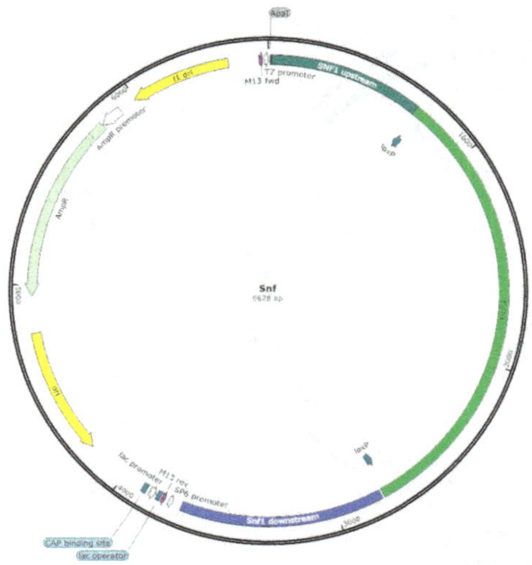	Uracil selection marker flanked by *SNF1* upstream and downstream sequences
FAA (FU)	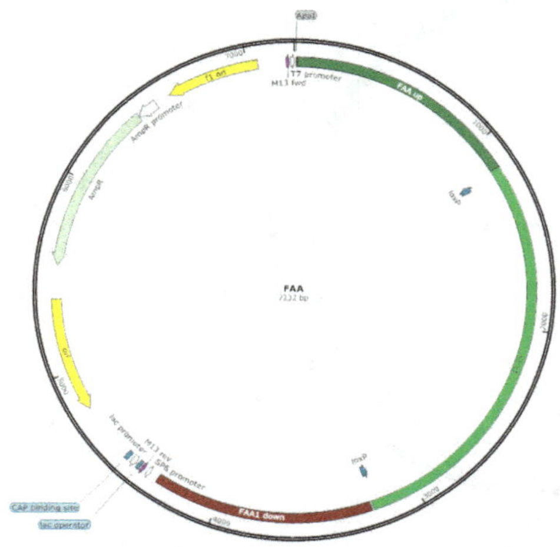	Uracil selection marker flanked by *FAA1* upstream and downstream sequences

The construction of the triple gene expression cassette was carried out by the amplification of *Yarrowia lipolytica* cytochrome P450 *YlALK5*, NADPH-P450 reductase *YlCPR*, and fatty alcohol oxidase *FAO1* gene segments. Cloning inserts were obtained from polymerase chain reaction (PCR) using Q5 high fidelity DNA polymerase (New England Biolabs, Ipswich, MA, USA), the gDNA of Po1f (ATCC MYA-2613) as the template, along with the primers presented in Table 3. The amplicons were individually digested and inserted into *Y. lipolytica* plasmid pGR12 (P_{FBA}-T_{lip1}), pJN44 (P_{TEFin}-T_{xpr2}), and pGR51 (P_{GPM1}-Tcyc1). These are the constitutive promoters with the following orders of strength: FBA1 > GPM1~TEF [59]. The segments of P_{GPM}-*YlCPR* Tcyc1 and P_{TEFin}-*FAO1*-T_{xpr2} were separated from the non-shuttle vectors by digestion with XbaI and SpeI. For a more distinguishable separation of the plasmid digestion products on the gel, the unique sites of XmnI and ScaI were also used in a triple digestion process. The target gel recovered fragments were consecutively inserted into the SpeI-digested and Fast Alkaline Phosphatase treated *ALK5*–pGR12 plasmid. Gene knock-out was carried out using the uracil maker containing vectors. The marker was surrounded by LoxP sites. For knock-out plasmid constructions, the 0.6–1.1 kb 5′- and 3′-flanking regions of the *Y. lipolytica LEU2*, *SNF1*, and *FAA1* genes were amplified with the primers listed in Table 3. The amplicons were inserted into the sites upstream and downstream of uracil gene upon digestion and purification. The triple gene expression cassette segment was double digested using XbaI and SpeI and purified on gel for subsequent insertion into the SpeI-digested and Fast Alkaline Phosphatase treated *LEU2* knock-out vector. The triple gene cassette was successfully integrated into the genomic *LEU2* locus, in a one step site-specific gene knock in/out. This was carried out to reach long-term genetic stability. *YlALK5* gene segment was also inserted into a vector under the control of the growth phase-dependent hp4d promoter. This promoter shows high quasi-constitutive activity [60] at the start of the stationary phase, and is relatively independent of medium composition and pH [61]. Application of the hp4d growth dependent promoter lessens the metabolic burden of *YlALK5* overexpression during the growth phase. The corresponding vector contains zeta sequences (Ylt1 retrotransposon LTRs), enabling its multiple random integration into the genome of strains devoid of Ylt1 retrotransposon, such as the P strain. This zeta-based vector bears the defective marker, ura3d4 allele, thus allowing a random ectopic integration of about ten copies into all H222 derivatives, which are devoid of Ylt1 retrotransposon [62].

Table 3. PCR primers used in this study.

No.	Name	Sequence (5′→3′, Underlined Restriction Site)
1	*Alk 5* F HindIII	GAGCGAAAGCTTATGCTACAACTCTTTGGCGTCC
2	*Alk 5* R PstI	CTTAGA CTGCAG CTACGCCTTCTCACCCTTATACA
3	*Alk 5* Zeta F A	AATGCTACAACTCTTTGGCGTCC
4	Alk 5 Zeta KpnI R	TTGCAAGGTACCCTACGCCTTCTCACCCTTATACATCT
5	*YlCPR* F HindIII	GAGCGAAAGCTT ATGGCTCTACTCGAC TCTC
6	*YlCPR* R SmaI	GTTAT CCCGGG CTACCACACATCTTCCTGG
7	*FAO1*F NdeI	CCTCA CATATGATGTCTGACGACAAGCACACT
8	*FAO1*R SmaI	GTTAT CCCGGG AGGATCTCCGACCTCGAATC
9	*LEU2* up F ApaI	CTATAGGGCCC ACCGGCAAGATCTCGTTAAGACAC
10	*LEU2* up R XbaI	GATCCTCTAGATGTGTGTGGTTGTATGTGTGATGTGG
11	*LEU2* down F SpeI	CTGGACTAGTCTCTATAAAAAGGGCCCAGCCCTG
12	*LEU2* down R NdeI	CCTCACATATG GACAGCCTTGACAACTTGGTTGTTG
13	*LEU2* F Ura	TACAGTTGTAACTATGGTGCTTATCTGGG
14	*LEU2* Ura R	CCTTGGGAACCACCACCGT
15	*LEU2* Ura F	ACTTCCTGGAGGCAGAAGAACTT
16	*LEU2* R Ura	ATAGCAAATTTAGTCGTCGAGAAAGGGTC
17	*SNF1* up F ApaI	CAATTGGGCCCGTGATCAAAGCATGAGATACTGTCAAGG
18	*SNF1* up R XbaI	GATCCTCTAGAGAGGTGGTGGAAGGAGTGGTATGTAGTC
19	*SNF1* down F SpeI	CTGGACTAGT TCATTAATACGTTTCCCTGGTG
20	*SNF1* down R NdeI	CCTCACATATGGGAATTCGTGCAGAAGAACA
21	*SNF1* F Ura	GCGGGAAATCAAGATTGAGA
22	*SNF1*Ura R	CGGTCCATTTCTCACCAACT
23	*SNF1* Ura F	CCTGGAGGCAGAAGAACTTG
24	*SNF1* R Ura	ACTACTGGCGGACTTTGTGG
25	*FAA1* up F ApaI	CAATT GGGCCC CCAGGTCTCAGTTGCACTTGC
26	*FAA1* up R XbaI	GATCC TCTAGA CAAATTATACCCCTCATCTCTCTAGGACA
27	*FAA1* down F SpeI	CT GGACTAGTTTGGTGAGCCCACCGC
28	*FAA1* down R NdeI	CCTCACATATGAACCTCCAGCAGACTAACTAGAACA
29	*FAA1* F Ura	ACTGTAGCTAGATGGGTGCC
30	*FAA1* Ura R	CGGTCCATTTCTCACCAACT
31	*FAA1* Ura F	CCTGGAGGCAGAAGAACTTG
32	*FAA1* R Ura	ACCAGCCCAGCCGG
33	*LEU2* up F Ura	TCATGTTCGTGGAGGGGAG
34	*LEU2* up Ura R	AAAACGCAGCTGTCAGACC
35	*LEU2* down F FAO	GGAGTCCAAGCCCTTCGA
36	*LEU2* down R	CACAAGACGTCAACTAAAGCGT

Site-specific gene knock in/out was achieved by transforming the linearized vectors containing homologous upstream and downstream sequences. *Y. lipolytica* was transformed with the linearized plasmids, consisting of NdeI-digested LU, ApaI-digested SU, NdeI-digested FU, and NsiI-digested L5CF. The zeta-containing vector was linearized using FD NotI enzyme. Transformation was performed using a Zymogen Frozen EZ yeast transformation kit II (Zymo Research, Irvine, CA, USA) according to the manufacturer's protocol. We modified this protocol to enhance the efficiency of multi-copy integration transformation. The *loxR–URA3–loxP* modules were rescued for subsequent genetic modification by the LoxP-Cre system as previously reported [63]. Gene deletions and expression cassette insertion were confirmed by PCR using the primers listed in Table 3.

Construction of vectors was carried out using standard cloning methods, with FastDigest restriction enzymes (Thermo Fisher Scientific, Waltham, MA, USA). Yeast genomic DNA was isolated as described previously [64] and was used for PCR amplification and verification. A DNA Clean &

Concentrator-5 Kit (Zymo Research, Irvine, CA, USA) was used for the purification of DNA products from PCR and digestion reactions. A GeneJET Gel Extraction Kit (Thermo Fisher Scientific) was used for the recovery of nucleic acid fragments from agarose gels.

2.4. Analytical Methods

2.4.1. Dry Biomass

Seven milliliters of culture broth were collected daily for dry cell weight, extracellular metabolites, fatty acids, and LCDCAs measurements. Five milliliters were centrifuged for 5 min at 13,300 rpm. The cell pellet underwent two stages of washing with a saline solution (0.9% NaCl) and distilled water. The biomass yield was measured gravimetrically by drying samples at 105 °C to reach a consistent weight. This was presented in grams of dry cell weight per liter (g·DCW/L).

2.4.2. Glycerol and Citric Acid Concentrations

High performance liquid chromatography (HPLC) with an Aminex HPX-87H column was used to measure the concentration of glycerol and citric acid. After centrifugation, supernatants were filtered using 0.22 μm pore-size membranes and eluted with 5 mM H_2SO_4 at a flow rate of 0.6 mL/min, at 65 °C. Refractive index (RI) and UV (210 nm) detectors were employed for the detection of target signals. Standards were used for the identification and quantification of glycerol and citric acid. Our methods did not adequately separate citric acid from its isomer, iso-citric acid. Thus, the reported concentration corresponds to the sum of these isomers, expressed as citric acid.

2.4.3. Qualitative and Quantitative Analysis of Fatty Acids and LCDCAs

Fatty acid methyl esters (FAME) and dicarboxylic dimethyl esters were made in hexane after lipid extraction and transesterification, in compliance with the previously described method [65]. Analysis of esterified compounds was conducted using an Agilent 7890A gas chromatography instrument coupled with a flame-ionization detector (FID) and a FAMEWAX column (30 m × 320 μm × 0.25 μm) (Restek Corporation, Bellefonte, PA, USA). The injection temperature and volume were set at 250 °C and 1 μL, respectively. The injection was performed with a split mode (ratio 20:1). The initial temperature of oven was 190 °C and progressed to 240 °C at a rate of 5 °C/min. The final temperature was maintained for 20 min. The FID temperature was 250 °C. FAME standards were used to identify the fatty acid peaks in the chromatograms. Quantification of fatty acids and LCDCAs were conducted by comparison with tridecanoic acid (C13:0) (0.5 mg/mL) (Sigma-Aldrich, St. Louis, MO, USA) as internal standard and by the calculation of response factor. The supernatant was filtered and used for possible extracellular metabolite extraction and quantification. Results show that the GC-FID method did not sufficiently separate C16 and C16:1 fatty acids, C18 and C18:1 fatty acids, LCDCA16 and LCDCA16:1, and LCDCA 18 and LCDCA 18:1. Therefore, the reported concentration represents the sum of these acid pairs, expressed as total fatty acid 16, total fatty acid 18, total LCDCA 16, and total LCDCA 18. LCDCA18:2 was also quantified and added to the category of LCDCA 18. An Agilent GC-MS was also used for the qualitative analysis of other dicarboxylic acids (DCAs), DCA12 and DCA14. The initial temperature of 170 °C with a holding time of 1 min, was increased to 250 °C at a rate of 3 °C/min, and held at the final temperature for 20 min. The split ratio was 30:1.

3. Results

3.1. Screening Process

We developed several single, double, and triple gene expression vectors with different promoters and terminators carrying homologous P450 monoxygenase (*YlAlk 2, 3, 5, 7,* and *10*), P450 reductase *YlCPR*, fatty alcohol oxidase *YlFAO1*, and fatty aldehyde dehydrogenase *YlHFD1* gene segments. We tested these vectors for their performance in the ω-oxidation of endogenous fatty acids to the corresponding LCDCA monomers. YlAlk gene products catalyze the rate-limiting step of the ω-oxidation pathway. Therefore, it is important to use strong FBA promoter for their overexpression when replicative vectors with low-copy numbers are used. Among YlALK 2, 3, 5, 7 and 10, we noted that *YlALK3*, *YlALK* 5, and *YlALK* 7 each have an effect on the production of LCDCA-16 with the following order of strength: YlALK5 > YlALK7 > YlALK3. We also observed that co-expression of *YlALK5* and *YlCPR* led to the generation of ω-hydroxy fatty acids of 16 carbons in the F strain. We selected *YlALK5* for single as well as multiple integration because it exhibited the highest activity toward the ω-oxidation of accumulated fatty acids among the tested monoxygenases. We constructed the F strain by deleting *YlLEU2*, *YlSNF1*, and *YlFAA1* in the HP-U (*POX*-deleted) strain. Then, the P strain was constructed by performing the site specific single-copy integration of the *YlALK5*, *YlCPR*, and *YlFAO1* expression cassette into the locus of the *LEU2* gene. It is notable that this single-copy integration improved the growth behavior of the resulting strain over the F strain in the minimal YNB-Ura media. Then, the M strain was constructed by random multiple integration of *YlALK5* into the P strain. Ten to twenty colonies were screened for their growth on selective YNB-Ura plate and LCDCA production. All of the F, P, and M strains were uracil prototrophic and leucine auxotrophic. The YNB-Ura culture of the M strain was additionally supplemented with an additional 400 mg/L leucine to study the possible stimulatory effect of this amino acid on the metabolite production by the M strain. This was labeled as the ML strain/treatment. Preliminary shake flask studies were conducted to find the best strain for a fermentation scale-up. However the production of LCDCAs in shake flask cultures is generally challenging due to insufficient aeration, agitation, and poor control over pH. The next section presents the data from the 6-day shake flask cultivations.

3.2. Preliminary Shake Flask Experiments for Fatty Acid and LCDCA Production

We tested four strains (wild type W and engineered F, P, and M) for fatty acids and monomer production in shake flask cultures. Glycerol was added to YNB-Ura at a concentration of 52 g/L to reach a C/N ratio of 50. This medium contained about 0.38 g/L of leucine and was supplemented with an additional 400 mg/L of leucine for the ML strain/treatment. Results of biomass production in the 6-day shake flask culture are summarized in Figure 1.

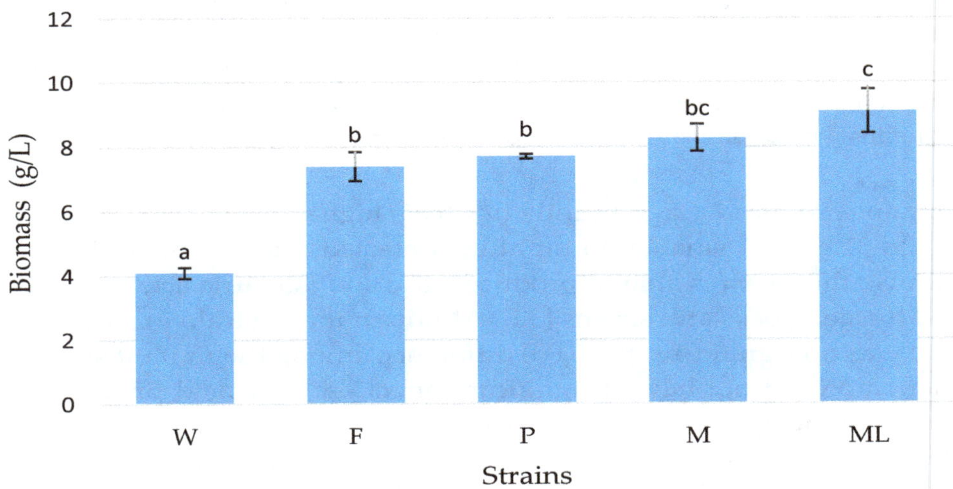

Figure 1. Study of 6-day shake flask cultures for biomass production by the W, F, P, M and ML strains/treatment. Glycerol based media under nitrogen-limiting conditions (C/N = 50). Error bars represent standard deviation for $n = 3$. W: H222 wild type strain, F: Δ*POX1-6* Δ*LEU2* Δ*SNF1* Δ*FAA1* Δ*SNF1::URA3*, P: F + Δ *LEU2:: URA3 YlALK5 YlCPR YlFAO1*, M: P + *YlALK5:: URA3*, ML: M + 400 mg/L leucine. abc Columns with dissimilar letters at the top are significantly different ($p < 0.05$).

The data present a significant ($p < 0.05$) difference between the wild type (W) strain and all mutant strains. This can be explained in part by the final pH of media, which can affect the $CaCO_3$ solubility. Although all shake flask media contained an equal amount of calcium carbonate (6 g/L), their final pH varied depending on the strain. For example, the final pH of the W strain was above 4.1, while it was in the range of 3.2 to 3.9 for other strains because of a higher citric acid production. The pH of media can affect $CaCO_3$ solubility and this interferes with dry biomass measurements. In addition to this explanation, we also noted that glycerol was almost completely consumed by the end of the 6-day period for all mutant strains. However, we observed a residual glycerol level of 15 g/L for the W strain. The integration of the fatty acid-oxidizing genes, did not significantly ($P > 0.05$) enhance the final biomass level under nitrogen limiting conditions. However, the supplementation of the YNB-Ura medium with an additional leucine caused the M strain to produce a significantly higher biomass level than F and P strains.

We also assessed the capability of the strains for citric acid production. Our methods could not distinguish between citric acid and isocitric acid, so the reported data in Figure 2 is the sum of these two isomers.

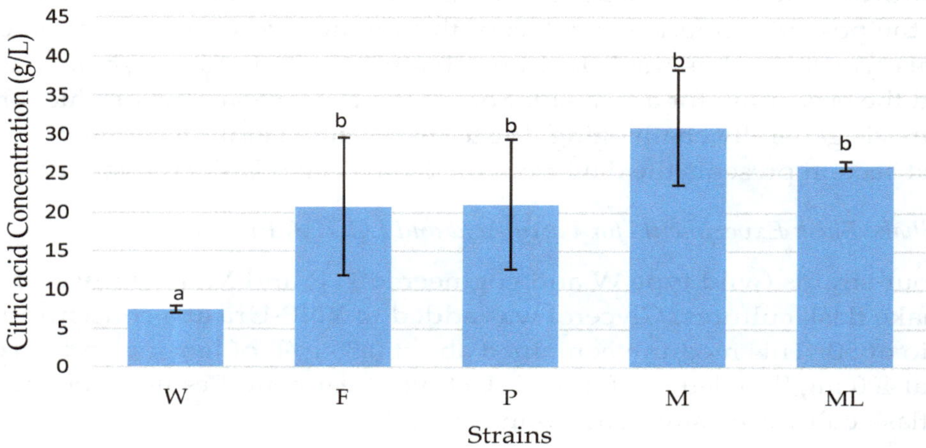

Figure 2. Study of 6-day shake flask cultures for citric acid production by the H, F, P, M and ML strains/treatment. Glycerol based media under nitrogen-limiting conditions (C/N = 50). Error bars represent a standard deviation for $n = 3$. W: H222 wild type strain, F: Δ*POX1-6* Δ*LEU2* Δ*SNF1* Δ*FAA1* Δ*SNF1::URA3*, P: F + Δ *LEU2:: URA3 YlALK5 YlCPR YlFAO1*, M: P + *YlALK5:: URA3*, ML: M + 400 mg/L leucine. ab Columns with dissimilar letters at the top are significantly different ($p < 0.05$).

All mutant strains produced a significantly ($p < 0.05$) higher level of citric acid than the W strain, in the range of 21 to 31 g/L. A small amount of the detected acid can be in the form of its isomer, iso-citric acid because this yeast is able to produce citric and iso-citric acid and their proportion is mainly influenced by feedstocks and strains [35]. Although the M strain produced the highest level of citric acid, there was no significant ($p > 0.05$) difference among the mutant strains. Moreover, the leucine supplementation did not have a significant effect on citric acid production. Citric acid is the major by-product of the lipid biosynthesis in *Y. lipolytica* under nitrogen-limiting conditions [66]. Higher re-direction of carbon flux from citric acid to fatty acid can be achieved by upregulating the first committed step of fatty acid biosynthesis, controlled by *ACC1* [67], and by removing the feedback inhibition effect of endogenous fatty acyl-CoA through thioesterase overexpression or acyl-CoA synthetase inactivation [49].

Figure 3 presents the total fatty acid and LCDCA monomer production by the strains from 6-day shake flask cultivation.

The wild-type W strain accumulated fatty acids at a low concentration of 340 mg/L with oleic acid (FA18:1) as the main constituent. The concentration of total fatty acids in the mutant strains was significantly ($p < 0.05$) higher than in the W strain. The deletion of *YlFAA1* in the F strain significantly ($p < 0.05$) increased the TFA level to 3300 mg/L in the F strain. These endogenous fatty acids were precursors for the LCDCA production. The LCDCAs were not produced by the W strain. The production of these monomers was initiated by the deletion of *YlFAA1* in the F strain. This shows the involvement of background ω-oxidation activity in LCDCA-16 and 18 production. This contribution was higher toward the end of 6-day shake flask cultivations. Single and multiple copy integration of the fatty acid-oxidizing genes significantly ($p < 0.05$) increased LCDCA production. This LCDCA level was similar for the P, M, and ML strains/treatment. Generally, LCDCA production in shake flasks is not stable and faces several challenges, including inadequate aeration, agitation, and pH control. Oxygen limitations are typical for shake flask cultures. Moreover, addition of 6 g/L $CaCO_3$ cannot sufficiently maintain the stability of pH value in shake flask cultures. These limitations undermine the effect of genetic or cultivation modification on LCDCA production. For instance, we observed the positive effect of leucine supplementation on biomass and LCDCA production by the M strain in bioreactor fermentations. However, application of similar treatment did not significantly change the LCDCA production level of the M strain in shake flask cultivations. Additionally, an extensive post-screening for choosing the best LCDCA-producing strain is beneficial to reach the strongest effect of multiple-integration. The preliminary shake flask studies were carried out, nonetheless, to select the appropriate strain and conditions for bioreactor fermentation. We obtained a LCDCA titer of 330 to 430 mg/L from the P, M and ML strains/treatment under nitrogen-limiting conditions. Chromatograms from the F and P culture indicate that a negligible amount of FA-17 was also produced by these strains.

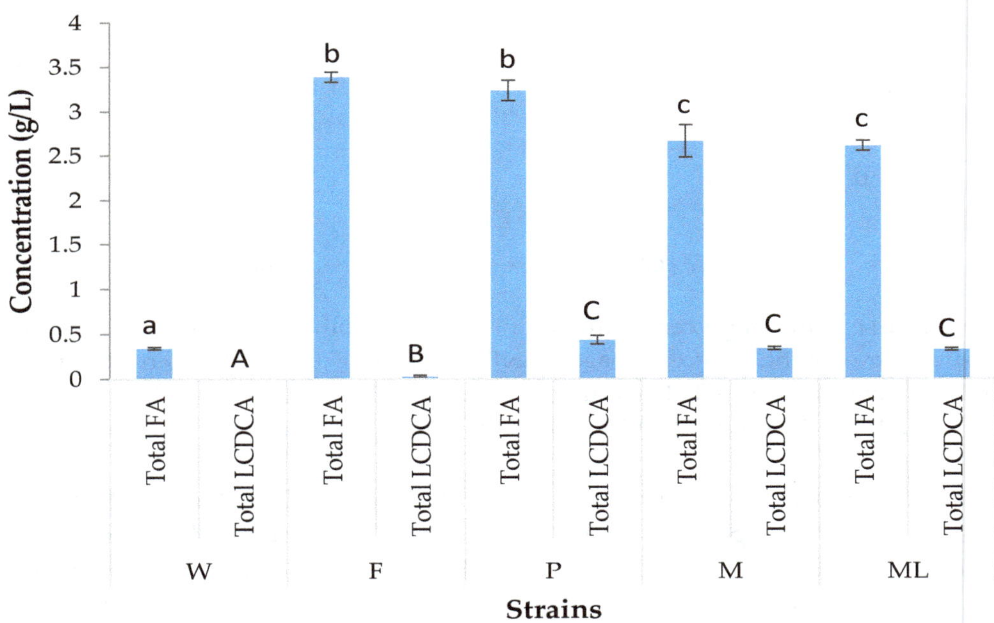

Figure 3. Study of 6-day shake flask cultures for fatty acid and LCDCA production by the H, F, P, M, and ML strains/treatment. Glycerol based media under nitrogen-limiting conditions (C/N = 50). Error bars a represent standard deviation of $n = 3$. W: H222 wild type strain, F: Δ*POX1-6* Δ*LEU2* Δ*SNF1* Δ*FAA1* Δ*SNF1::URA3*, P: F + Δ *LEU2*:: *URA3 YlALK5 YlCPR YlFAO1*, M: P + *YlALK5*:: *URA3*, ML: M + 400 mg/L leucine. Total FA: total fatty acids of 16 and 18 carbons, Total LCDCA: total long chain dicarboxylic acids of 16 and 18 carbons. ABC/abc Columns with dissimilar letters at the top are significantly different ($p < 0.05$).

In general, the deletion of *POX* genes prevents the degradation of fatty acid precursors and monomers. Additionally, the inactivation of *YlFAA1* promotes the generation of free fatty acids, which do not have the feedback inhibitory effect of acyl-CoA on the fatty acid biosynthesis pathway. The combination of these genetic modifications with the deletion of *YlSNF1* can promote the re-direction of the carbon flux towards malonyl-CoA and eventually free fatty acid accumulation. In addition to background ω-oxidation activity, the upregulation of the ω-oxidation pathway can enhance the transformation of free fatty acid pool into the monomers of less toxicity. Therefore, our genetic engineering served the dual purpose of enhancing fatty acid accumulation and producing LCDCA monomers. The ML strain and treatment was selected for further bioreactor fermentation experiments.

3.3. Batch Fermentation

The batch bioreactor fermentation was carried out using YNB-Ura medium with a C/N of 50. Glycerol was added to the medium at a concentration of 52 g/L. The production medium was also supplemented with 400 mg/L leucine. The time course study of biomass and citric acid production as well as glycerol utilization during the 4-day fermentation is presented in Figure 4.

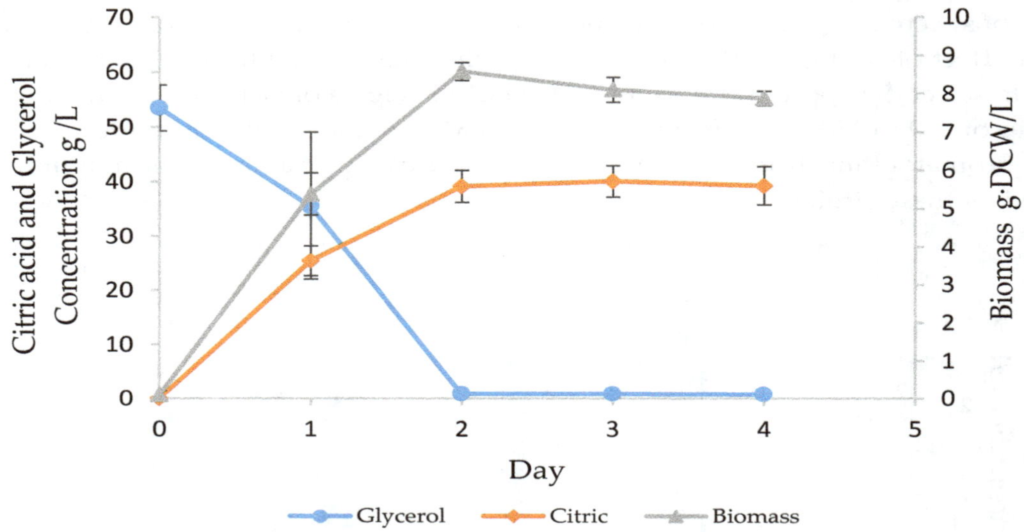

Figure 4. Time course study of glycerol utilization, biomass proliferation, and citric acid production by the ML strain/treatment. Glycerol based YNB-Ura media under nitrogen-limiting conditions (C/N = 50). Data from 2 rounds of batch fermentation. *ML: ΔPOX1-6 ΔLEU2 ΔSNF1 ΔFAA1 + Δ LEU2:: YlALK5 YlCPR YlFAO1 + YlALK5:: URA3* + 400 mg/L leucine.

Glycerol was consumed rapidly within 2 days of fermentation. During this period, the citric acid and biomass reached their peak at about 39 and 8.6 g/L, respectively. Afterwards, there was a slight reduction in the concentration of biomass, possibly due to some cell attachment to the fermenter vessel. The ML strain and treatment produced citric acid at a yield of more than 0.75 g/g of consumed glycerol. During the last two days of fermentation, the citric acid concentration did not undergo a marked change. It is notable that the mutant strains of this study are derived from H222 stain, which has a strong potential for citric acid production. Additionally, nitrogen-limiting conditions reroute carbon flux to lipid accumulation for storage and to citric acid for secretion. Both citric acid and fatty acids compete for the building units of acetyl-CoA [68]. To gain insight into monomers production conditions, we analyzed the concentration of fatty acids and monomers during the 4-day fermentation. Results are shown in Figure 5.

Batch fermentation in the 1-L bioreactor significantly enhanced the titer of LCDCA from 0.33 g/L, which was obtained in the shake flask culture, to 3.49 g/L for the ML strain and treatment. This corresponds to a minimum yield of 0.067 g/g of consumed glycerol and to a productivity of 0.04 g/L·h. This significant improvement was mainly due to adequate aeration, agitation, and precise control over the pH. The increase in LCDCA level continued until the last day of the 4-day fermentation at the expense of fatty acid pool.

The LCDCAs are more soluble than their fatty acid counterparts. This feature facilitates their transportation across the cell. We also analyzed the concentration of extracellular LCDCA monomers in cell-free production media to find more about the spatial distribution of LCDCAs. Figure 6 presents the results from this analysis.

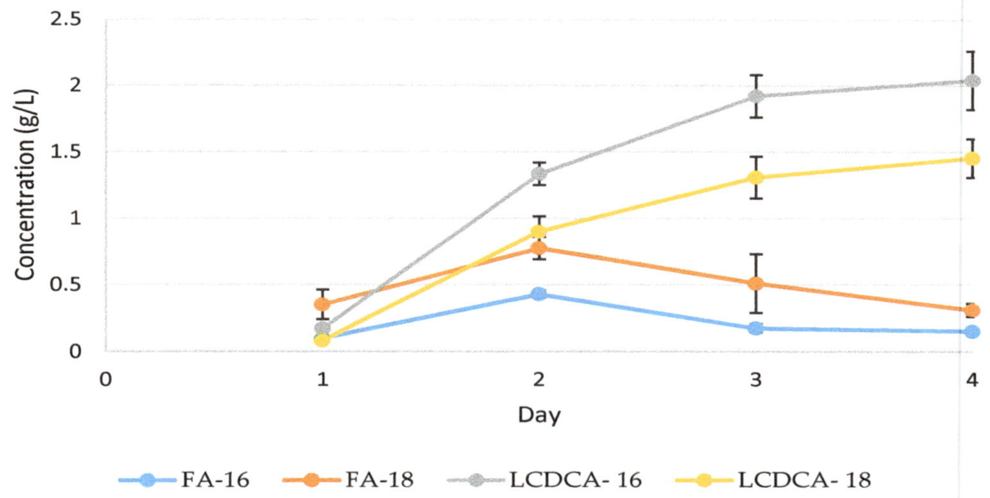

Figure 5. Time course study of fatty acids and LCDCA production by the ML strain/treatment. Glycerol based media under nitrogen-limiting conditions (C/N = 50). ML: Δ*POX1-6* Δ*LEU2* Δ*SNF1* Δ*FAA1* Δ*LEU2:: YlALK5 YlCPR YlFAO1* + *YlALK5:: URA3* + 400 mg/L leucine. FA-16: fatty acid of 16 carbons, FA-18: fatty acid of 18 carbons, LCDCA-16: long chain dicarboxylic acids of 16 carbons, LCDCA-18: long chain dicarboxylic acids of 18 carbons. Data from two rounds of batch fermentation.

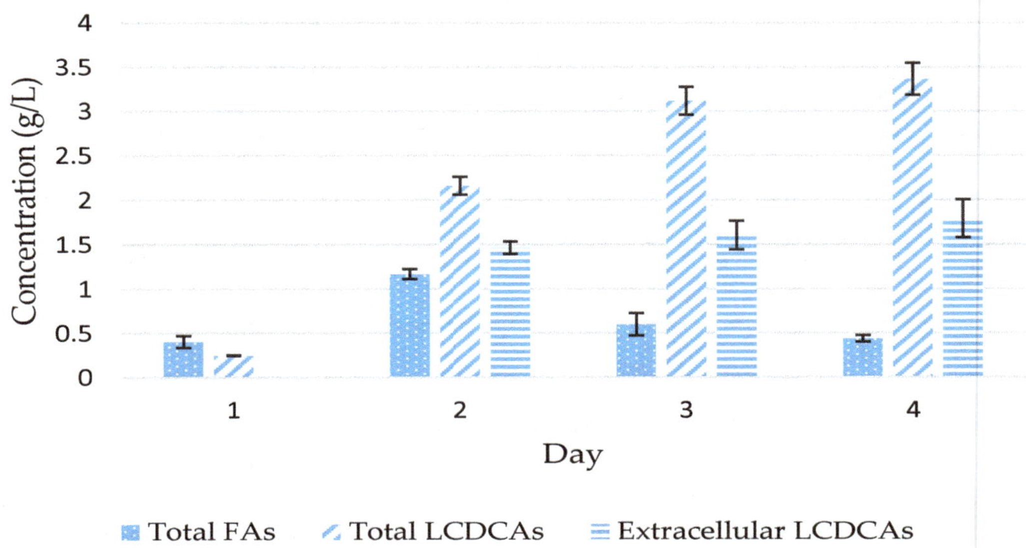

Figure 6. Time course study of total fatty acid, total LCDCA, and extracellular LCDCA production by the ML strains/treatment. Glycerol based media under nitrogen-limiting conditions (C/N = 50). Data from 2 rounds of batch fermentation. ML: Δ*POX1-6* Δ*LEU2* Δ*SNF1* Δ*FAA1* Δ*LEU2::YlALK5 YlCPR YlFAO1* +*YlALK5:: URA3* + 400 mg/L leucine. Total FAs: fatty acids of 16 and 18 carbons, Total LCDCAs: long chain dicarboxylic acids of 16 and 18 carbons. Data from 2 rounds of batch fermentation.

According to these results, more than half of the LCDCAs were present in the extracellular medium while the fatty acids were intracellularly used as precursors for the monomer production. Figure 6 also indicates that the availability of intracellular free fatty acids is the rate limiting step for *de novo* LCDCA production in the 1-L bioreactor fermentation. It should be noted that a small amount of dicarboxylic acids of 12 and 14 carbons was also found in the extracellular medium (less than 0.2 g/L). This can be due to the activity of *YlALK5* that was integrated into the genomes at multiple copies. The composition of metabolites was also determined. Results are presented in Figure 7.

Figure 7 shows that LCDCAs became predominant fatty acid-based metabolites at the end of the 4-day fermentation. Also, LCDCA-16 accounted for more than 50% of LCDCAs, meaning that the ω-oxidation rate of 16 carbon-fatty acid was higher than that of 18 carbon-fatty acids. This rate is influenced by the background ω-oxidation activity, as well as the activity of those genes that have been overexpressed. Table 4 presents the summary of data from biomass, citric acid, and LCDCAs production.

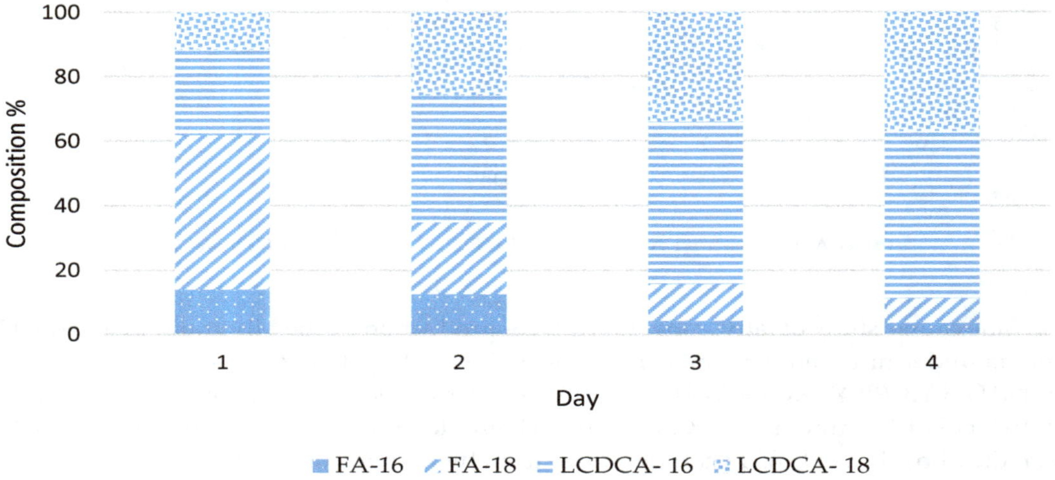

Figure 7. Percentage composition of fatty acids and LCDCAs produced by the ML strain/treatment. Glycerol based media under nitrogen-limiting conditions (C/N = 50). ML: Δ*POX1-6* Δ*LEU2* Δ*SNF1* Δ*FAA1* Δ*LEU2::YlALK5 YlCPR YlFAO1* +*YlALK5:: URA3* + 400 mg/L leucine. FA-16: fatty acid of 16 carbons, FA-18: fatty acid of 18 carbons, LCDCA-16: long chain dicarboxylic acids of 16 carbons, LCDCA-18: long chain dicarboxylic acids of 18 carbons. Data from 2 rounds of batch fermentation.

Table 4. Summary of the production level for biomass, citric acid, and LCDCAs.

DCW (g/L)	Citric Acid (g/L)	LCDCA Titer (g/L)	LCDCA Yield (g/g)	LCDCA Productivity (g/L·h)
8.58 ± 0.23	39.2 ± 3.5	3.49 ± 0.14	0.06	0.04

4. Discussion

The *de novo* production of fatty acid-derived monomeric compounds from unrelated feedstock has been previously demonstrated. For example, the overexpression of a plant thioesterase and of an engineered self-sufficient monooxygenase fusion protein in a β-oxidation deficient *E. coli* strain resulted in 234 μM ω-hydroxy octanoic acid production in a fed-batch fermentation [69]. In another study, *E. coli* was engineered for sugar-based hydroxyl fatty acid production [70]. That was achieved by the coexpression of acetyl-CoA carboxylase and acyl-CoA thioesterase in an inactivated acyl-CoA synthetase background, followed by the expression of a heterologous fatty acid hydroxylase. This resulted in the production of different hydroxyl fatty acids with chain lengths of 10 to 18 carbons at a final titer of 58.7 and 548 mg/L in culture broth and fed-batch fermentation, respectively. The engineered reversal of the β-oxidation cycle, in combination with the expression of heterologous thioesterase and monooxygenase systems, has led to the biosynthesis of medium chain ω-hydroxy fatty acids and dicarboxylic acids at a titer of 0.8 and 0.5 g/L, respectively, from glycerol in *E. coli* [71]. This bacterial host has also been engineered for the biosynthesis of long chain hydroxylated fatty acids at a titer of 117 mg/L [72]. In that study, both thioesterase and fatty acid hydroxylase $P450_{BM3}$ were involved in the production of free fatty acids and in their conversion into oxygenated monomers. Bowen et al. (2015) also engineered *E. coli* for sugar-based C12 and C14 ω-hydroxy FAs and α,ω-DCAs production [6]. This was achieved by expressing high-specificity acyl ACP thioesterases (TEs) to detach free fatty acids from the fatty acid synthase (FAS II). The application of p450 (CYP) and alcohol/aldehyde dehydrogenases led to the generation of ω-hydroxy FAs and subsequently α,ω-DCAs at a titer and productivity of 600 mg/L and 0.026 g/L·h, respectively. Figure 8 presents a summary of aforementioned strategies.

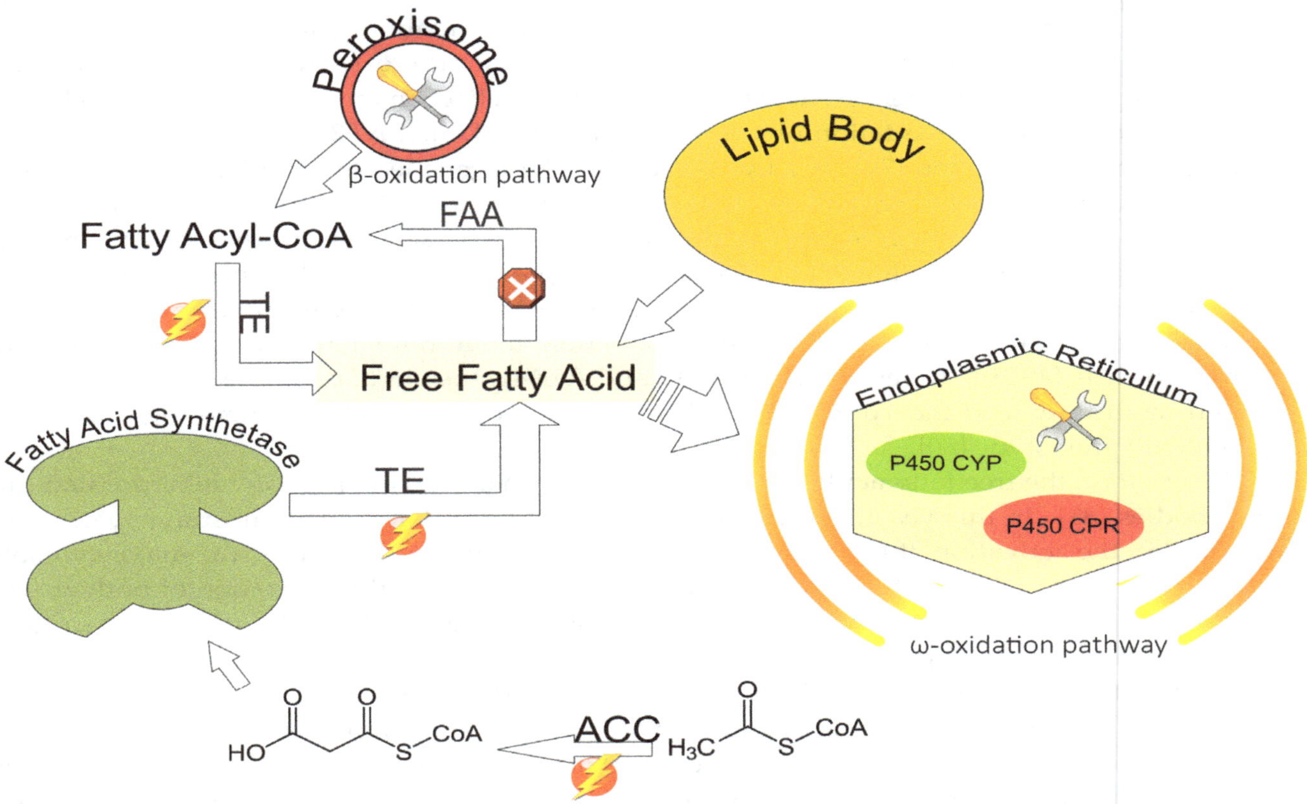

Figure 8. Genetic engineering strategies for the *de novo* production of oxygenated fatty acid monomers: overexpression of acetyl-CoA carboxylase (ACC) for higher carbon flux re-direction toward fatty acid biosynthesis pathway, overexpression of length specific thioesterases (TE) for the hydrolysis of acyl-CoA and consequently mitigation of their feedback inhibition effect, inactivation of the fatty acyl-CoA synthase (FAA) for the prevention of free fatty acid re-activation, inactivation or reversal of the β-Oxidation cycle for enhancing fatty acid pool and preventing degradation of fatty acid derivatives, and selective upregulation of the ω-oxidation pathway for the terminal oxidation of endogenous fatty acids to their oxygenated counterparts.

The above-mentioned yields and titers are notably less than those reported for whole-cell biotransformation of exogenously added long chain fatty acids to the corresponding monomers [8,73]. For instance, engineering *Y. lipolytica* for the production of LCDCA from C18 oil allowed to reach a titer of 16 g/L. This was accomplished through the blockage of β-oxidation and overexpression of *ALK1,2* together with the *CPR* gene under the control of p*POX2* [46].

This study reports the construction of *Y. lipolytica* as an oleaginous yeast cell platform for the *de novo* production of LCDCA monomers from renewable hydrophilic feedstock. Our LCDCA-producing platform is independent of oily feedstock whose utilization has potential disadvantages such as toxicity, low solubility, and weaker cellular uptake [7]. Instead, this sustainability-driven biosynthetic route utilizes hydrophilic substrates for the dual purpose of energy production and LCDCA biosynthesis under nitrogen-limiting conditions.

We developed our platform through a combinatorial genetic engineering strategy in a β-oxidation deficient strain. Our strategy involved the inactivation of *SNF1* for enhancing malonyl-CoA biosynthesis, disruption of the major cytosolic acyl-CoA synthetase (*FAA1*) for generating free fatty acid precursors, and upregulation of native ω-oxidation pathway for improving oxidation of the endogenous free fatty acid precursors into the corresponding long chain monomers. This strategy was developed based on the functions of the targeted genes.

In general, it is necessary to delete the *POX* genes in order to inactivate the β-oxidation pathway as the major degradation route. This prevents the degradation of fatty acid precursors and LCDCAs [45,47]. However, complete disruption of the peroxisome creates unwanted consequences. Also, defects in peroxisome biogenesis may have an inhibitory effect on the native P450 system [23,74], and may interfere with the oxidation of various metabolites [75]. *SNF2* deletion has been found to enhance lipid accumulation [76]. Inactivation of the *SNF1* gene led to an enhanced constitutive fatty acid accumulation and citric acid production in *Y. lipolytica* [66,77]. An alternative strategy to enhance carbon flux re-direction from citric acid production to fatty acid biosynthesis is to overexpress *ACC1* or use hyperactive Acc1p [67,78]. This strategy can prevent the possible unintended metabolic consequences resulting from *SNF1* deletion. Inactivation of *FAA1* can interfere with the re-activation of released fatty acids that could serve in energy production, membrane maintenance, or triglyceride (TAG) homeostasis [49,79]. Deletion of the *FAA1* and *FAA4* genes in *S. cerevisiae* have been found to interfere with fatty acid incorporation in phospholipids or TAG, while generating a free fatty acid-producing phenotype [80].

In *Y. lipolytica*, the free fatty acids that are activated by YlFaa1p are partially incorporated into the lipid bodies [81]. In this yeast, several gene products that differ in their substrate specificities are involved in the activation of fatty acids in different organelles [82]. Faa1p is the major cytosolic long chain fatty acid activator [83]. This gene plays a crucial role in the utilization of both *ex novo* fatty acids and those obtained from the metabolism of *n*-alkanes [84]. *FAA1* also plays a role in the storage of fatty acids in lipid bodies [81,83]. Besides YlFat1p, which is mainly involved in fatty acid remobilization, there are at least 10 *AAL* genes involved in the activation of fatty acids in the peroxisome and lipid bodies. The inactivation of these genes can prevent the cell from storing fatty acids in lipid bodies [82]. The deficiency of fatty acid activation systems can also negatively affect yeast growth in batch cultures with glucose [85], due to either the detergent-like properties of free fatty acids or metabolic disorder [86].

We observed some residual fatty acids at the end of the 4-day batch fermentation. Since these fatty acids may occur in the esterified form in lipid bodies, they do not act as substrates for the ω-oxidation reactions. Our genetic engineering did not involve the disruption of TAG biosynthesis. This is because the generation of free fatty acids, particularly unsaturated ones, causes toxicity in the cell with deficient lipid body formation [87]. This occurs for several reasons: the destabilizing effect of fatty acid overproduction on the cellular membrane [80,88], the harmful effects of oxidative species derived from the oxidation of unsaturated fatty acids [89], and a lack of protection against excess lipid intermediates [90].

Free fatty acid accumulation or starvation can also trigger the ω-hydroxylation of fatty acids [91,92]. The resulting hydroxylated fatty acid monomers may also interfere with the TAG homeostasis [93]. We propose that ω-oxidation serves as a rescue route to detoxify free acids in the β-oxidation-disrupted strains, especially when *FAA1* is inactivated. The stress relieving effect of fatty acid oxidizing enzyme has also been observed in plants [23]. In *Y. lipolytica*, *ALK* genes are involved in the detoxification of fatty alcohol, dodecanol [36]. Thus, this background ω-oxidative activity processes free fatty acids to oxygenated monomers of higher solubility and less toxicity. This can explain the stimulatory effect of ω-oxidation upregulation on biomass proliferation in the F strain with deficient FAA1p activity. In fact, the natural enzymatic machinery of the host cell plays an influential role in fatty acid over-oxidation [18].

Results show that our combinatorial genetic engineering for the biosynthesis of free fatty acids and their terminal oxidation to the corresponding LCDCAs served multiple purposes. These included the reduction of free fatty acid toxicity, increase of product solubility, facilitation of product transportation across the membrane, and the creation of a pull for further fatty acid flux direction toward ω-oxidation.

We observed that free fatty acids, generated from the *FAA1* inactivation, are suitable substrates for the native ω-oxidation pathway. This accords with a previous study, which reported the hydroxylation of the free fatty acid pool by a fatty acid hydroxylase (CYP102A1) from *Bacillus megaterium* [70]. Free fatty acids do not have the feedback inhibition regulatory effect of fatty acyl-CoAs on some genes of lipid synthesis [48,86]. To obtain free fatty acids, the inactivation of native fatty acid activation systems or overexpression of heterologous/homologous thioesterases has been suggested in the literature. For example, inactivation of *FAA1* and *FAA4* genes in *S. cerevisiae* was found to mitigate the feedback inhibition of acyl-CoA on fatty acid synthesis and undermine the β-oxidation upregulating-effect of fatty acid overproduction [86]. Overexpression of thioesterases has also been employed to release the free fatty acids of specific chain lengths [6].

Studies show that *YlALK3*, *YlALK5*, and *YlALK7* exhibit ω-oxidation activity against dodecanoic acid [36,41]. However none could result in a detactable amount of products from ω-oxidation of hexadecanoic acid in a Δ*alk1-12* strain [36]. It was postulated that Alk3, 5, and 7p may prefer shorter fatty acids, while larger fatty acids are preferrably used for cell maintenance. In this study, the DCAs of 12 and 14 carbons were detected as minor products. Their production can be due to the overexpression of *YlALK5*. Additionally, the overexpression of this gene, using the replicative vector with the strong constitutive promoter, contributed toward LCDCA-16 and ω-hydroxy fatty acid-16 production in our preliminary shake flask cultivations. In fact, a high gene dosage is necessary to achieve sufficient P450 expression in *Y. lipolytica* [94]. This contribution was in addtiton to the background ω-oxidation activity in the F strain toward LCDCA-16 and 18 production. We also observed the contribution of *YlCPR* and *YlFAO1* to the multiple-step oxidation of endogenous long chain free fatty acids precursors to the monomers. In the same fashion, amplification of the P450 reductase gene was found to enhance LCDCA production from *C. tropicalis* [14]. Similarly, it was reported that the overexpression of *YlFAO1* improves LCDCA production from exogenous hydrophobic substrates [42]. This accords with the contribution of *FAO1* to diacid production observed in *C. tropicalis* [95]. Various fatty aldehyde dehydrogenases have been identified in *Y. lipolytica* that are involved in the detoxification of fatty acid aldehyde intermediates by oxidizing them to dicarboxylic acids [96–98]. These native genes likely contribute to the background ω-oxidation activity.

A bioreactor is much more efficient than a shake flask for DCA production [7]. Adequate aeration, agittion, and control over pH play major roles in the P450-based LCDCA production. The inner pH of the cell, as well as aeration, both affect the induction of yeast CYP monooxygenase and CYP reductase [10]. Studies show that a higher dissolved oxygen (DO) level can remarkably improve P450 activity [99,100]. This level was kept at 80% saturation for LCDCA production from *C. ropiclais* [14]. In fact, the native P450 sytem utilizes molecular oxygen that is abundant in a bioreactor fermentation. In our study, fermentation scale-up to 1-L bioreactor under the DO level of at least 50% significantly enhanced the rate of free fatty acid oxidation to LCDCAs. The majority of free fatty acids were converted to LCDCA during the four days of bioreactor fermentation, reaching a titer of 2 g/L for LCDCA-16 and 1.45 g/L for LCDCA-18. Adjustment of the pH is also required for the biosynthesis of LCDCAs. During the production phase, the pH was adjusted to 8 since the LCDCAs are completely diasscoiated at this level. For instance, the diassociation constant of DCA13 is less than 5.8. A higher extracellular pH level increases the solubility of LCDCAs, improves their transportation efficiency, and consequently mitigates the corresponding product inhibition [101]. Adjustment of pH as well as metabolite production in the production media increases the demand for antifoam addition. Thus, the antfoam was also periodically added to the media.

Transport of LCDCAs out of the cell can improve P450 activity [100] and simplify recovery for an enhanced downstream process at a lower cost. Our engineered strain showed good secretion efficiency, since more than half of the synthetized monomers were found in the media [102]. In fact, DCA of 13 carbons has been found to undergo passive diffusion due to the concentration difference between the inside and outside cells [101]. At a low intracellular pH, the undissociated form of the carboxylic acid is expected, which can mitigate their transportation across the cell membrane [103]. In this study, we showed that LCDCAs are extracellular products. Secretion allows LCDCA production to exceed the lipid accumulation capacity of the cell, and promotes their recovery from media after cell harvest at a lower cost. This also can alleviate their inhibitory effect on P450-based hydroxylation reaction due to their competitive binding to P450 [104].

We supplemented the production media with 400 mg/L leucine to take the advantage of its possible stimulatory effect on the metabolite produciton, under nitrogen-limiting conditions. Supplementation of media with leucine had significant effect on biomass level in shake flask cutures. In addition to that, our preliminary bioreactor fermentation using YNB-Ura, without additional leucine supplementation, resulted in less biomass and LCDCA production from the M strain (Biomass up to 7 g/L and maximum LCDCA of 3 g/L). Similalrly, addition of leucine was reported to increase the growth and lipid accumulation in *S. cerevisiae* devoid of active *SNF2* [105].

Although nitrogen-limiting conditions hinder growth, they induce lipid and citric acid production. The resting cell production of LCDCAs has several advantages, including the dedication of a larger proportion of supplied oxygen to metabolism and the targeted P450 reaction [106], the repeated utilization of LCDCA-exporting cells, the use of high cell concentration, particularly for the multi-copy transformants [107], as well as more strain robustness against fermentation media. In fact, resting cells consume nearly half of what they typically utilize when they have access to a nitrogen source [106]. We used this strategy for the *de novo* production of free fatty acids. This strategy also caused our engineered cell factory to produce more than 36 g/L of citric acid as the major bioproduct which was nearly stable during the 4-day bioreactor fermentation. The production of this byproduct, which is extensively used in pharmaceutical industries [108] raises more revenue and offsets the total production cost.

This *Y. lipolytica* platform can be tailored to produce LCDCA of higher specificity with respect to chain length and saturation. This yeast accumulates a higher proportion of unsaturated fatty acids, oleic and linoleic acids, when it is grown on sugar, due to the strong expression of endogenous desaturases [53]. Overexpression of these native desaturase genes, including Δ-9 and Δ-12 desaturases, can increase the proportion of desaturated fatty acids in lipid bodies. Together with *FAA1* inactivation, this can result in a larger pool of unsaturated free fatty acids for higher unsaturated LCDCA production. In fact, *Y. lipolytica* predominantly accumulate fatty acids in the form of C16:0 and C18:1. These fatty acids are more susceptible to remobolization [109]. Disruption of their re-activation in the Δ*FAA1* background, enbales their availibiltiy for the ω-oxidation. Our study confirmed that this oxidation does not change the configuration of double bonds. As a result, LCDCA16 and LCDCA18:1 were the predominant LCDCA products of our platform. Moreover, according to the chromatogram of GC-FID, a smaller portion of LCDCA-18 had two double bonds (less than 10%). These bonds are dervied from the linolenic acid, FA 18:2, whose concntration was less than oleic acid. Regarding chain length specificity and selectivity, there is room for improvement. For example, manipulation of *YlYAS* genes [110], and selective overexpression of elongase, monooxygenase, and thioesterase can be employed to modulate background activity and enhance the length-specificity of the terminal oxidation.

Y. lipolytica has several advantages as a LCDCA-producing platform. This yeast platform has its own native P450 system that catalyzes terminal oxidation of alkanes and fatty acids [36]. Moreover, this platform possesses an endogenous redox partner (cytochrome P450 reductase), has the capability for large free fatty acid accumulation [111], and can efficiently generates NADPH cofactor mainly through the pentose phosphate pathway (PPP) [112,113]. This pathway is not tightly regulated by nitrogen concentration [114]. Together with glycolysis, it can contribute significantly to the energy supply of resting cells [115]. At the stationary phase, a higher expression of cytochrome P450 monoxygenases [104] and lipid remobilization [40] are also expected. These features enable the large-scale application of engineered *Y. lipolytica* cells without the need for cofactor addition. Regeneration of NADPH is crucial in whole-cell P450-based biotransformation [116], particularly in the NADPH- and adenosine triphosphate (ATP)-consuming process of fatty acid biosynthesis and oxidation. Moreover, whole cells also provide a protected environment for the increased stability of the P450 enzymes [51]. A whole cell biocatalyst is a cheaper option than the isolated or immobilized enzyme [106]. The pathogenic yeast, *C. tropicalis*, has been commonly engineered for whole cell biotransformation of aliphatic feedstock to LCDCAs [14,25]. However, *Y. lipolytica* has safe status, and has been categorized as a GRAS microorganism by the FDA (Food and Drug Administration) [56,88]. This status, as well as its probiotic properties [117], can pave the way toward the biosynthesis of LCDCAs that meet the safety requirements of the food and pharmaceutical industries.

Despite some chemical based-production of LCDCA at large scale, the high price of the LCDCAs, and their limited commercial availability are still major constraints in expanding the spectrum of their application [25]. Therefore, researchers should use various renewable resources for the sustainable production of these multi-purpose building blocks. An alternative sustainable solution is to establish a P450-based biological approach. However, commercial application of a P450-based platform has also its own advantages and limitations. For example, the scalability of modified cytochrome P450 for multi-kilogram scale synthesis remains a challenge [118]. The majority of P450-based biocatalysts have not met a minimum space-time yield of 0.1 g/L·h in previous studies. A solution to this limitation would be the production of high-value pharmaceuticals. In this scenario, a minimum process productivity of 0.001 g/L·h can justify large-scale bioprocessing from an economic standpoint [19]. The LCDCAs have relatively high selling price, depending on their purity, and therefore are outstanding targets for white biotechnology [12]. Our study is the first report on engineering *Y. lipolytica* for the efficient *de novo* production of LCDCAs from glycerol. Results from this study showed a minimum productivity of 0.04 g/L·h, which is more than the foregoing required productivity. However, the provision of adequate agitation and oxygen transfer in larger fermenters will be challenging to maintain this minimum productivity. Our biological-based approach has several advantages over chemically-based approaches, including a broad range of renewable feedstock, milder process conditions, a more environmentally friendly process, and higher selectivity, particularly toward long chain unsaturated fatty acids [10,119].

These monomers have superior properties, promising future, and a large global market that is expected to reach USD $300.3 millions by 2025 [120]. These highly demanding monomers are used in manufacturing various high-performance polymeric compounds such as nylon and other polyamides. They have a wide scope of application in the industrial manufacture of automobiles, medicines, fragrances, adhesives, and macrolide antibiotics [10]. However, pharmaceutical applications exhibit the quickest growth in terms of revenue [120].

Compared to the shorter diacids, these building blocks and their LCDCA-based polymers have superior bio-based properties such as improved biodegradability, hydrolytic stability, crystallinity, melting temperature, optical transparency, dispersion, miscibility, material toughness, and flexibility; lower moisture uptake, dielectric constant, and surface tension [5]. These long chain monomers are also the building blocks of naturally occurring polyesters [121]. Additionally, the double bonds of unsaturated monomers, which are maintained during terminal oxidation, show promise for additional functional groups and linkages. These features also allow for the production of novel unsaturated LCDCAs [54] to synthesize cross-linked polymer networks with adjustable properties. Synthesizing these unsaturated long chain monomers is difficult via chemical routes due to the reactivity and participation of the double bonds in unwanted reactions [122].

Our findings suggest that the tunable high-level expression of selected P450 genes, by high copy and stable integration or by high copy number expression plasmids, can foster LCDCA overproduction. Application of a lipid responsive strong promoter may provide additional benefit for P450 gene expression. The multiple-copy integration of a selected gene at random loci is influenced by the integration site underlining the need for an intensive screening afterwards [62,123]. We recommend an intensive screening process for choosing the most productive multiple-copy transformants. We selected the *YlALK5* for multiple-copy integration since it exhibited the highest ω-oxidation activity, among the tested *YlALKSs* toward the endogenously synthesized fatty acids of 16 carbons. The *YlALK5* gene product is a good candidate for protein engineering, allowing higher selectivity and turn-over rates for long chain fatty acids. The combination of the *YlALK5* and *YlFAO1* overexpression can create a push and pull of fatty acid precursors and oxygenated fatty acids in the ω-oxidation pathway. Further studies are recommended to identify the major native cytochome monoxygenese member involved in the temrinal oxidation of 18 carbon-fatty acids. We also suggest modulating the ratio of P450 monoxygenase/reductase overexpression for better performance. Our engineered cell factory can be further optimized to produce a higher proportion of LCDCAs with specific chain lengths or degrees of desaturation. This can be achieved in two ways: by introducing heterologous thioesterase-coding genes that have long chain fatty acid specificity (e.g., *Cocos nucifera* fatty acyl-ACP thioesterase CnFatB2 and *Iris germanica* acyl-ACP thioesterase AAG43857 [124]) or by manipulating the native desaturase and elongase systems. The *LEU2* gene was disrupted for enabling a successful site-specific multiple gene integration and testing various expression systems. However, we suggest targeting other sites for future multiple-gene integrations, due to the positive role of the *LEU2* gene on biomass and metabolite production. We used a standard minimal medium for production. However, the availability of rich nutrients, heme containing compounds, and low cost substrates, including lignocellulosic hydrolysates and crude glycerol, could enhance production level and economy of this bioprocess. The overexpression of key genes from PPP can promote NADPH availability [113], and this may serve the NADPH-consuming process of endogenous fatty acid and LCDCA production. We also recommend the application of immobilized cells and a biofilm reactor for integrated production and recovery and for efficient use of multi-copy transformants. The *de novo* production of LCDCA can be further improved by generating a larger pool of free fatty acids [125], which have stronger stimulatory effect on the induction of the background ω-oxidation activity. This could be achieved by overexpressing the upstream and downstream targets of the fatty acid biosynthesis pathway. Acetyl-CoA carboxylase (*ACC1*) and thioesterase (*TE*) are examples of these targets. Disruption of other acyl-CoA synthetases (e.g., *YlFat1*) [81] and overexpression of intracellular lipases (e.g., *TGL3* and *TGL4*) [126] may also contribute to a larger pool of free fatty acids. Microscopic examination of resting cells may reveal more information about the status of remaining fatty acids. Moreover, time course qualitative analysis of free fatty acids can also provide further insight into their threshold for the induction of background ω-oxidation activity. Despite the good LCDCA transport efficiency of our platform, the overexpression of potential heterologous transporters such as CmCDR1 [127], or homologous transporter like ABC2 [128] could improve LCDCA export to minimize product inhibition. Optimization of culture conditions was beyond the scope of this study. However, production titer and yield can be enhanced by optimizing production media and developing fed-batch or continuous bioprocess. For example, providing a reduced aeration rate [129] or nitrogen as co-feed [130] might lessen citric acid production in favor of more fatty acid accumulation. Summary of elements for further improvement is shown in Figure 9.

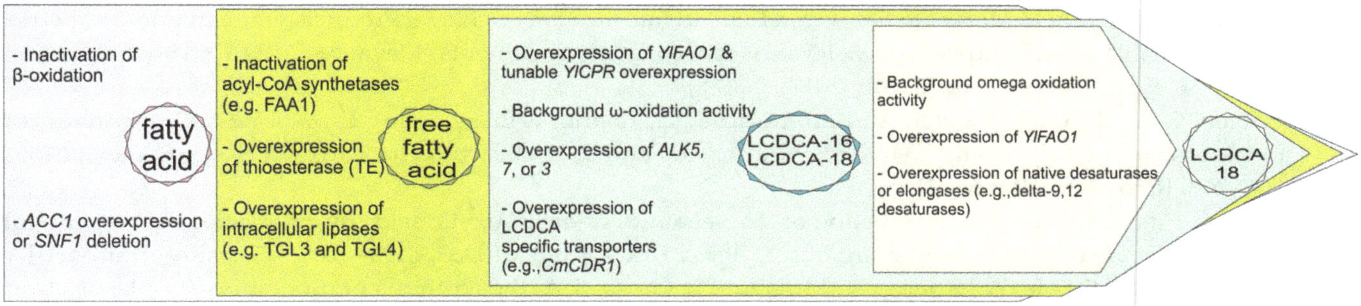

Figure 9. Summary of elements for further improvement.

5. Conclusions

Our study presents the construction of *Y. lipolytica* as a cell factory platform for the *de novo* production of LCDCAs from glycerol. We accomplished this by harnessing the native capabilities of this yeast for endogenous fatty acid accumulation and ω-oxidation. Our combinatorial genetic engineering mainly involved inactivating the peroxisomal fatty acid degradation, disrupting cytosolic fatty acid activation, and upregulating fatty acid ω-oxidation. Further enhancement was obtained through fermentation scale-up to the 1-L nitrogen-limited bioreactor culture. Our engineering strategies led *Y. lipolytica* strains to produce about 3.49 g/L LCDCA, with a yield of 0.06 g/g. To the best of our knowledge, this is the highest *de novo* production level ever reported for LCDCA monomers. We envision that this study constitutes a foundational work in developing *Y. lipolytica* as a safe oleaginous yeast platform to upgrade low-value, unrelated feedstock to high-value multi-purpose LCDCAs at a larger scale in the foreseeable future. This work represents a key advance in yeast strain manipulation and fermentation for the biosynthesis of LCDCAs from renewable non-oily feedstock. Last, this research paves the way for an industrial process that can supplement the global supply of LCDCA from a second source, independent of the non-renewable hydrophobic substrates such as crude oil.

Acknowledgments: We would like to thank Washington State University for funding this project, Michael Gatter from the Institute of Microbiology TU Dresden, Germany, for providing strains H222 and H222ΔP. We also greatly appreciate valuable technical advices from Scott E. Baker and Amir H. Ahkami (Pacific Northwest National Laboratory); our colleagues in Washington State University: Birgitte Ahring (Bioproducts, Sciences, & Engineering Laboratory), Jonathan Lomber (Central Analytical Chemistry Laboratory), James G Wallis (Browse lab, Institute of Biological Chemistry), Anna Berim (Gang Lab), Xiaochao Xiong, and Yuxiao Xie (Bioprocessing and Bioproduct Engineering Laboratory).

Author Contributions: Ali Abghari developed the concept, designed the experiments, performed the experiments, analyzed the data, and drafted this paper. Catherine Madzak provided advice about the use of pINA1291 vector, analysis of results, and revision of the draft. Shulin Chen revised the manuscript and approved the final version for publication. All authors read and approved the final manuscript.

References

1. Pellis, A.; Herrero Acero, E.; Gardossi, L.; Ferrario, V.; Guebitz, G.M. Renewable building blocks for sustainable polyesters: New biotechnological routes for greener plastics. *Polym. Int.* **2016**, *65*, 861–871. [CrossRef]
2. Liao, J.C.; Mi, L.; Pontrelli, S.; Luo, S. Fuelling the future: Microbial engineering for the production of sustainable biofuels. *Nat. Rev. Microbiol.* **2016**, *14*, 288–304. [CrossRef] [PubMed]
3. Chung, H.; Yang, J.E.; Ha, J.Y.; Chae, T.U.; Shin, J.H.; Gustavsson, M.; Lee, S.Y. Bio-based production of monomers and polymers by metabolically engineered microorganisms. *Curr. Opin. Biotechnol.* **2015**, *36*, 73–84. [CrossRef] [PubMed]
4. Díaz, A.; Katsarava, R.; Puiggalí, J. Synthesis, properties and applications of biodegradable polymers derived from diols and dicarboxylic acids: From polyesters to poly (ester amide) s. *Int. J. Mol. Sci.* **2014**, *15*, 7064–7123. [CrossRef] [PubMed]

5. Beuhler, A. C18 Diacid Market to Grow and Expand into an Array of Novel Products with Superior Properties. 2013. Available online: http://www.elevance.com/images/documents/Elevance-ODDA-C18-white-paper_20130916_F.pdf (accessed on 22 July 2017).
6. Bowen, C.H.; Bonin, J.; Kogler, A.; Barba-Ostria, C.; Zhang, F. Engineering *Escherichia coli* for conversion of glucose to medium-chain ω-hydroxy fatty acids and α, ω-dicarboxylic acids. *ACS Synth. Biol.* **2015**, *5*, 200–206. [CrossRef] [PubMed]
7. Sathesh-Prabu, C.; Lee, S.K. Production of long-chain α, ω-dicarboxylic acids by engineered escherichia coli from renewable fatty acids and plant oils. *J. Agric. Food Chem.* **2015**, *63*, 8199–8208. [CrossRef] [PubMed]
8. Lu, W.; Ness, J.E.; Xie, W.; Zhang, X.; Minshull, J.; Gross, R.A. Biosynthesis of monomers for plastics from renewable oils. *J. Am. Chem. Soc.* **2010**, *132*, 15451–15455. [CrossRef] [PubMed]
9. Ngo, H.L.; Foglia, T.A. Synthesis of long chain unsaturated-α, ω-dicarboxylic acids from renewable materials via olefin metathesis. *J. Am. Oil Chem. Soc.* **2007**, *84*, 777–784. [CrossRef]
10. Huf, S.; Kruegener, S.; Hirth, T.; Rupp, S.; Zibek, S. Biotechnological synthesis of long-chain dicarboxylic acids as building blocks for polymers. *Eur. J. Lipid Sci. Technol.* **2011**, *113*, 548–561. [CrossRef]
11. Cho, C.; Choi, S.Y.; Luo, Z.W.; Lee, S.Y. Recent advances in microbial production of fuels and chemicals using tools and strategies of systems metabolic engineering. *Biotechnol. Adv.* **2015**, *33*, 1455–1466. [CrossRef] [PubMed]
12. Weiss, A. Selective microbial oxidations in industry: Oxidations of alkanes, fatty acids, heterocyclic compounds, aromatic compounds and glycerol using native or recombinant microorganisms. *Mod. Biooxid. Enzym. React. Appl.* **2007**, 193–209. [CrossRef]
13. Kroha, K. Industrial biotechnology provides opportunities for commercial production of new long-chain dibasic acids. *Inform* **2004**, *15*, 568–571.
14. Picataggio, S.; Rohrer, T.; Deanda, K.; Lanning, D.; Reynolds, R.; Mielenz, J.; Eirich, L.D. Metabolic engineering of candida tropicalis for the production of long–chain dicarboxylic acids. *Nat. Biotechnol.* **1992**, *10*, 894–898. [CrossRef]
15. Maurer, S.C.; Schmid, R.D. Biocatalysts for the epoxidation and hydroxylation of fatty acids and fatty alcohols. *Handb. Ind. Biocatal.* **2005**. [CrossRef]
16. Kogure, T.; Horiuchi, H.; Matsuda, H.; Arie, M.; Takagi, M.; Ohta, A. Enhanced induction of cytochromes p450alk that oxidize methyl-ends of n-alkanes and fatty acids in the long-chain dicarboxylic acid-hyperproducing mutant of candida maltosa. *FEMS Microbiol. Lett.* **2007**, *271*, 106–111. [CrossRef] [PubMed]
17. Kim, D.; Cryle, M.J.; De Voss, J.J.; de Montellano, P.R.O. Functional expression and characterization of cytochrome p450 52a21 from candida albicans. *Arch. Biochem. Biophys.* **2007**, *464*, 213–220. [CrossRef] [PubMed]
18. Scheps, D.; Honda Malca, S.; Richter, S.M.; Marisch, K.; Nestl, B.M.; Hauer, B. Synthesis of ω-hydroxy dodecanoic acid based on an engineered cyp153a fusion construct. *Microb. Biotechnol.* **2013**, *6*, 694–707. [CrossRef] [PubMed]
19. Girhard, M.; Bakkes, P.J.; Mahmoud, O.; Urlacher, V.B. P450 biotechnology. In *Cytochrome p450*; Springer: Berlin, Gemany, 2015; pp. 451–520.
20. Li, H.; Pinot, F.; Sauveplane, V.; Werck-Reichhart, D.; Diehl, P.; Schreiber, L.; Franke, R.; Zhang, P.; Chen, L.; Gao, Y. Cytochrome p450 family member cyp704b2 catalyzes the ω-hydroxylation of fatty acids and is required for anther cutin biosynthesis and pollen exine formation in rice. *Plant Cell* **2010**, *22*, 173–190. [CrossRef] [PubMed]
21. Malca, S.H.; Scheps, D.; Kühnel, L.; Venegas-Venegas, E.; Seifert, A.; Nestl, B.M.; Hauer, B. Bacterial cyp153a monooxygenases for the synthesis of omega-hydroxylated fatty acids. *Chem. Commun.* **2012**, *48*, 5115–5117. [CrossRef] [PubMed]
22. Höfer, R.; Briesen, I.; Beck, M.; Pinot, F.; Schreiber, L.; Franke, R. The arabidopsis cytochrome p450 cyp86a1 encodes a fatty acid ω-hydroxylase involved in suberin monomer biosynthesis. *J. Exp. Bot.* **2008**, *59*, 2347–2360. [CrossRef] [PubMed]
23. Kandel, S.; Sauveplane, V.; Compagnon, V.; Franke, R.; Millet, Y.; Schreiber, L.; Werck-Reichhart, D.; Pinot, F. Characterization of a methyl jasmonate and wounding-responsive cytochrome p450 of arabidopsis thaliana catalyzing dicarboxylic fatty acid formation in vitro. *FEBS J.* **2007**, *274*, 5116–5127. [CrossRef] [PubMed]

24. Benveniste, I.; Tijet, N.; Adas, F.; Philipps, G.; Salaün, J.-P.; Durst, F. Cyp86a1 fromarabidopsis thalianaencodes a cytochrome p450-dependent fatty acid omega-hydroxylase. *Biochem. Biophys. Res. Commun.* **1998**, *243*, 688–693. [CrossRef] [PubMed]
25. Craft, D.L.; Madduri, K.M.; Eshoo, M.; Wilson, C.R. Identification and characterization of the cyp52 family of candida tropicalis atcc 20336, important for the conversion of fatty acids and alkanes to alpha,omega-dicarboxylic acids. *Appl. Environ. Microbiol.* **2003**, *69*, 5983–5991. [CrossRef] [PubMed]
26. Eschenfeldt, W.H.; Zhang, Y.; Samaha, H.; Stols, L.; Eirich, L.D.; Wilson, C.R.; Donnelly, M.I. Transformation of fatty acids catalyzed by cytochrome p450 monooxygenase enzymes of candida tropicalis. *Appl. Environ. Microbiol.* **2003**, *69*, 5992–5999. [CrossRef] [PubMed]
27. Hoffmann, S.M.; Danesh-Azari, H.R.; Spandolf, C.; Weissenborn, M.J.; Grogan, G.; Hauer, B. Structure-guided redesign of CYP153AM. aq for the improved terminal hydroxylation of fatty acids. *ChemCatChem* **2016**, *8*, 3234–3239. [CrossRef]
28. Schreiber, L.; Durst, F.; Pinot, F. Cyp94a5, a new cytochrome p450 from nicotiana tabacum is able to catalyze the oxidation of fatty acids to the v-alcohol and to the corresponding diacid. *FEBS J.* **2001**, *268*, 3083–3090.
29. McLean, K.J.; Luciakova, D.; Belcher, J.; Tee, K.L.; Munro, A.W. Biological diversity of cytochrome p450 redox partner systems. In *Monooxygenase, Peroxidase and Peroxygenase Properties and Mechanisms of Cytochrome p450*; Springer: Berlin, Gemany, 2015; pp. 299–317.
30. Girhard, M.; Tieves, F.; Weber, E.; Smit, M.S.; Urlacher, V.B. Cytochrome p450 reductase from candida apicola: Versatile redox partner for bacterial p450s. *Appl. Microbiol. Biotechnol.* **2013**, *97*, 1625–1635. [CrossRef] [PubMed]
31. Theron, C.W. *Hetrologous Expression of Cytochrome p450 Monoxygenase by the Yeast Yarrowia Lipolytica*; University of the Free State: Bloemfontein, South Africa, 2007.
32. Iida, T.; Sumita, T.; Ohta, A.; Takagi, M. The cytochrome p450alk multigene family of an n-alkane-assimilating yeast, yarrowia lipolytica: Cloning and characterization of genes coding for new CYP52 family members. *Yeast* **2000**, *16*, 1077–1087. [CrossRef]
33. Abghari, A.; Chen, S. *Yarrowia lipolytica* as an oleaginous cell factory platform for the production of fatty acid-based biofuel and bioproducts. *Front. Energy Res.* **2014**, *2*. [CrossRef]
34. Juretzek, T.; Mauersberger, S.; Barth, G. Recombinant Haploid or Diploid Yarrowia Lipolytica Cells for the Functional Heterologous Expression of Cytochrome p450 Systems. Patent WO2000003008 A3, 27 April 2000.
35. Fickers, P.; Benetti, P.H.; Waché, Y.; Marty, A.; Mauersberger, S.; Smit, M.S.; Nicaud, J.M. Hydrophobic substrate utilisation by the yeast yarrowia lipolytica, and its potential applications. *FEMS Yeast Res.* **2005**, *5*, 527–543. [CrossRef] [PubMed]
36. Iwama, R.; Kobayashi, S.; Ishimaru, C.; Ohta, A.; Horiuchi, H.; Fukuda, R. Functional roles and substrate specificities of twelve cytochromes p450 belonging to cyp52 family in n-alkane assimilating yeast yarrowia lipolytica. *Fungal Genet. Biol.* **2016**, *91*, 43–54. [CrossRef] [PubMed]
37. Braun, A.; Geier, M.; Buehler, B.; Schmid, A.; Mauersberger, S.; Glieder, A. Steroid biotransformations in biphasic systems with yarrowia lipolytica expressing human liver cytochrome p450 genes. *Microb. Cell Factories* **2012**, *11*, 106. [CrossRef] [PubMed]
38. Mauersberger, S. Cytochromes p450 of the alkane-utilising yeast yarrowia lipolytica. In *Yarrowia Lipolytica*; Springer: Berlin, Gemany, 2013; pp. 227–262.
39. Takai, H.; Iwama, R.; Kobayashi, S.; Horiuchi, H.; Fukuda, R.; Ohta, A. Construction and characterization of a yarrowia lipolytica mutant lacking genes encoding cytochromes p450 subfamily 52. *Fungal Genet. Biol.* **2012**, *49*, 58–64. [CrossRef] [PubMed]
40. Thevenieau, F.; Beopoulos, A.; Desfougeres, T.; Sabirova, J.; Albertin, K.; Zinjarde, S.S.; Nicaud, J.-M. Uptake and assimilation of hydrophobic substrates by the oleaginous yeast yarrowia lipolytica. In *Handbook of Hydrocarbon and Lipid Microbiology*; Timmis, K.N., Ed.; Springer: Berlin/Heidelberg, Gemany, 2010; pp. 1513–1527.
41. Hanley, K.; Nguyen, L.V.; Khan, F.; Pogue, G.P.; Vojdani, F.; Panda, S.; Pinot, F.; Oriedo, V.B.; Rasochova, L.; Subramanian, M. Development of a plant viral-vector-based gene expression assay for the screening of yeast cytochrome p450 monooxygenases. *Assay Drug Dev. Technol.* **2003**, *1*, 147–160. [CrossRef] [PubMed]
42. Gatter, M.; Förster, A.; Bär, K.; Winter, M.; Otto, C.; Petzsch, P.; Ježková, M.; Bahr, K.; Pfeiffer, M.; Matthäus, F. A newly identified fatty alcohol oxidase gene is mainly responsible for the oxidation of long-chain ω-hydroxy fatty acids in yarrowia lipolytica. *FEMS Yeast Res.* **2014**, *14*, 858–872. [CrossRef] [PubMed]

43. Iwama, R.; Kobayashi, S.; Ohta, A.; Horiuchi, H.; Fukuda, R. Alcohol dehydrogenases and an alcohol oxidase involved in the assimilation of exogenous fatty alcohols in yarrowia lipolytica. *FEMS Yeast Res.* **2015**, *15*, fov014. [CrossRef] [PubMed]
44. Beopoulos, A.; Chardot, T.; Nicaud, J.-M. *Yarrowia lipolytica*: A model and a tool to understand the mechanisms implicated in lipid accumulation. *Biochimie* **2009**, *91*, 692–696. [CrossRef] [PubMed]
45. Smit, M.S.; Mokgoro, M.M.; Setati, E.; Nicaud, J.-M. A, ω-dicarboxylic acid accumulation by acyl-coa oxidase deficient mutants of yarrowia lipolytica. *Biotechnol. Lett.* **2005**, *27*, 859–864. [CrossRef] [PubMed]
46. Wache, Y. *Yarrowia lipolytica* biotechnological applications: Production of dicarboxylic acids and flagrances by yarrowia lipolytica. *Microbiol. Monogr.* **2013**, *25*, 151–170.
47. Nicaud, J.-M.; Thevenieau, F.; Le Dall, M.-T.; Marchal, R. Production of Dicarboxylic Acids by Improved Mutant Strains of Yarrowia Lipolytica. U.S. Patent 20100041115 A1, 18 February 2010.
48. Foo, J.L.; Susanto, A.V.; Keasling, J.D.; Leong, S.S.J.; Chang, M.W. Whole-cell biocatalytic and de novo production of alkanes from free fatty acids in saccharomyces cerevisiae. *Biotechnol. Bioeng.* **2017**, *114*, 232–237. [CrossRef] [PubMed]
49. Ledesma-Amaro, R.; Dulermo, R.; Niehus, X.; Nicaud, J.-M. Combining metabolic engineering and process optimization to improve production and secretion of fatty acids. *Metab. Eng.* **2016**, *38*, 38–46. [CrossRef] [PubMed]
50. Cintolesi, A.; Rodríguez-Moyá, M.; Gonzalez, R. Fatty acid oxidation: Systems analysis and applications. *Wiley Interdiscip. Rev. Syst. Biol. Med.* **2013**, *5*, 575–585. [CrossRef] [PubMed]
51. O'Reilly, E.; Köhler, V.; Flitsch, S.L.; Turner, N.J. Cytochromes p450 as useful biocatalysts: Addressing the limitations. *Chem. Commun.* **2011**, *47*, 2490–2501. [CrossRef] [PubMed]
52. Endoh-Yamagami, S.; Hirakawa, K.; Morioka, D.; Fukuda, R.; Ohta, A. Basic helix-loop-helix transcription factor heterocomplex of yas1p and yas2p regulates cytochrome p450 expression in response to alkanes in the yeast yarrowia lipolytica. *Eukaryot. Cell* **2007**, *6*, 734–743. [CrossRef] [PubMed]
53. Beopoulos, A.; Mrozova, Z.; Thevenieau, F.; Le Dall, M.T.; Hapala, I.; Papanikolaou, S.; Chardot, T.; Nicaud, J.M. Control of lipid accumulation in the yeast yarrowia lipolytica. *Appl. Environ. Microbiol.* **2008**, *74*, 7779–7789. [CrossRef] [PubMed]
54. Goldbach, V.; Roesle, P.; Mecking, S. Catalytic isomerizing ω-functionalization of fatty acids. *ACS Catal.* **2015**, *5*, 5951–5972. [CrossRef]
55. Isikgor, F.; Becer, C.R. Lignocellulosic biomass: A sustainable platform for production of bio-based chemicals and polymers. *Polym. Chem.* **2015**, *6*, 4497–4559. [CrossRef]
56. Groenewald, M.; Boekhout, T.; Neuveglise, C.; Gaillardin, C.; Dijck, P.W.; Wyss, M. *Yarrowia lipolytica*: Safety assessment of an oleaginous yeast with a great industrial potential. *Crit. Rev. Microbiol.* **2014**, *40*, 187–206. [CrossRef] [PubMed]
57. Sambrook, J.; Russell, D.W. *Molecular Cloning: A Laboratory Manual*, 3rd ed.; Coldspring-Harbour Laboratory Press: Plymouth, UK, 2001.
58. Tan, M.-J.; Chen, X.; Wang, Y.-K.; Liu, G.-L.; Chi, Z.-M. Enhanced citric acid production by a yeast yarrowia lipolytica over-expressing a pyruvate carboxylase gene. *Bioprocess Biosyst. Eng.* **2016**, *39*, 1–8. [CrossRef] [PubMed]
59. Xie, D.; Jackson, E.N.; Zhu, Q. Sustainable source of omega-3 eicosapentaenoic acid from metabolically engineered yarrowia lipolytica: From fundamental research to commercial production. *Appl. Microbiol. Biotechnol.* **2015**, *99*, 1599–1610. [CrossRef] [PubMed]
60. Chuang, L.-T.; Chen, D.-C.; Nicaud, J.-M.; Madzak, C.; Chen, Y.-H.; Huang, Y.-S. Co-expression of heterologous desaturase genes in yarrowia lipolytica. *New Biotechnol.* **2010**, *27*, 277–282. [CrossRef] [PubMed]
61. Madzak, C.; Nicaud, J.M.; Gaillardin, C. Yarrowia lipolytica. In *Production of Recombinant Proteins: Novel Microbial and Eukaryotic Expression Systems*; Gellissen, G., Ed.; Wiley-Blackwell: Hoboken, NJ, USA, 2005; pp. 163–189.
62. Madzak, C. *Yarrowia lipolytica*: Recent achievements in heterologous protein expression and pathway engineering. *Appl. Microbiol. Biotechnol.* **2015**, *99*, 4559–4577. [CrossRef] [PubMed]
63. Fickers, P.; Le Dall, M.T.; Gaillardin, C.; Thonart, P.; Nicaud, J.M. New disruption cassettes for rapid gene disruption and marker rescue in the yeast yarrowia lipolytica. *J. Microbiol. Methods* **2003**, *55*, 727–737. [CrossRef] [PubMed]

64. Lõoke, M.; Kristjuhan, K.; Kristjuhan, A. Extraction of genomic DNA from yeasts for pcr-based applications. *Biotechniques* **2011**, *50*, 325–328. [CrossRef] [PubMed]
65. O'fallon, J.; Busboom, J.; Nelson, M.; Gaskins, C. A direct method for fatty acid methyl ester synthesis: Application to wet meat tissues, oils, and feedstuffs. *J. Anim. Sci.* **2007**, *85*, 1511–1521. [CrossRef] [PubMed]
66. Abghari, A.; Chen, S. Engineering *yarrowia lipolytica* for enhanced production of lipid and citric acid. *Fermentation* **2017**, *3*, 34. [CrossRef]
67. Tai, M.; Stephanopoulos, G. Engineering the push and pull of lipid biosynthesis in oleaginous yeast yarrowia lipolytica for biofuel production. *Metab. Eng.* **2013**, *15*, 1–9. [CrossRef] [PubMed]
68. Morin, N.; Cescut, J.; Beopoulos, A.; Lelandais, G.; Le Berre, V.; Uribelarrea, J.-L.; Molina-Jouve, C.; Nicaud, J.-M. Transcriptomic analyses during the transition from biomass production to lipid accumulation in the oleaginous yeast yarrowia lipolytica. *PLoS ONE* **2011**, *6*, e27966. [CrossRef] [PubMed]
69. Kirtz, M.; Klebensberger, J.; Otte, K.B.; Richter, S.M.; Hauer, B. Production of ω-hydroxy octanoic acid with escherichia coli. *J. Biotechnol.* **2016**, *230*, 30–33. [CrossRef] [PubMed]
70. Cao, Y.; Cheng, T.; Zhao, G.; Niu, W.; Guo, J.; Xian, M.; Liu, H. Metabolic engineering of *Escherichia coli* for the production of hydroxy fatty acids from glucose. *BMC Biotechnol.* **2016**, *16*, 26. [CrossRef] [PubMed]
71. Clomburg, J.M.; Blankschien, M.D.; Vick, J.E.; Chou, A.; Kim, S.; Gonzalez, R. Integrated engineering of β-oxidation reversal and ω-oxidation pathways for the synthesis of medium chain ω-functionalized carboxylic acids. *Metab. Eng.* **2015**, *28*, 202–212. [CrossRef] [PubMed]
72. Wang, X.; Li, L.; Zheng, Y.; Zou, H.; Cao, Y.; Liu, H.; Liu, W.; Xian, M. Biosynthesis of long chain hydroxyfatty acids from glucose by engineered escherichia coli. *Bioresour. Technol.* **2012**, *114*, 561–566. [CrossRef] [PubMed]
73. Picataggio, S.; Rohrer, T.; Eirich, L.D. Method for Increasing the Omega-Hydroxylase Activity in Candida Tropicalis. U.S. Patent 5620878 A, 15 April 1997.
74. Sumita, T.; Iida, T.; Yamagami, S.; Horiuchi, H.; Takagi, M.; Ohta, A. Ylalk1 encoding the cytochrome p450alk1 in yarrowia lipolytica is transcriptionally induced by n-alkane through two distinct cis-elements on its promoter. *Biochem. Biophys. Res. Commun.* **2002**, *294*, 1071–1078. [CrossRef]
75. Titorenko, V.I.; Rachubinski, R.A. Dynamics of peroxisome assembly and function. *Trends Cell Biol.* **2001**, *11*, 22–29. [CrossRef]
76. Runguphan, W.; Keasling, J.D. Metabolic engineering of saccharomyces cerevisiae for production of fatty acid-derived biofuels and chemicals. *Metab. Eng.* **2014**, *21*, 103–113. [CrossRef] [PubMed]
77. Seip, J.; Jackson, R.; He, H.; Zhu, Q.; Hong, S.-P. Snf1 is a regulator of lipid accumulation in yarrowia lipolytica. *Appl. Environ. Microbiol.* **2013**, *79*, 7360–7370. [CrossRef] [PubMed]
78. Hofbauer, H.F.; Schopf, F.H.; Schleifer, H.; Knittelfelder, O.L.; Pieber, B.; Rechberger, G.N.; Wolinski, H.; Gaspar, M.L.; Kappe, C.O.; Stadlmann, J. Regulation of gene expression through a transcriptional repressor that senses acyl-chain length in membrane phospholipids. *Dev. Cell* **2014**, *29*, 729–739. [CrossRef] [PubMed]
79. Thevenieau, F.; Le Dall, M.T.; Nthangeni, B.; Mauersberger, S.; Marchal, R.; Nicaud, J.M. Characterization of yarrowia lipolytica mutants affected in hydrophobic substrate utilization. *Fungal Genet. Biol.* **2007**, *44*, 531–542. [CrossRef] [PubMed]
80. Scharnewski, M.; Pongdontri, P.; Mora, G.; Hoppert, M.; Fulda, M. Mutants of saccharomyces cerevisiae deficient in acyl-coa synthetases secrete fatty acids due to interrupted fatty acid recycling. *FEBS J.* **2008**, *275*, 2765–2778. [CrossRef] [PubMed]
81. Dulermo, R.; Gamboa-Meléndez, H.; Ledesma-Amaro, R.; Thévenieau, F.; Nicaud, J.-M. Unraveling fatty acid transport and activation mechanisms in yarrowia lipolytica. *Biochim. Biophys. Acta (BBA)-Mol. Cell Biol. Lipids* **2015**, *1851*, 1202–1217. [CrossRef] [PubMed]
82. Dulermo, R.; Gamboa-Meléndez, H.; Ledesma, R.; Thevenieau, F.; Nicaud, J.-M. *Yarrowia lipolytica* aal genes are involved in peroxisomal fatty acid activation. *Biochim. Biophys. Acta (BBA)-Mol. Cell Biol. Lipids* **2016**, *1861*, 555–565. [CrossRef] [PubMed]
83. Fukuda, R.; Ohta, A. Utilization of hydrophobic substrate by yarrowia lipolytica. In *Yarrowia Lipolytica Genetics, Genomics, and Physiology*; Barth, G., Ed.; Springer: Berlin/Heidelberg, Gemany, 2013; pp. 111–119.
84. Park, J.S.; Iwama, R.; Kobayashi, S.; Ohta, A.; Horiuchi, H.; Fukuda, R. Involvement of acyl-coa synthetase genes in n-alkane assimilation and fatty acid utilization in yeast yarrowia lipolytica. *FEMS Yeast Res.* **2015**, *15*, fov031. [CrossRef]
85. Krivoruchko, A.; Zhang, Y.; Siewers, V.; Chen, Y.; Nielsen, J. Microbial acetyl-coa metabolism and metabolic engineering. *Metab. Eng.* **2015**, *28*, 28–42. [CrossRef] [PubMed]

86. Chen, L.; Zhang, J.; Lee, J.; Chen, W.N. Enhancement of free fatty acid production in saccharomyces cerevisiae by control of fatty acyl-coa metabolism. *Appl. Microbiol. Biotechnol.* **2014**, *98*, 6739–6750. [CrossRef] [PubMed]
87. Petschnigg, J.; Wolinski, H.; Kolb, D.; Zellnig, G.; Kurat, C.F.; Natter, K.; Kohlwein, S.D. Good fat, essential cellular requirements for triacylglycerol synthesis to maintain membrane homeostasis in yeast. *J. Biol. Chem.* **2009**, *284*, 30981–30993. [CrossRef] [PubMed]
88. Stefan, A.; Ugolini, L.; Lazzeri, L.; Conte, E.; Hochkoeppler, A. The expression of the cuphea palustris thioesterase cpfatb2 in yarrowia lipolytica triggers oleic acid accumulation. *Biotechnol. Prog.* **2015**, *32*, 26–35. [CrossRef] [PubMed]
89. Xu, P.; Qiao, K.; Stephanopoulos, G. Engineering oxidative stress defense pathways to build a robust lipid production platform in yarrowia lipolytica. *Biotechnol. Bioeng.* **2017**, *114*, 1521–1530. [CrossRef] [PubMed]
90. Kohlwein, S.D. Triacylglycerol homeostasis: Insights from yeast. *J. Biol. Chem.* **2010**, *285*, 15663–15667. [CrossRef] [PubMed]
91. Hardwick, J.P. Cytochrome p450 omega hydroxylase (cyp4) function in fatty acid metabolism and metabolic diseases. *Biochem. Pharmacol.* **2008**, *75*, 2263–2275. [CrossRef] [PubMed]
92. Björkhem, I. On the mechanism of regulation of omega oxidation of fatty acids. *J. Biol. Chem.* **1976**, *251*, 5259–5266. [PubMed]
93. Wanders, R.J.; Komen, J.; Kemp, S. Fatty acid omega-oxidation as a rescue pathway for fatty acid oxidation disorders in humans. *FEBS J.* **2011**, *278*, 182–194. [CrossRef] [PubMed]
94. Geier, M.; Braun, A.; Emmerstorfer, A.; Pichler, H.; Glieder, A. Production of human cytochrome p450 2d6 drug metabolites with recombinant microbes–a comparative study. *Biotechnol. J.* **2012**, *7*, 1346–1358. [CrossRef] [PubMed]
95. Eirich, L.D.; Craft, D.L.; Steinberg, L.; Asif, A.; Eschenfeldt, W.H.; Stols, L.; Donnelly, M.I.; Wilson, C.R. Cloning and characterization of three fatty alcohol oxidase genes from candida tropicalis strain atcc 20336. *Appl. Environ. Microbiol.* **2004**, *70*, 4872–4879. [CrossRef] [PubMed]
96. Kang, W.-R.; Seo, M.-J.; An, J.-U.; Shin, K.-C.; Oh, D.-K. Production of δ-decalactone from linoleic acid via 13-hydroxy-9 (z)-octadecenoic acid intermediate by one-pot reaction using linoleate 13-hydratase and whole yarrowia lipolytica cells. *Biotechnol. Lett.* **2016**, *38*, 817–823. [CrossRef] [PubMed]
97. Iwama, R.; Kobayashi, S.; Ohta, A.; Horiuchi, H.; Fukuda, R. Fatty aldehyde dehydrogenase multigene family involved in the assimilation of n-alkanes in yarrowia lipolytica. *J. Biol. Chem.* **2014**, *289*, 33275–33286. [CrossRef] [PubMed]
98. Gatter, M.; Matthaus, F.; Barth, G. Yeast Strains and Method for the Production of Omega-Hydroxy Fatty Acids and Dicarboxylic Acids. U.S. Patent 20160304913 A1, 20 October 2016.
99. Vanhanen, S.; West, M.; Kroon, J.T.; Lindner, N.; Casey, J.; Cheng, Q.; Elborough, K.M.; Slabas, A.R. A consensus sequence for long-chain fatty-acid alcohol oxidases from candida identifies a family of genes involved in lipid ω-oxidation in yeast with homologues in plants and bacteria. *J. Biol. Chem.* **2000**, *275*, 4445–4452. [CrossRef] [PubMed]
100. Liu, S.; Li, C.; Fang, X.; Cao, Z.A. Optimal ph control strategy for high-level production of long-chain α, ω-dicarboxylic acid by candida tropicalis. *Enzym. Microb. Technol.* **2004**, *34*, 73–77. [CrossRef]
101. Lin, R.; Cao, Z.; Zhu, T.; Zhang, Z. Secretion in long-chain dicarboxylic acid fermentation. *Bioprocess Eng.* **2000**, *22*, 391–396. [CrossRef]
102. Bankar, A.V.; Kumar, A.R.; Zinjarde, S.S. Environmental and industrial applications of yarrowia lipolytica. *Appl. Microbiol. Biotechnol.* **2009**, *84*, 847–865. [CrossRef] [PubMed]
103. Casal, M.; Paiva, S.; Queirós, O.; Soares-Silva, I. Transport of carboxylic acids in yeasts. *FEMS Microbiol. Rev.* **2008**, *32*, 974–994. [CrossRef] [PubMed]
104. Liu, S.; Li, C.; Xie, L.; Cao, Z.A. Intracellular ph and metabolic activity of long-chain dicar□ylic acid-producing yeast*Candida tropicalis*. *J. Biosci. Bioeng.* **2003**, *96*, 349–353. [CrossRef]
105. Kamisaka, Y.; Tomita, N.; Kimura, K.; Kainou, K.; Uemura, H. DGA1 (diacylglycerol acyltransferase gene) overexpression and leucine biosynthesis significantly increase lipid accumulation in the Δ*snf*2 disruptant of saccharomyces cerevisiae. *Biochem. J.* **2007**, *408*, 61–68. [CrossRef] [PubMed]
106. Lundemo, M.T.; Woodley, J.M. Guidelines for development and implementation of biocatalytic p450 processes. *Appl. Microbiol. Biotechnol.* **2015**, *99*, 2465–2483. [CrossRef] [PubMed]
107. Wang, Z.; Zhao, F.; Chen, D.; Li, D. Biotransformation of phytosterol to produce androsta-diene-dione by resting cells of mycobacterium in cloud point system. *Process Biochem.* **2006**, *41*, 557–561. [CrossRef]

108. Vandenberghe, L.P.S.; Soccol, C.R.; Pandey, A.; Lebeault, J.-M. Microbial production of citric acid. *Braz. Arch. Biol. Technol.* **1999**, *42*, 263–276. [CrossRef]
109. Papanikolaou, S.; Aggelis, G. Selective uptake of fatty acids by the yeast yarrowia lipolytica. *Eur. J. Lipid Sci. Technol.* **2003**, *105*, 651–655. [CrossRef]
110. Hirakawa, K.; Kobayashi, S.; Inoue, T.; Endoh-Yamagami, S.; Fukuda, R.; Ohta, A. Yas3p, an opi1 family transcription factor, regulates cytochrome p450 expression in response to n-alkanes in yarrowia lipolytica. *J. Biol. Chem.* **2009**, *284*, 7126–7137. [CrossRef] [PubMed]
111. Beopoulos, A.; Haddouche, R.; Kabran, P.; Dulermo, T.; Chardot, T.; Nicaud, J.-M. Identification and characterization of dga2, an acyltransferase of the dgat1 acyl-coa: Diacylglycerol acyltransferase family in the oleaginous yeast yarrowia lipolytica. New insights into the storage lipid metabolism of oleaginous yeasts. *Appl. Microbiol. Biotechnol.* **2012**, *93*, 1523–1537. [CrossRef] [PubMed]
112. Wasylenko, T.M.; Ahn, W.S.; Stephanopoulos, G. The oxidative pentose phosphate pathway is the primary source of nadph for lipid overproduction from glucose in yarrowia lipolytica. *Metab. Eng.* **2015**, *30*, 27–39. [CrossRef] [PubMed]
113. Silverman, A.M.; Qiao, K.; Xu, P.; Stephanopoulos, G. Functional overexpression and characterization of lipogenesis-related genes in the oleaginous yeast yarrowia lipolytica. *Appl. Microbiol. Biotechnol.* **2016**, *100*, 3781–3798. [CrossRef] [PubMed]
114. Zhang, H.; Wu, C.; Wu, Q.; Dai, J.; Song, Y. Metabolic flux analysis of lipid biosynthesis in the yeast yarrowia lipolytica using 13 c-labled glucose and gas chromatography-mass spectrometry. *PLoS ONE* **2016**, *11*, e0159187. [CrossRef]
115. Fang, F.; Dai, B.; Zhao, G.; Zhao, H.; Sun, C.; Liu, H.; Xian, M. In depth understanding the molecular response to the enhanced secretion of fatty acids in saccharomyces cerevisiae due to one-step gene deletion of acyl-coa synthetases. *Process Biochem.* **2016**, *51*, 1162–1174. [CrossRef]
116. Jin, Z.; Wong, A.; Foo, J.L.; Ng, J.; Cao, Y.X.; Chang, M.W.; Yuan, Y.J. Engineering saccharomyces cerevisiae to produce odd chain-length fatty alcohols. *Biotechnol. Bioeng.* **2016**, *113*, 842–851. [CrossRef] [PubMed]
117. Vohra, A.; Syal, P.; Madan, A. Probiotic yeasts in livestock sector. *Anim. Feed Sci. Technol.* **2016**, *219*, 31–47. [CrossRef]
118. Martinez, C.A.; Rupashinghe, S.G. Cytochrome p450 bioreactors in the pharmaceutical industry: Challenges and opportunities. *Curr. Top. Med. Chem.* **2013**, *13*, 1470–1490. [CrossRef] [PubMed]
119. Holtmann, D.; Hollmann, F. The oxygen dilemma: A severe challenge for the application of monooxygenases? *ChemBioChem* **2016**, *17*, 1391–1398. [CrossRef] [PubMed]
120. Grand View Research, Inc. *Long Chain Dicarboxylic Acid Market Analysis*; Grand View Research: Maharashtra, India, 2016.
121. Stempfle, F.; Ortmann, P.; Mecking, S. Long-chain aliphatic polymers to bridge the gap between semicrystalline polyolefins and traditional polycondensates. *Chem. Rev.* **2016**, *116*, 4597–4641. [CrossRef] [PubMed]
122. De Montellano, P.R.O. *Cytochrome p450: Structure, Mechanism, and Biochemistry*; Springer Science & Business Media: Berlin, Gemany, 2005.
123. Cambon, E.; Piamtongkam, R.; Bordes, F.; Duquesne, S.; Laguerre, S.; Nicaud, J.-M.; Marty, A. A new yarrowia lipolytica expression system: An efficient tool for rapid and reliable kinetic analysis of improved enzymes. *Enzym. Microb. Technol.* **2010**, *47*, 91–96. [CrossRef]
124. Xu, P.; Gu, Q.; Wang, W.; Wong, L.; Bower, A.G.W.; Collins, C.H.; Koffas, M.A.G. Modular optimization of multi-gene pathways for fatty acids production in *E. coli*. *Nat. Commun.* **2013**, *4*, 1409. [CrossRef] [PubMed]
125. Teixeira, P.G.; Ferreira, R.; Zhou, Y.J.; Siewers, V.; Nielsen, J. Dynamic regulation of fatty acid pools for improved production of fatty alcohols in saccharomyces cerevisiae. *Microb. Cell Factories* **2017**, *16*, 45. [CrossRef] [PubMed]
126. Dulermo, T.; Tréton, B.; Beopoulos, A.; Gnankon, A.P.K.; Haddouche, R.; Nicaud, J.-M. Characterization of the two intracellular lipases of y. Lipolytica encoded by tgl3 and tgl4 genes: New insights into the role of intracellular lipases and lipid body organisation. *Biochim. Biophys. Acta (BBA)-Mol. Cell Biol. Lipids* **2013**, *1831*, 1486–1495. [CrossRef] [PubMed]
127. Sagehashi, Y.; Horiuchi, H.; Fukuda, R.; Ohta, A. Identification and characterization of a gene encoding an abc transporter expressed in the dicarboxylic acid-producing yeast candida maltosa. *Biosci. Biotechnol. Biochem.* **2013**, *77*, 2502–2504. [CrossRef] [PubMed]

128. Chen, B.; Ling, H.; Chang, M.W. Transporter engineering for improved tolerance against alkane biofuels in saccharomyces cerevisiae. *Biotechnol. Biofuels* **2013**, *6*, 21. [CrossRef] [PubMed]
129. Kavšček, M.; Bhutada, G.; Madl, T.; Natter, K. Optimization of lipid production with a genome-scale model of yarrowia lipolytica. *BMC Syst. Biol.* **2015**, *9*, 72. [CrossRef] [PubMed]
130. Cescut, J. *Accumulation D'acylglycérols par des Espèces Levuriennes à Usage Carburant Aéronautique: Physiologie et Performances de Procédés*; INSA Toulouse: Toulouse, France, 2009. (In French) Available online: http://www.theses.fr/2009ISAT0022 (accessed on 22 July 2017).

13

Characterization of Old Wine Yeasts Kept for Decades under a Zero-Emission Maintenance Regime

Katrin Matti [1], Beatrice Bernardi [1,2], Silvia Brezina [1], Heike Semmler [1], Christian von Wallbrunn [1], Doris Rauhut [1] and Jürgen Wendland [1,*]

[1] Department of Microbiology and Biochemistry, Hochschule Geisenheim University, Von-Lade-Strasse 1, D-65366 Geisenheim, Germany; Katrin.Matti@hs-gm.de (K.M.); Beatrice.Bernardi@hs-gm.de (B.B.); Silvia.Brezina@hs-gm.de (S.B.); Heike.Semmler@hs-gm.de (H.S.); Christian.Wallbrunn@hs-gm.de (C.v.W.); Doris.Rauhut@hs-gm.de (D.R.)

[2] Research Group of Microbiology (MICR)—Functional Yeast Genomics, Vrije Universiteit Brussel, Pleinlaan 2, BE-1050 Brussels, Belgium

* Correspondence: juergen.wendland@hs-gm.de

Abstract: All laboratories dealing with microbes have to develop a strain maintenance regime. While lyophilization based on freeze-drying may be feasible for large stock centers, laboratories around the world rely on cryopreservation and freezing of stocks at −80 °C. Keeping stocks at these low temperatures requires investments of several thousand kW/h per year. We have kept yeast stocks for several decades at room temperature on agar slants in glass reagent tubes covered with vaspar and sealed with cotton plugs. They were part of the Geisenheim Yeast Breeding Center stock collection that was started in the 19th century, well before −80 °C refrigeration technology was invented. Of these stocks, 60 tubes were analyzed and around one-third of them could be regrown. The strains were typed by sequencing of rDNA PCR fragments. Based on BlastN analyses, twelve of the strains could be assigned to *Saccharomyces cerevisiae*, two to *S. kudriavzevii*, and the others to *Meyerozyma* and *Candida*. The strains were used in white wine fermentations and compared to standard wine yeasts Uvaferm/GHM (Geisenheim) and Lalvin EC1118. Even with added nitrogen, the strains exhibited diverse fermentation curves. Post-fermentation aroma analyses and the determination of residual sugar and organic acid concentrations indicated that some strains harbor interesting flavor characteristics, surpassing current standard yeast strains. Thus, old strain collections bear treasures for direct use either in wine fermentations or for incorporation in yeast breeding programs aimed at improving modern wine yeasts. Furthermore, this provides evidence that low-cost/long-term culture maintenance at zero-emission levels is feasible.

Keywords: strain collection; aroma profiling; gas chromatography; wine yeast; *Saccharomyces*; fermentation; volatile aroma compounds

---

1. Introduction

At the end of the 19th century, Emil Christian Hansen at the Carlsberg Laboratory in Copenhagen, Denmark, established the first pure culture lager yeast strain, Unterhefe No. 1 [1]. This strain then became known as *Saccharomyces carlsbergensis*. The finding that one yeast strain was sufficient to generate a fermented beverage of high quality started a new era and lead to new developments in the beer and dairy industry. It was soon recognized by Julius Wortmann at the Geisenheim Research Center in Germany that Hansen's findings were also applicable to wine making [2]. This started efforts in collecting wine yeast strains from different vineyards and wineries in the Rheingau area. These strains were characterized for their fermentation capacity and flavor attributes. At the "Geisenheimer Hefe-Reinzuchtstation" (Geisenheim Yeast Breeding Center), founded in 1894, these strains were produced as liquid starter cultures and dispatched to the wineries upon request.

With the isolation of pure yeast cultures came the responsibility to maintain stocks of these cultures. Wine making requires yeast starter cultures only once a year just after the grape harvest. By contrast, yeasts for beer production are in constant use throughout the year. Even before the isolation of pure cultures, there was an interest in generating dry yeast cakes for longer term storage. The history of both the patents and literature in this field has been covered in depth by a recent excellent review [3]. A solid supply of dehydrated yeast became a necessity for long distance shipments, which came around 1940 when the Fleischman Co. produced active dry yeast. This yeast required 'reactivation', i.e., rehydration prior to use. In the 1970s, Lesaffre introduced an instant dry yeast which could be used directly without reactivation. In microbiological laboratories, however, bacterial and yeast cultures are nowadays generally preserved by storing at −80 °C which, since the 1970s, has become technically feasible on a larger scale [4].

Thus, for decades after the 1880s, yeast cultures had to be kept by other means. Two techniques used were water stocks and yeast slants covered with vaspar (a mix of paraffin and Vaseline) [5,6]. The method of storing yeasts in distilled water at room temperature, as proposed by Castellani, is not only a cheap way of preserving cultures but is also a very effective way of culturing a collection over many years without the need for constant propagation [7]. This method is particularly useful for yeasts [8]. Storage of fungal cells in distilled water can be extended for 20 years [9]. Therefore, it was stated that storing yeast cultures in distilled water may reach similar efficiencies as freezing at −80 °C [10]. The use of a paraffin or vaspar overlay is also a very cheap way of yeast culturing, although the viability may be reduced when compared to the other methods.

There are only a few long-term studies describing yeast viability, one of which used the traditional method of yeasts grown on slants and covered by paraffin oil and found cells to be viable after a seven-year incubation period [11]. At the Geisenheim Yeast Breeding Center, we have a large collection of wine yeasts and non-conventional yeasts dating back to the 1890s. Samples were routinely stored with a vaspar overlay. Of course, over time, the strain collection was transferred to either storage in liquid nitrogen or in freezers at −80 °C. Nevertheless, we still stored a few samples for over 30 years at room temperature in the old way. In this study, we examined 60 tubes containing these decade-old samples. The yeast were restreaked, and those strains that could be regrown were subjected to fermentation studies and volatile aroma analyses. Our results show that strain collections can safely be stored at room temperature. Such a strain maintenance regime could contribute to energy conservation and the reduction of CO_2 emissions.

2. Materials and Methods

2.1. Strains and Media

The yeast strains used in this study are shown in Table 1, including standard wine yeast strains used for comparison. Yeast strains were subcultured in YPD (1% yeast extract, 2% peptone, 2% glucose).

Table 1. Strains used in this study.

Original Label	Sequence-Based Assignment
Zell 1895	***Saccharomyces cerevisiae***
Valdepenas Criptana 1909	***Saccharomyces cerevisiae***
Brettanomyces claussenii IHG Berlin 1959	*Meyerozyma guilliermondii*
Ungstein 1892	***Saccharomyces cerevisiae***
Riesling Krim 1896	***Saccharomyces cerevisiae***
Olewig II 1896	*Meyerozyma guilliermondii*
Dürkheim 1892	*Meyerozyma guilliermondii*
Rüdesheimer Hinterhaus 1893	***Saccharomyces cerevisiae***
Alpiarca II 1896	***Saccharomyces cerevisiae***
Candida tropicalis	*Candida sanyaensis*
Heimersheimer Ruth 1895	***Saccharomyces cerevisiae***
Steinberg 1893	***Saccharomyces cerevisiae***
Rüdesheimer Berg	***Saccharomyces cerevisiae***
Würzburg (Stein)	*Saccharomyces kudriavzevii*
Winningen 1892	*Saccharomyces kudriavzevii*
Scy 1892	***Saccharomyces cerevisiae***
Bordeaux 1892	***Saccharomyces cerevisiae***
Geisenheimer (Mäuerchen) 1893	***Saccharomyces cerevisiae***

Bold: to indicate true *S. cerevisiae* strains. These strains will be important for winemaking. The others are non-conventional yeasts.

2.2. Molecular Analysis of Yeast Strains

Typing of the strains was determined by performing ITS-PCR (ITS, internal transcribed spacer) using standard ITS1F-fungal specific-(5′-CTTGGTCATTTAGAGGAAGTAA-3′) and ITS4-universal-(5′-TCCTCCGCTTATTGATATGC-3′) primers and sequencing of the PCR products as previously described [12]. Sequencing was conducted by Starseq, Mainz, Germany.

2.3. Fermentation Conditions

Lab-scale fermentations were carried out in duplicate with a standard pasteurized white wine must with a sugar concentration of 72 °Oechsle. The must was supplemented by the addition of Fermaid E (inactivated yeast product; according to the supplier's instructions; Lallemand, Vienna, Austria). Cells were inoculated at a density of OD_{600} = 0.5. The fermentation temperature was set to 18 °C and cultures were incubated with constant stirring at 300 rpm.

2.4. Analytical Methods

At the end of the fermentations, several compounds including fructose, glucose, ethanol, and organic acids were analyzed by high-performance liquid chromatography (HPLC) with an Agilent 1100 Series (Agilent Technologies, Waldbronn, Germany). Quantitative analyses were done as described in [13]. The HPLC equipment was equipped with a variable wavelength detector (UV/VIS) and a refractive index detector (RID). The column for separation is an Allure Organic Acids (Restek GmbH, Bad Homburg, Germany) with a 5 µm particle size, 60 A pore size and dimensions of 250 mm x i.d. 4.6 mm. A water-based solution of sulfuric acid (0.0139% *v*/*v*) and ethanol (0.5% *v*/*v*) was used as eluent. Gas chromatography was conducted using a GC 7890A (Agilent, Santa Clara, CA, USA), coupled with a MSD 5977B mass spectrometer (Agilent, Santa Clara, CA, USA). The determination of aroma compounds followed the analytical approach described in Belda et al. [14] according to the method of Camara et al. [15].

3. Results

3.1. Regrowing of Dormant Strains Kept under Vaspar for Decades

At the Geisenheim Yeast Breeding Center (GYBC), we stored two racks containing 60 reagent tubes with yeast strains. The tubes contained slants on which yeast strains were spread and grown and then overlaid with vaspar (Figure 1). This was the standard procedure for maintaining strains at the GYBC. The collections were generated before the introduction of −80 °C refrigeration. These samples represent an even older stock as younger samples (20+ years of age) were kept in reagent tubes with screw cap closures. The apparent age of the old samples is, therefore, estimated to be over 30 years but less than 60 years, as one isolate carried a label with the year 1959 indicative of the year of isolation (Table 1).

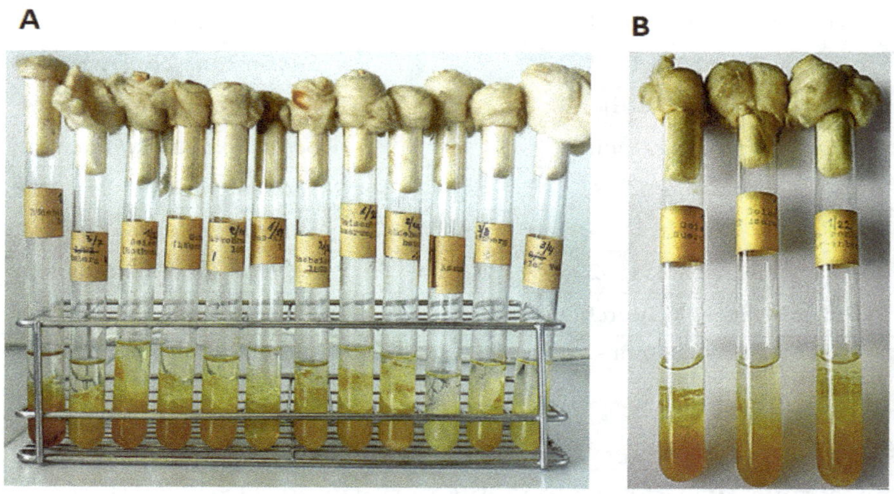

Figure 1. (**A**,**B**). Old samples from the Geisenheimer Yeast Breeding Center. Samples were generated as agar slants with yeasts grown and covered in vaspar. Tubes were plugged by a cotton ball. Samples were labeled according to the location of the isolate and the year of isolation (see Table 1).

We wanted to find out if these strains were still alive and, once propagated, what their fermentation behavior would be like. To this end, we either took samples with an inoculation loop and restreaked them on full medium YPD or inoculated them in liquid YPD. Astonishingly, about one-third (18 out of 60) of the strains could be regrown and cultivated under these conditions.

The strain labels often indicated the area where these strains had been isolated and did not necessarily identified the species. Thus, we went on to type the strains by PCR amplification of a region of the ribosomal DNA using a standard primer pair designed for fungal species (ITS1 and ITS4). These primers are located at the end of the 18S and start of the 28S rDNA and thus amplify the internal transcribed spacer (ITS) region including the 5.8S rDNA. This region is highly variable and allows strain determination to the species level. PCR products were sequenced, and the sequences compared to the NCBI non-redundant database using BlastN. The sequence comparisons indicated that most strains could indeed be assigned to *Saccharomyces cerevisiae*. Two of the strains were found to be *S. kudriavzevii* while three strains matched *Meyerozyma guilliermondii* and one strain could be assigned to a newly described species of *Candida sanyaensis* (Table 1).

3.2. Fermentation Performance

We went on to study the fermentation characteristics of the *Saccharomyces* strains. To this end, strains were used to ferment a standard white wine must of 72 °Oechsle to which additional amino nitrogen was added via an inactivated yeast product (Fermaid E). Fermentations were carried out at 18 °C with stirring over a period of 12 days and fermentation rates were followed by daily measurements of CO_2 release. Strains were compared to the standard wine strain Lalvin EC1118 (Figure 2). It turned out that half of the strains generated a weight loss slightly larger than Lalvin EC1118, while the other half performed less well than Lalvin EC1118 in this respect. The largest weight loss was found with *S. cerevisiae* strain Steinberg 1893 and the *S. kudriavzevii* strain Würzburg (Stein). Lalvin EC1118 required a short lag phase of one day to enter alcoholic fermentation. Several of the tested yeast strains exhibited an extended lag phase of 3–4 days, particularly the strains that later showed the greatest weight loss and also Geisenheimer Mäuerchen, Winningen 1892, and Bordeaux 1892. Thus, even given the extended lag phase, these strains managed a complete fermentation within the 12-day fermentation window.

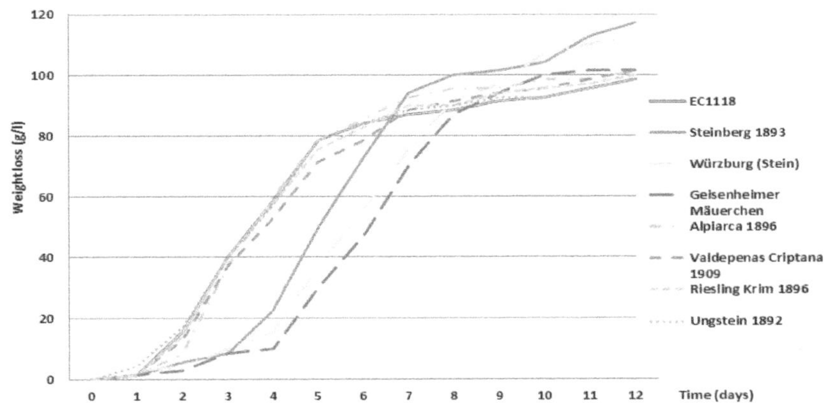

(A)

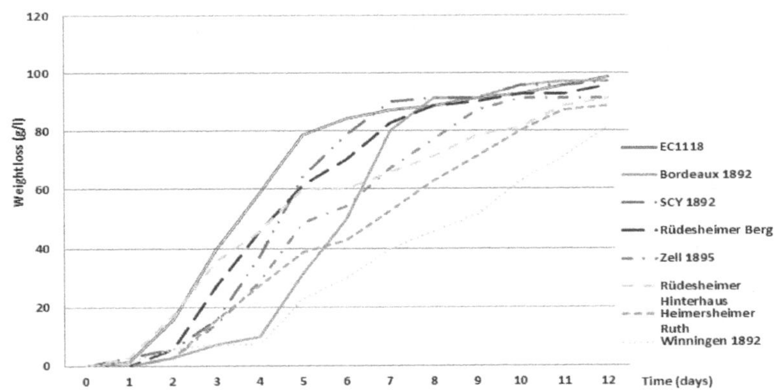

(B)

Figure 2. Fermentation curves based on CO_2 release/weight loss of *Saccharomyces* cultures derived from isolates of the old collection. Release of CO_2 was measured daily. (**A**) Strains are shown that released more CO_2 than the EC1118 control wine yeast strain. (**B**) Strains are shown that released less CO_2 than the EC1118 control wine yeast strain.

To analyze the fermented liquids in more detail, the residual sugars, organic acids, and final ethanol content were determined (Table 2). Most of the *Saccharomyces* strains reached complete fermentation with around 7% alcohol content. Two glucophilic strains, however, failed to utilize all of the fructose in the 12-day fermentation time. The *S. kudriavzevii* strains produced less alcohol than the *S. cerevisiae* strains. The two *S. cerevisiae* strains Rüdesheimer Hinterhaus 1893 and Heimersheimer Ruth, while using up glucose, did not utilize fructose completely during the 12-day fermentation. All strains showed a similar organic acid profile, with malate being the pronounced acid. Rüdesheimer Hinterhaus 1893, on the other hand, showed a surprising amount of shikimic acid (Table 2).

Table 2. HPLC analyses of residual sugars, organic acids, and total alcohol of *Saccharomyces* wine yeasts.

Strains	*Glucose*	*Fructose*	*Total Sugar*	*Malate*	*Shikimic Acid*	*Lactate*	*Acetate*	*Citric acid*	*Ethanol*	*Ethanol*
	[g/L]	[g/L]	[g/L]	[g/L]	[mg/L]	[g/L]	[g/L]	[g/L]	[g/L]	[%]
EC1118 wine yeast	<1	<1		4.6	38	0.2	0.1	0.2	60.7	7.7
GHM wine yeast	<1	<1		4.4	40	0.3	0.2	0.2	56.3	7.1
Saccharomyces cerevisiae										
Zell 1895	<1	<1		3.7	41	0.4	0.1	0.2	60.2	7.6
Valdepenas Criptana 1909	<1	<1		3.5	39	0.2	0.1	0.2	54.9	7.0
Ungstein 1892	<1	<1		2.9	28	0.2	0.2	0.2	56.6	7.2
Riesling Krim 1896	<1	<1		2.2	20	0.1	0.2	0.2	57.1	7.3
Rüdesheimer Hinterhaus 1893	<1	4.1	4.1	3.9	297	0.9	0.2	0.4	53.8	6.9
Alpiarca II 1896	<1	<1		4.1	38	0.2	0.1	0.2	56.8	7.2
Heimersheimer Ruth 1895	<1	6.9	6.9	3.4	34	0.3	0.2	0.2	56.4	7.2
Steinberg 1893	<1	<1		3.8	32	0.1	0.1	0.2	56.4	7.1
Rüdesheimer Berg	<1	<1		4.0	39	0.5	0.2	0.2	55.9	7.1
Scy 1892	<1	<1		3.6	31	0.1	0.2	0.2	58.7	7.4
Bordeaux 1892	<1	<1		2.9	29	0.2	0.1	0.2	56.9	7.2
Geisenheimer (Mäuerchen)	<1	<1		3.8	27	0.1	0.4	0.2	57.1	7.2
S. kudriavzevii										
Würzburg (Stein)	<1	<1		3.7	26	<0.1	0.2	0.2	54.2	6.9
Winningen 1892	2.1	19.2	20.8	4.1	37	0.5	0.2	0.2	49.7	6.3

3.3. *Production of Aroma Compounds*

We routinely examined 28 aroma compounds, specifically alcohols and esters (Table S1). A comparison within strains using a selection of eight major compounds is shown in Figure 3 and Table 3. While all species produced a range of compounds, it was interesting to see that a major current wine production strain, EC1118, was actually not the highest producer of certain aroma compounds in our assay. The three strains that produced most fruity esters were Rüdesheimer Hinterhaus 1893, Alpiarca 1896, and Valdepenas Criptana 1909. Additionally, Rüdesheimer Hinterhaus 1893 championed the production of isoamyl acetate (acetic acid 3-methyl butyl ester) and 2-phenylethyl acetate (acetic acid 2-phenylethylester), which is the acetate ester of 2-phenylethanol. This is apparently a consequence of Ehrlich pathway output as regarding the production of alcohols, particularly i-butanol, isoamylalcohol (3-methyl-butanol), and 2-phenlyethanol, the strain that came on top for each of the compounds was also the Rüdesheimer Hinterhaus 1893 yeast (see Table S1).

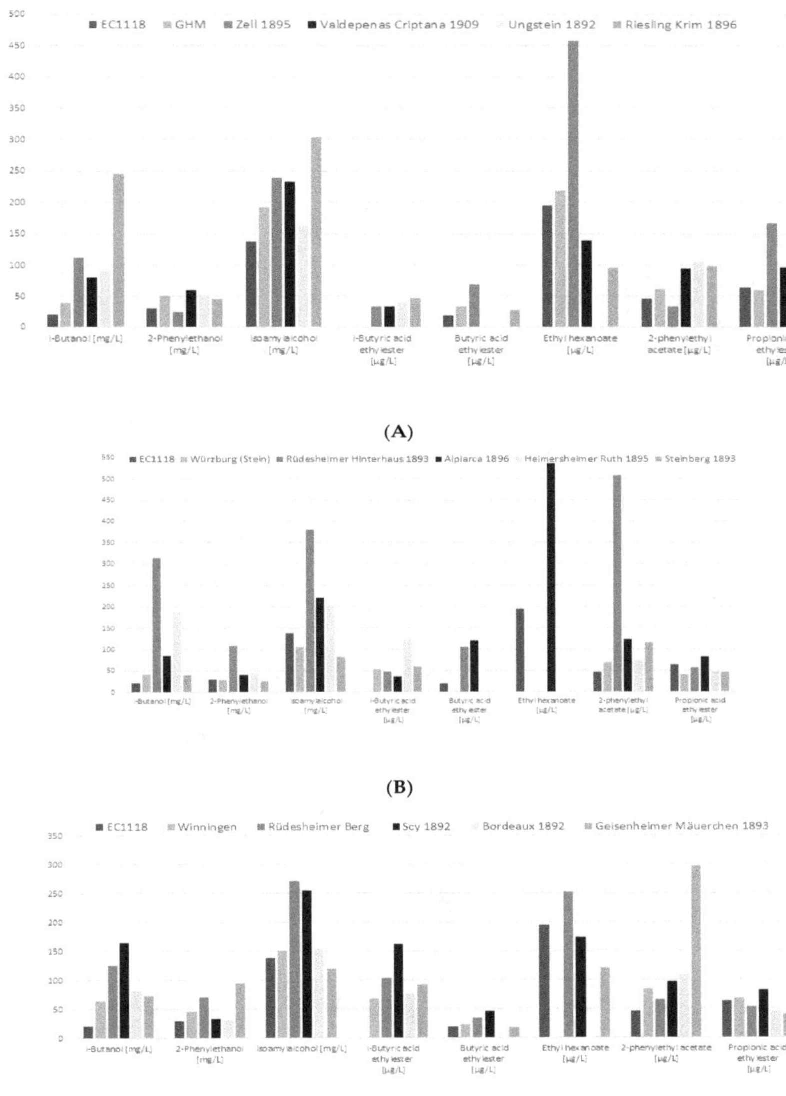

Figure 3. (**A**–**C**). Bar charts with selected alcohol (in mg/L) and ester (in µg/L) aroma compounds of *Saccharomyces* strains compared to the EC1118 wine yeast. Flavor compounds were measured using gas chromatography at the end of fermentation. The full list of aroma compounds is shown in Table S1.

Table 3. Aroma compound generation determined at the end of fermentation.*

	i-Butanol [mg/L]	2-Phenyl-Ethanol [mg/L]	Isoamyl Alcohol [mg/L]	i-Butyric Acid Ethyl-Ester [µg/L]	Butyric Acid Ethylester [µg/L]	Hexanoic Acid Ethyl-Ester [µg/L]	2-Phenylethyl Acetate [µg/L]	Propionic Acid Ethyl-Ester [µg/L]
EC1118	21	30	138	nq	20	195	46	64
GHM	40	51	192	nq	34	218	62	60
Saccharomyces cerevisiae								
Zell 1895	112	24	238	34	69	**457**+	34	**167**
Valdepenas Criptana 1909	81	60	233	33	nq	138	94	**95**
Ungstein 1892	90	52	163	41	nq	nq	106	23
Riesling Krim 1896	**246**	46	303	48	28	96	98	25
Rüdesheimer Hinterhaus 1893	**314**	**108**	**380**	47	**106**	nq	**507**	57
Alpiarca 1896	85	41	220	35	**120**	**536**	123	82
Heimersheimer Ruth 1895	187	44	202	**122**	nq	nq	74	46
Steinberg 1893	40	25	83	59	nq	nq	116	46
Rüdesheimer Berg	126	72	**271**	104	35	252	67	54
Scy 1892	164	34	255	**162**	46	174	98	82
Bordeaux 1892	83	31	155	77	nq	nq	110	46
Geisenheimer Mäuerchen 1893	74	**95**	120	92	18	121	**297**	41
S. kudriavzevii								
Würzburg (Stein)	42	29	106	54	nq	nq	68	40
Winningen	65	46	151	68	23	nq	85	68

* nq = not quantified; + = pronounced values are shown in bold and highlighted in orange.

What was interesting to note is that some strains produced either high levels of hexanoic acid ethylester (ethyl hexanoate) or high levels of acetic acid phenylethylester, but not of both substances. This was observed when comparing Rüdesheimer Hinterhaus 1893 with Zell 1895 and Alpiarca 1896. The first was low on hexanoic acid ethylester but produced high levels of 2-phenylethyl acetate, while the latter strains produced high levels of hexanoic acid ethylester and much lower levels of 2-phenylethyl acetate (Table 3, Figure 3; see Section 4).

4. Discussion

Alcoholic beverage production has been carried out by spontaneous fermentation throughout the ages, and today, often still the preferred method of fermentation for some. In beer and baking enterprises, it was already realized in the middle ages that there are special properties in the slurry that leavens bread and makes beer ferment and foam. It was the traditional occupation of a 'Hefner' in Germany to maintain and provide sufficient supplies of this leavening activity [16]. Yet, as it was not clear what the causal activity in the slurry was, the German 'Reinheitsgebot' (purity law) from 1516 only stated that beer should be brewed using barley malt, hops, and water. With the work of Pasteur, published in his Etudes sur le vin and Etudes sur la biere, and the work of others, it became evident that yeast, *Saccharomyces cerevisiae*, was responsible for the observed fermenting power.

The pure yeast strain isolation procedure worked out by Hansen was transferred to the wine industry by the German Julius Wortmann, who founded the 'Geisenheim Yeast Breeding Center' in 1894 [1,2]. The isolation of pure yeast strains in Geisenheim had been initiated in the early 1890s and these strains are still preserved today. Today's strain maintenance relies on deep freezing of culture collections at −80 °C, while supplies for the industry are generally provided as instant dry yeasts [17,18].

Yet, at the Geisenheim Yeast Breeding Center, stocks were originally maintained as slants with a paraffin overlay. Cultures were stored either in this manner or, even simpler, in plain water [5,6]. A recent report analyzed long-term storage (12 years) of >1000 stocks and provided evidence that water storage yielded survival rates of 98.9%, closely resembling that of frozen stocks (99.5%), while survival rates under a mineral oil layer were a bit less with 88.2% [10]. Two racks of Geisenheim yeast stocks were kept over the years, more as display items than out of necessity. Such a long-term storage has not been previously reported. In fact, it is not entirely clear anymore when exactly these stocks were generated (besides the fact that they are very old and between 30 and 60 years of age). Collectively, these studies show, however, that yeast strain collections can be routinely kept under a zero-emission regime, which could be used as an incentive to reduce the energy-demanding storage of cultures at −80 °C.

The 'old' Geisenheim yeast strains can be both a heritage and a source of new strains for yeast breeding programs. As a heritage, they could be used to strengthen the local character of wines produced with them and thus contribute to the terroir of these wines [19–21]. To be useful as a breeding stock, these old strains need to be characterized in more detail, preferably including their genomes [22,23]. In our study, the yeast with the most pronounced flavor production capability of alcohols and esters was the *S. cerevisiae* strain Rüdesheimer Hinterhaus 1893 ('Hinterhaus' refers to backyard). It was apparent that this strain is a remarkable producer of acetate esters, but not so of medium chain fatty esters. The former are produced by the acetyl transferases Atf1 and Atf2, while the latter are generated by the acyl-coenzymeA:ethanol O-acyltransferases Eeb1 and Eht1 [24,25]. This suggests strong activity of the ATF alcohol acetyl-coA transferases in the Rüdesheimer Hinterhaus strain isolated already in 1893, with reduced formation of medium chain fatty acids. It will be interesting to explore these properties through a full-scale wine fermentation trial and use in-depth genomics and transcriptomics to generate molecular markers for yeast breeding.

The strive for more volatile aromas has spurred the search for alternative yeasts and renewed the interest in spontaneous fermentations [26]. Yet, due to the unpredictable and inconsistent outcomes of spontaneous fermentations, the development of improved starter cultures or consortia may provide better alternatives [27,28].

5. Conclusions

In conclusion, our study of yeast isolates stored for a very long period opens several new research avenues in the use of these strains, either directly as starter cultures or as stocks for the Geisenheim Yeast Breeding Center; both will be interesting to exploit in the future. For general use, our data and previous work, e.g., on water cultures show that zero-emission strain-keeping of yeast cultures is feasible and should be more generally exploited.

Author Contributions: Conceptualization, J.W., D.R., and C.v.W.; methodology, K.M., B.B., S.B., H.S.; validation, J.W., K.M., B.B., S.B., H.S.; formal analysis, J.W., K.M., S.B., H.S.; investigation, J.W., K.M., B.B., S.B., H.S.; resources, J.W.; D.R., C.v.W. data curation, J.W., K.M., S.B., H.S.; writing—original draft preparation, J.W.; writing—review and editing, J.W., D.R., C.v.W., K.M., B.B., S.B., H.S.; visualization, K.M. and J.W. supervision, J.W. and D.R.; project administration, J.W. and D.R.; funding acquisition, J.W. All authors have read and agreed to the published version of the manuscript.

Acknowledgments: We thank Judith Muno-Bender for taking photographs of the old yeast samples; Stefanie Fritsch for calibrations of the analytical equipment and Bettina Mattner for excellent technical service and all Department members for input and support of this project.

References

1. Hansen, E.C. Recherches sur la physiologie et la morphologie des ferments alcooliques V. Methodes pour obtenir des cultures pures de *Saccharomyces* et de mikroorganismes analogues. *C R Trav. Lab. Carlsberg* **1888**, *2*, 143–167.
2. Wortmann, J. *Anwendung und Wirkung Reiner Hefen*; Verlagsbuchhandlung Paul Parey: Berlin, Germany, 1895.
3. Gélinas, P. Active dry yeast: Lessons from patents and science. *Compr. Rev. Food Sci. Saf.* **2019**, *18*, 1227–1255. [CrossRef]
4. Wellman, A.M.; Stewart, G.G. Storage of brewing yeasts by liquid nitrogen refrigeration. *Appl. Microbiol.* **1973**, *26*, 577–583. [CrossRef] [PubMed]
5. Castellani, A. Viability of some pathogenic fungi in distilled water. *J. Trop. Med. Hyg.* **1939**, *42*, 225–226.
6. Hartsell, S.E. The longevity of bacterial cultures under paraffin oil. *J. Bacteriol.* **1947**, *53*, 801. [PubMed]
7. Hartung de Capriles, C.; Mata, S.; Middelveen, M. Preservation of fungi in water (castellani): 20 years. *Mycopathologia* **1989**, *106*, 73–79. [CrossRef] [PubMed]
8. Rodrigues, E.G.; Lírio, V.S.; Lacaz, C.d.S. Preservation of fungi and actinomycetes of medical importance in distilled water. *Rev. Inst. Med. Trop. Sao Paulo* **1992**, *34*, 159–165. [CrossRef]
9. Richter, D.L. Revival of saprotrophic and mycorrhizal basidiomycete cultures after 20 years in cold storage in sterile water. *Can. J. Microbiol.* **2008**, *54*, 595–599. [CrossRef]
10. Karabicak, N.; Karatuna, O.; Akyar, I. Evaluation of the viabilities and stabilities of pathogenic mold and yeast species using three different preservation methods over a 12-year period along with a review of published reports. *Mycopathologia* **2016**, *181*, 415–424. [CrossRef]
11. Henry, B.S. Viability of yeast cultures preserved under mineral oil. *J. Bacteriol.* **1947**, *54*, 264. [CrossRef]
12. Hamelin, R.C.; Berube, P.; Gignac, M.; Bourassa, M. Identification of root rot fungi in nursery seedlings by nested multiplex pcr. *Appl. Environ. Microbiol.* **1996**, *62*, 4026–4031. [CrossRef] [PubMed]
13. Schneider, A.; Gerbi, V.; Redoglia, M. A Rapid HPLC method for separation and determination of major organic acids in grape musts and wines. *Am. J. Enol. Vitic* **1987**, *38*, 151–155.
14. Belda, I.; Ruiz, J.; Esteban-Fernández, A.; Navascués, E.; Marquina, D.; Santos, A. Microbial contribution to wine aroma and its intended use for wine quality improvement. *Molecules* **2017**, *22*, 189. [CrossRef] [PubMed]
15. Camara, J.S.; Alves, M.A.; Marques, J.C. Development of headspace solid-phase microextraction-gas chromatography–mass spectrometry methodology for analysis of terpenoids in Madeira wines. *Anal. Chim. Acta* **2006**, *555*, 191–200. [CrossRef]
16. Meusdoerffer, F.G. A Comprehensive History of Beer Brewing. In *Handbook of Brewing: Processes, Technology, Markets*; Eßlinger, H.M., Ed.; WILEY-VCH: Weinheim, Germany, 2009.

17. Querol, A.; Barrio, E.; Huerta, T.; Ramon, D. Molecular monitoring of wine fermentations conducted by active dry yeast strains. *Appl. Environ. Microbiol.* **1992**, *58*, 2948–2953. [CrossRef]
18. Beltran, G.; Torija, M.J.; Novo, M.; Ferrer, N.; Poblet, M.; Guillamon, J.M.; Rozes, N.; Mas, A. Analysis of yeast populations during alcoholic fermentation: A six year follow-up study. *Syst. Appl. Microbiol.* **2002**, *25*, 287–293. [CrossRef]
19. Bokulich, N.A.; Thorngate, J.H.; Richardson, P.M.; Mills, D.A. Microbial biogeography of wine grapes is conditioned by cultivar, vintage, and climate. *Proc. Natl. Acad. Sci. USA* **2014**, *111*, E139–E148. [CrossRef]
20. Gilbert, J.A.; van der Lelie, D.; Zarraonaindia, I. Microbial terroir for wine grapes. *Proc. Natl. Acad. Sci. USA* **2014**, *111*, 5–6. [CrossRef]
21. Capozzi, V.; Garofalo, C.; Chiriatti, M.A.; Grieco, F.; Spano, G. Microbial terroir and food innovation: The case of yeast biodiversity in wine. *Microbiol. Res.* **2015**, *181*, 75–83. [CrossRef]
22. Peter, J.; De Chiara, M.; Friedrich, A.; Yue, J.X.; Pflieger, D.; Bergstrom, A.; Sigwalt, A.; Barre, B.; Freel, K.; Llored, A.; et al. Genome evolution across 1,011 Saccharomyces cerevisiae isolates. *Nature* **2018**, *556*, 339–344. [CrossRef]
23. Langdon, Q.K.; Peris, D.; Baker, E.P.; Opulente, D.A.; Nguyen, H.V.; Bond, U.; Goncalves, P.; Sampaio, J.P.; Libkind, D.; Hittinger, C.T. Fermentation innovation through complex hybridization of wild and domesticated yeasts. *Nat. Ecol. Evol.* **2019**, *3*, 1576–1586. [CrossRef] [PubMed]
24. Saerens, S.M.; Delvaux, F.R.; Verstrepen, K.J.; Thevelein, J.M. Production and biological function of volatile esters in saccharomyces cerevisiae. *Microb. Biotechnol.* **2010**, *3*, 165–177. [CrossRef]
25. Saerens, S.M.; Verstrepen, K.J.; Van Laere, S.D.; Voet, A.R.; Van Dijck, P.; Delvaux, F.R.; Thevelein, J.M. The *Saccharomyces cerevisiae EHT1* and *EEB1* genes encode novel enzymes with medium-chain fatty acid ethyl ester synthesis and hydrolysis capacity. *J. Biol. Chem.* **2006**, *281*, 4446–4456. [CrossRef] [PubMed]
26. Bougreau, M.; Ascencio, K.; Bugarel, M.; Nightingale, K.; Loneragan, G. Yeast species isolated from Texas High Plains vineyards and dynamics during spontaneous fermentations of Tempranillo grapes. *PLoS ONE* **2019**, *14*, e0216246. [CrossRef] [PubMed]
27. Jolly, N.P.; Varela, C.; Pretorius, I.S. Not your ordinary yeast: Non-*Saccharomyces* yeasts in wine production uncovered. *FEMS Yeast Res.* **2014**, *14*, 215–237. [CrossRef] [PubMed]
28. Bagheri, B.; Bauer, F.F.; Setati, M.E. The impact of *Saccharomyces cerevisiae* on a wine yeast consortium in natural and inoculated fermentations. *Front. Microbiol.* **2017**, *8*, 1988. [CrossRef] [PubMed]

14

Fluorinated Phenylalanine Precursor Resistance in Yeast

Ian S. Murdoch, Samantha L. Powers and Aaron Z. Welch *

Biomolecular Sciences Institute, Florida International University, Miami 33199, FL, USA; imurd001@fiu.edu (I.S.M.); spowe009@fiu.edu (S.L.P.)

* Correspondence: aawelch@fiu.edu

Abstract: Development of a counter-selection method for phenylalanine auxotrophy could be a useful tool in the repertoire of yeast genetics. Fluorinated and sulfurated precursors of phenylalanine were tested for toxicity in *Saccharomyces cerevisiae*. One such precursor, 4-fluorophenylpyruvate (FPP), was found to be toxic to several strains from the *Saccharomyces* and *Candida* genera. Toxicity was partially dependent on *ARO8* and *ARO9*, and correlated with a strain's ability to convert FPP into 4-fluorophenylalanine (FPA). Thus, strains with deletions in *ARO8* and *ARO9*, having a mild phenylalanine auxotrophy, could be separated from a culture of wild-type strains using FPP. Tetrad analysis suggests FPP resistance in one strain is due to two genes. Strains resistant to FPA have previously been shown to exhibit increased phenylethanol production. However, FPP resistant isolates did not follow this trend. These results suggest that FPP could effectively be used for counter-selection but not for enhanced phenylethanol production.

Keywords: fluorinated compounds; counter selection; phenylalanine; phenylethanol; yeast

1. Introduction

The phenylalanine biosynthetic pathway is a route to production of the high-value chemical 2-phenylethanol that is used for its rosy scent in cosmetics, foods, and cleaning supplies [1]. Phenylethanol is produced mainly by synthetic chemical processes but also naturally by yeast, as a degradation product of phenylalanine, moving through the intermediate metabolites of phenylpyruvate then phenylacetaldehyde (Figure 1). In *Saccharomyces cerevisiae,* one method of increasing the production of these chemicals is to select for resistance to the fluorinated phenylalanine analog 4-fluorophenylalanine (FPA) [2]. One facet of the mechanism of resistance is enhanced endogenous production of phenylalanine, thus requiring less dependence on exogenously provided phenylalanine [3]. Hence, FPA can be used to increase production of phenylethanol.

There are numerous counter-selection methods for a variety of genes. One method, the use of toxic metabolites to counter-select for auxotrophies such as uracil or tryptophan, has been a boon in yeast genetics, enabling numerous techniques such as the plasmid shuffle and Synthetic Genetic Array analysis [4–6]. These techniques rely on adding a fluorine to a precursor of the final biosynthetic pathway product, such that use of the fluorinated precursor results in a fluorinated end-product, which is toxic to the cell. There are many potential mechanisms of resistance to toxic precursors. One such mechanism is the prohibition of toxic metabolite import, as in the case of canavanine resistance via abrogation of the arginine transporter [4]. Another is to decrease endogenous production of the precursor while relying on exogenous end-product for growth, as in the case of deletion of *URA3* and reliance upon external uracil for 5-fluoroorotic acid resistance. In the case of resistance to a toxic end-product, one mechanism of resistance is via increasing endogenous production of the end-product so as to decrease exogenous nutrient dependence, such as in the case of increased phenylalanine production in response to fluorophenylalanine [3].

We sought to identify toxic precursors to the phenylalanine biosynthetic pathway that could be used for auxotrophic counter-selection, and that may lead to increased production of phenylethanol.

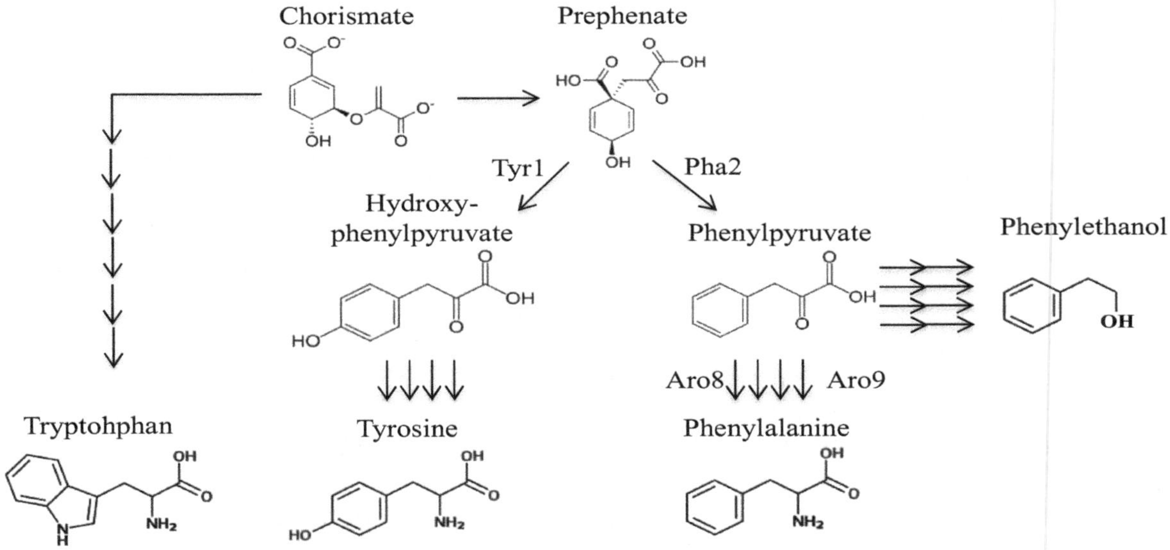

Figure 1. Diagram of the yeast pathway for biosynthesis of tryptophan, tyrosine, and phenylalanine. Number of enzymes and steps are indicated by number of arrows.

2. Materials and Methods

2.1. Media and Growth Conditions

Rich medium, or YPD, was made with BactoTM Yeast Extract at 10 g L^{-1} (Becton Dickinson, Franklin Lakes, NJ, USA), BactoTM Peptone at 20 g L^{-1} (Becton Dickinson), and dextrose at 20 g L^{-1} (Fisher Scientific, Waltham, MA, USA). Minimal medium contains 1.8 g L^{-1} yeast nitrogen base without amino acid or ammonium sulfate and 6.2 g L^{-1} ammonium sulphate and dextrose at 20 g L^{-1}. Synthetic complete medium contains minimal medium with 20 mg L^{-1} of all the following compounds: uracil, lysine, tyrosine, tryptophan, phenylalanine, adenine, leucine, histidine, and methionine. For clarification, the phenylalanine-related chemicals used in this paper are described in Table 1, their molecular structures are shown in Figure 2.

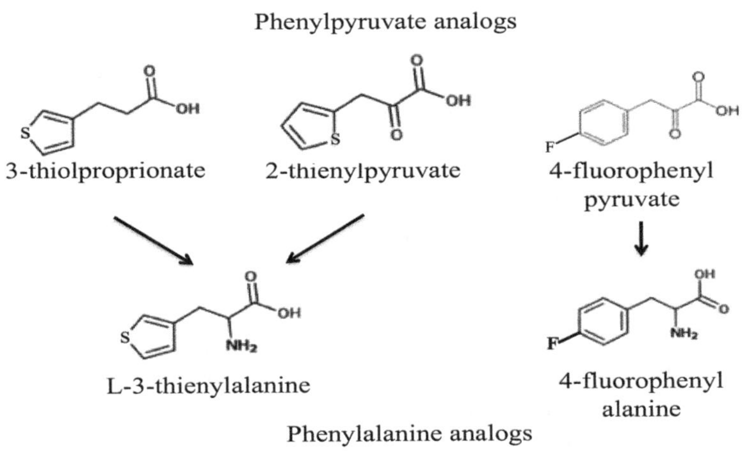

Figure 2. Analogs of phenylpyruvate and phenylalanine tested.

Table 1. List of phenylalanine-related drugs used.

Chemical Name	Abbreviation	CAS #	Source
4-fluorophenylalanine	FPA	1132-68-9	Chem-Impex
4-fluorophenylpyruvate	FFP	7761-30-0	Enamine
2-thienylpyruvate	2TP	15504-41-3	Enamine
3-thiolproprionate	3TP	16378-06-6	Enamine
L-3-thienylalnine	L3T	3685-51-6	Sigma

Chem-Impex (Wood Dale, IL, USA);Enamine (Monmouth Jct., NJ, USA); Sigma (St. Louis, MO, USA). CAS# refers to Chemical Abstracts Service registry number for indicated compound.

Experiments using FPA, FPP, 2TP, 3TP, or L3T were performed in MM2 medium, which consists of minimal medium with 2 mg L^{-1} of all the following compounds: uracil, tyrosine, phenylalanine, leucine, histidine, and methionine. Chemicals FPA, 2TP, 3TP, or L3T were suspended in water, while FPP and FPA were suspended in dimethyl sulfoxide (DMSO).

Culture density was determined by a spectrophotometer reading of the optical density at 600 nm, which was correlated to cell density via hemocytometer counts. Strains used are listed in Table 2. Strains mentioned with a dash (e.g., S288c-3) are derivatives of the parental strain named preceding the dash. Mutation cultures were obtained by exposure to ultraviolet light at 230 nm for 10 s that resulted in viability between 30–70% as measured by methylene blue.

Table 2. List of yeast strains used.

Strain	Genotype	Source
26704c	*MATα ura3 aro8-2 aro9-1*	[7]
BY4700	*MATα S288c ura3Δ0*	ATCC
BY4742	*MATα S288c his3Δ1 leu2Δ0 lys2Δ0 ura3Δ0*	ATCC
A364A	*MATa ade1 ade2 ura1 his7 lys2 tyr1 gal1*	ATCC
AW051	*MATa ura3 aro8-2 aro9-1*	This study
AW052	*MATa ura3 aro8-2 aro9-1*	This study
AW063	*MATα ade2 pha2 arg8 lys2*	This study
AW077	*MATα ade2 ura1*	This study
AW081	*MATa ura1 his7*	This study
AW083	*MATα lys2 ade1 ade2*	This study
AW102	*MATa/MATα wild-type*	This Study
AW108	*MATα S288c pha2::G418*	This study
AW109	*MATα S288c pha2::G418*	This study
AW258	*MATα ura1*	This study
AW259	*MATα his7*	This study
AW261	*MATα met15*	This study
IM006	*MATα lys2*	This study
Calb	*Candida albicans ATCC 10231*	Gift from Darlene Miller
Ctrop	*Candida tropicalis*	Gift from Darlene Miller
S288c	*MATα SUC2 mal mel CUP1 flo1 flo8-1 hap1*	ATCC
SPara	*lys2 Saccharomyces paradoxus*	Gift from Doug Koshland
SBaya	*ura3 Saccharomyces bayanus*	Gift from Doug Koshland

American Type Culture Collection (ATCC)

Drug resistance was measured by calculating the relative viability of cells, which is the number of viable cells resulting from a drug treated culture divided by the number of viable cells resulting from a carrier treated (water or DMSO) culture. Viability was measured as colony forming units on a petri dish by directly plating cells to a plate or by first serially diluting in water, then pinning to drug and carrier plates (frogging method) or by using the Tadpoling method, discussed below. For timecourse experiments, relative viability was calculated by using the number of viable cells at timepoint zero as the denominator in all subsequent timepoints.

Briefly, the Tadpoling method consists of inoculating 20 µL of yeast containing culture into 180 µL YPD in a well of a 96-well plate and serially diluting 20 µL from this well to subsequent wells [8]. This plate is incubated until individual colonies can be counted in the most dilute well, yielding a quantifiable number used to calculate cell viability in the original well by multiplying by the dilution factor.

$$hours \times \log_2(\frac{\text{ODin}}{\text{ODfin}})$$

Growth rate was calculated using the following expression: where ODin is the initial A600, ODfin is the final A600, hours is the number of hours in between A600 readings.

2.2. Determination of Phenylethanol and Phenylalanine

After growth in 2 mg L^{-1} uracil, tyrosine, phenylalanine, leucine, histidine, and methionine with 1.8 g L^{-1} yeast nitrogen base without amino acid or ammonium sulfate with 20 g L^{-1} dextrose and 6 g L^{-1} phenylalanine for 6–8 days at 30 °C, culture supernatant was removed and selected ion monitoring (SIM) was performed to detect phenylethanol and phenylalanine using liquid chromatography-tandem mass spectrometry (LC-MS/MS). A gradient was run from 100% 50 mM ammonium acetate to 100% methanol with 0.1% formic acid over the course of 20 min. Phenylethanol retention time was 12.5 min while that of phenylalanine was 4.5.

2.3. Determination of Fluorophenylalanine

Yeast samples were grown in medium containing FPP at a concentration of 200 mg L^{-1} for 16 h then yeast samples were resuspended in 70% ethanol, boiled for 10 min, then the supernatant was transferred to a new tube and after evaporated in a spinning-vacuum centrifuge. Samples were fully dissolved in 1 mL of 3:1:1 water:acetronitrile:isopropanol, with 2 min of vortexing and 20 min sonication. The resulting solutions were 5× diluted (200/1000) in 10 mM ammonium formate in water (mobile phase A) and was injected in the LC-MS/MS system, an AB Sciex QTRAP 5500 Triple-Quadrupole mass spectrometer, equipped with a Turbospray ESI source. A binary gradient separation program was employed using a reverse phase HPLC column (Dionex Acclaim 120 C18 Column, 250 × 2.1 mm, 5 µm). Sample injection (20 µL) and LC separation was performed by a Shimadzu Prominence LC-20AD Ultra-Fast Liquid Chromatograph.

3. Results

First, we sought to identify a toxic phenylalanine analog and work backward to design toxic phenylalanine precursors. Two promising toxic phenylalanine analogs were identified from the literature: L-3-thienylalanine (L3T), consisting of phenylalanine with the benzene ring substituted for thiophene, and fluorophenylalanine (FPA), consisting of phenylalanine with a fluorine attached to the benzene ring (Figure 2, Table 1). A 2.6 mg L^{-1} concentration of L-3-thienylalanine has been shown to inhibit growth of *S. cerevisiae* by 50% [9]. First, the toxicity of L3-thienylalanine was tested by plating wild-type strains S288c and AW077 to medium with L3-thienylalanine at 0 or 300 mg L^{-1} and determining the relative viability. L3T killed all AW077 cells and nearly all of S288c cells (Figure 3a). Next the toxicity of FPA was tested by plating strains S288c and AW077 onto medium containing FPA at 0 or 300 mg L^{-1} and determining the relative viability (Figure 3b). FPA at both of these concentrations killed almost all cells of S288c and AW077. Thus, we reconfirmed that L3T and FPA are toxic to *S. cerevisiae*.

Next, we sought to design a molecule that may be enzymatically converted to a toxic phenylalanine analog. By identifying a toxic phenylalanine precursor, we could use this compound to select for strains that have mutations in the phenylalanine biosynthetic pathway. We considered several precursors of phenylalanine; chorismate (this precursor is used to synthesize not only phenylalanine, but also tryptophan, tyrosine, folate, and ubiquinone). Thus, a strain with a chorismate deletion would have numerous other auxotrophies besides phenylalanine. Next, prephenate—however, this is not a good candidate for this particular intervention as adding fluorine or sulfur on the cyclohexadiene ring could interfere with transformation to a benzene ring. Additionally, there seems to be at least two mechanisms to convert prephenate to phenylalanine, as we found that deletion of *PHA2* does not confer complete phenylalanine requirement. To test this, strains lacking *PHA2* were inoculated alongside a strain with functioning *PHA2* in minimal medium with or without phenylalanine added (Figure 4, Table 3). The A600 was measured over four days, and maximum growth rate (lowest doubling time) was calculated. Strains AW108 and AW109 grew much better with phenylalanine added (Table 3), showing the dependency on phenylalanine; however, they still grew over time. This is in contrast to previously described observations of strains lacking *PHA2*. However, it is likely that the extremely slow growth rate explains this conclusion (Figure 4) [10]. The next candidate is phenylpyruvate, which does not require modification of its benzene ring prior to becoming phenylalanine, and it is the substrate for only two essentials molecules: phenylalanine and tyrosine. Use of phenylpyruvate has one drawback: there are two primary enzymes that can complete the transamination reaction converting phenylpyruvate to phenylalanine (*ARO8* and *ARO9*) and several others that are predicted to be capable of performing this reaction (Bat1, Bat2, His5) [11].

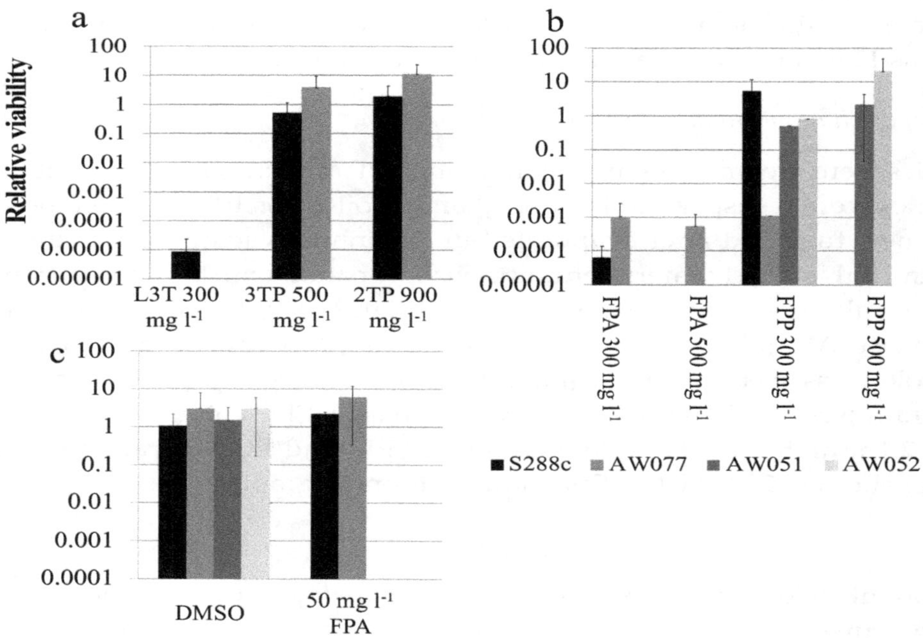

Figure 3. Viability of yeast in response to toxic phenylalanine or phenylpyruvate analogs. Cultures of yeast strains were serially diluted in water then plated onto medium containing either (**a**) L3T, 3TP, 2TP or (**b**) FPA, FPP, or (**c**) dimethyl sulfoxide, FPA. Relative viability was obtained by dividing the number of colony forming units (CFUs) from drug plates by CFUs from YPD plates. Plates were incubated for 2–3 days and CFUs counted. The average and standard deviation of three independent trials is shown.

Table 3. Density of two-day-old cultures.

Strain	Genotype	MM + U [a]	MM + UP [b]
AW108	*pha2*	0.06	3.8
AW109	*pha2*	0.08	3.6
AW051	*aro8 aro9 ura3*	0.37	1.1
AW052	*aro8 aro9 ura3*	1.2	2.2
BY4700	*ura3*	3.6	3.7

MM: Minimal Medium [a] Minimal medium with 20 mg L^{-1} uracil, [b] Minimal medium with 20 mg L^{-1} of uracil and phenylalanine.

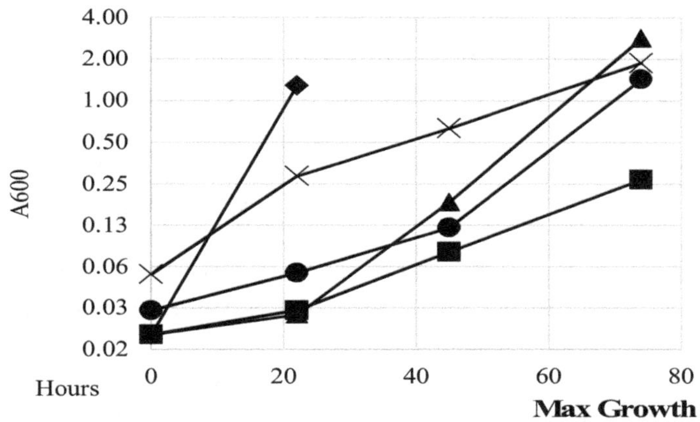

Strain	Genotype	Max Growth (hrs/doubling)
AW108	MATα S288c pha2::G418	16.1
AW109	MATα S288c pha2::G418	8.1
AW051	MATa ura3 aro8-2 aro9-1	6.8
AW052	MATa ura3 aro8-2 aro9-1	5.3
BY4700	MATα S288c ura3Δ0	3.4

Figure 4. Growth rate of strains in medium lacking phenylalanine. The indicated strains were grown overnight in synthetic complete medium then washed and inoculated to minimal medium with uracil and rotated at 23 °C. The A600 was measured over subsequent hours. Max growth was the lowest growth rate calculated for all points tested.

This presents a problem for selection, as strains would need to have deletions in two genes, rather than just one, in the case of uracil (URA3), lysine (LYS2), or arginine (CAN1) [6,12,13]. To test if *ARO8* and *ARO9* were required for phenylalanine production, strains lacking *ARO8* and *ARO9* (AW051, AW052) were constructed and inoculated alongside a wild-type strain in minimal medium with or without phenylalanine added (Figure 4, Table 3).

The A600 was measured over four days, and maximum growth rate (lowest doubling time) was calculated. Strains AW051 and AW052 grew much better with phenylalanine added (Table 3), showing their dependency on phenylalanine. However, they still grew over time. These strains do grow faster than the *PHA2* mutants but are not completely rescued solely by phenylalanine, as they also require tyrosine, hence the lower A600 on the second day of growth with phenylalanine added.

Furthermore, to verify there were no residual nutrients in the minimal medium, strains with various individual auxotrophies were inoculated to minimal medium and were found to not grow (Table 4).

Table 4. Density of three-day-old cultures.

Strain	Genotype	MM	MM + Aux [a]
AW108	*pha2*	3.0	5.8
AW109	*pha2*	2.8	5.6
AW258	*ura1*	0.1	3.6
AW259	*his7*	0.04	0.3
AW261	*met15*	0.08	5.2
IM006	*lys2*	0.07	1.4

[a] Minimal medium with 20 mg L^{-1} of auxotrophy required for strain.

Nonetheless, multiple enzymes producing phenylalanine could be viewed as a boon, as cells are less likely to develop spontaneous resistance to a toxic phenylpyruvate analog, as they would have to delete multiple processing enzymes. Thus we set out to design phenylpyruvate-like derivatives that may be catalytically converted into toxic phenylalanine analogs so that these can be used for counter-selection of phenylalanine auxotrophies.

The first two phenylpyruvate-like derivatives of L-3-thienylalanine tested were 3-thiolproprionate (3TP) and 2-thienylpyruvate (2TP). Wild-type yeast strains S288c and AW077 were pinned to plates containing these analogs at 500 mg L^{-1} and 900 mg L^{-1}, respectively, and relative viability was calculated (Figure 3a). Despite higher concentrations of the phenylpyruvate derivatives compared to L3T, S288c and AW077 strains exhibited no decrease in viability on 3TP and 2TP containing medium. Thus, while the phenylalanine analog was toxic to these strains, the phenylpyruvate analogs were not. Differences in toxicity could be due to adsorption to the cell wall, decreased uptake, or incompatibility with enzymes.

The toxicity of the FPA derivative fluorophenylpyruvate (FPP) was next examined. Strains AW077 and S288c were serially diluted in water, then pinned to plates containing 0, 300 mg L^{-1}, or 500 mg L^{-1} FPP and relative viability was calculated (Figure 3b). At 300 mg L^{-1} FPP strain S288c showed mild growth inhibition but no lethality, while strain AW077 was more growth-inhibited. At 500 mg L^{-1}, no growth was detected from S288c and AW077 strains (Figure 3b). These results indicate that the phenylpyruvate analog FPP is toxic to S288c and AW077 strains.

Phenylpyruvate is converted to phenylalanine primarily by the proteins Aro8p and Aro9p. However, there are other proteins that can catalyze this reaction. To test if the mechanism of FPP toxicity was due to its conversion to FPA, the toxicity of FPP was tested on strains lacking genes *ARO8* and *ARO9* (Figure 3b). Strains lacking *ARO8* and *ARO9* (AW051, AW052) are able to grow on this medium, showing approximately equal growth on YPD. However, these strains are hypersensitive to FPA, which may reflect the cells increased import of phenylalanine, due to a phenylalanine auxotrophy (Figure 3c). This result suggests that *ARO8* and *ARO9* are primarily responsible for affecting the toxicity of FPP.

For FPP to be of use as an auxotrophic marker, the natural resistance for it should be low. For example, the frequency of spontaneous resistance to 5-fluoroanthranilic acid or 5-fluoroorotic acid, for selection of tryptophan or uracil auxotrophs, respectively, was 6×10^{-7} [5]. To test the natural resistance frequency of S288c and AW077 strains to FPP, we plated 5×10^{6} cells on plates containing 500 mg L^{-1} FPP. We found no resistant cells grew on this medium, indicating natural resistance of wild-type strains to FPP is below a frequency of 2×10^{-7}. As a comparison to this, we also plated these strains to medium containing 500 mg L^{-1} FPA. Strains S288c and AW077 formed 240 and 86 colonies giving a frequency of resistance of 3×10^{-5} and 5×10^{-5}, respectively. This indicates that the natural frequency of resistance to FPP is quite low, while that of FPA is higher. This indicates that resistance to FPP toxicity requires either more mutations or more specific mutations, as compared to FPA, and that perhaps these molecules do not induce toxicity via the same mechanism.

To further investigate the difference between FPP and FPA resistance, strains that were FPA resistant were tested for FPP resistance. Isolates that were resistant to FPA were plated onto medium with or without FPP and relative viability was calculated (Figure 5). Three isolates derived from strain AW083 and one from AW077 showed resistance to FPA, but no resistance to FPP. This indicates that FPA resistance is not sufficient for FPP resistance. However, an isolate from a wild-type yeast wine strain (AW102) and an S288c mutant isolate (S288c-3) were recovered that showed resistance to FPP and FPA. When the strain S288c-3 was mated to a non-resistant strain (AW081) and sporulated, the resulting haploid segregants generally showed a pattern of 1:3 resistance to FPP, while resistance to FPA showed a pattern of 2:2 (Table 5).

Table 5. Tetrad analysis of FPP and FPA resistance.

Segregation	FPA Resistance	FPP Resistance
2:2	20	5
1:3	6	21

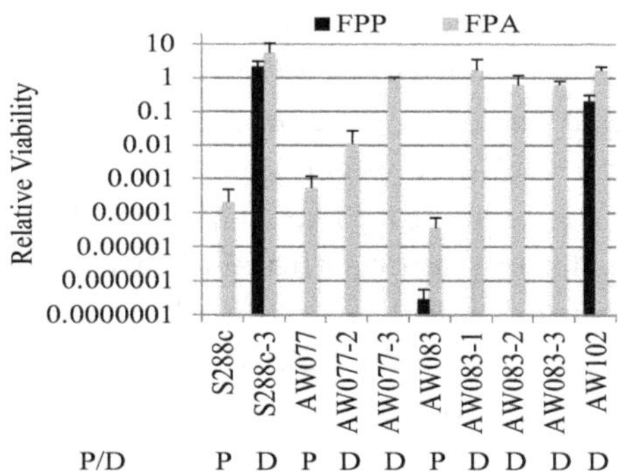

Figure 5. Viability of selected strains in response to FPP or FPA. Cultures of parental and derivative yeast strains were serially diluted in water then plated onto medium containing either 500 mg L^{-1} of FPA, FPP. Plates were incubated for 2–3 days and colony forming units (CFUs) counted. Relative viability was obtained by dividing the number of CFUs from drug plates by CFUs from YPD plates. P indicates the parental strain while D indicates a derivative of that strain. The average and standard deviation of three independent trials is shown.

A predominance of 1:3 segregation (tetratype) indicates that two genes are segregating independently, following the pattern of 1:1:4 PD:NPD:TT, while a 2:2 pattern indicates a single gene. All haploid segregants that were resistant to FPP were also resistant to FPA. The haploid segregants that were only resistant to FPA generally showed slow growth on FPA. One interpretation of these results is that detoxification of FPP requires the activity of two genes, either of which will suffice to detoxify FPA.

To determine if cells converted FPP into FPA or some other compound, we evaluated the presence of FPA in cells treated with FPP. Strains were inoculated to medium containing FPP and allowed to incubate for > 12 h, enzymatically digested to recover individual amino acids, and then tested for the presence of FPA using LC-MS/MS (Figure 6). Strains sensitive to FPP, AW077, and S288c were observed to accumulate the highest amounts of FPA, while the less sensitive strain AW051 accumulated less. The modest decrease in FPP conversion of AW051 is not too surprising, as there are other enzymes capable of converting phenylpyruvate to phenylalanine (Bat1, Bat2, His5), and this strain is capable of growth without phenylalanine supplementation. Strains S288c-3 and AW102, both of which are resistant to FPP, had the lowest levels of FPA. This suggests that their mechanism of resistance involves decreased conversion of FPP to FPA. Thus cells can convert FPP into FPA, and this may be the cause of some of the toxicity of FPP.

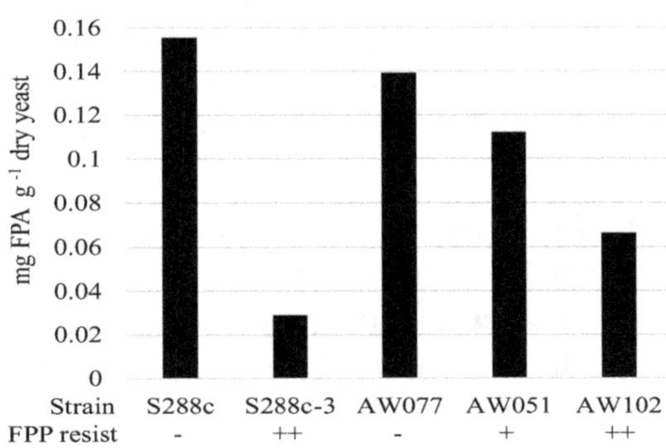

Figure 6. FPP is converted by cells to FPA. Yeast cells were inoculated to Synthetic Complete medium overnight at 30 °C, then inoculated to MM2 medium with 200 mg L^{-1} FPP and incubated at 30 °C for 16 h. Cultures were isolated, dried, weighed, and analyzed for the quantity of FPA. FPP resistance is indicated below each strain corresponding to a qualitative measure of their resistance to FPP with - signifying little or no resistance, + indicating moderate resistance, and ++ indicating strong resistance.

For FPP to be an effective selective agent, it should exert its toxic effect amongst diverse genre of yeast. To test the evolutionary conservation of FPP toxicity, strains from diverse species such as *Candida tropicalis, Candida albicans, Saccharomyces bayanus*, and *Saccharomyces paradoxus* were exposed to 0 or 500 mg L^{-1} FPP or FPA and their viability was measured over several days, and relative viabilities were calculated (Figure 7). To test these strains, the FPP and FPA sensitivity test was performed by inoculating cells into media with said compounds and evaluating culture viability using the Tadpoling method [8]. All the strains tested retained viability or slightly increased in cell number when inoculated into media without FPP or FPA, except for AW083. When inoculated to media containing FPA, *C. tropicalis* exhibited high resistance while most of the other strains lost viability over the course of four days. Both *C. albicans* and *S. bayanus* exhibited enhanced sensitivity to FPA, as these lost almost all culture viability within one day. Surprisingly, in media containing FPP, all strains showed greater sensitivity than in FPA and lost at least four orders of magnitude culture viability within two days. This result would suggest that FPP may be transported more effectively than FPA, or that FPP can exert its toxicity through a mechanism distinct from that of FPA. *S.paradoxus* and AW051 (*aro8-2 aro9-1*) showed the least sensitivity to FPP of all the strains tested, but even these cultures succumbed after four days. This data shows that FPP can act through diverse yeast genre to exert its toxic effect and may be a more effective fungicide than FPA.

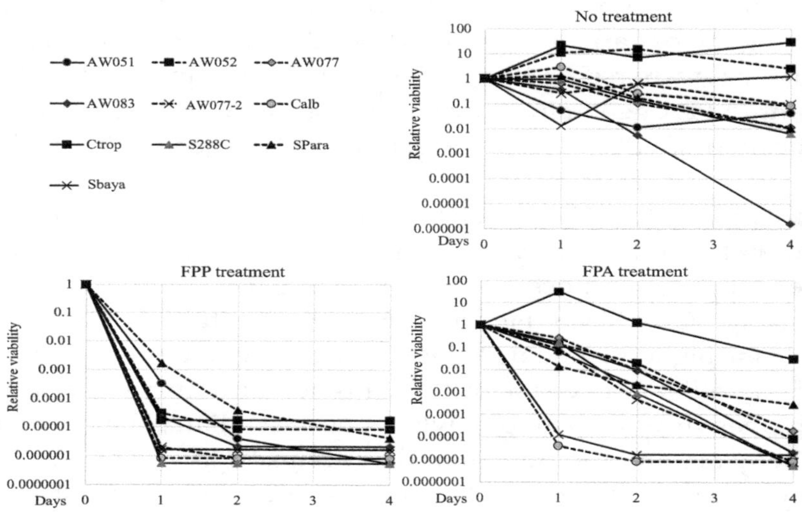

Figure 7. Toxicity of FPP and FPA are well conserved. Yeast cells were inoculated to Synthetic Complete medium overnight, then inoculated to MM2 medium with 500 mg L^{-1} FPP, FPA, or no drug, and incubated at 37 °C in an immobile 96-well plate. Culture viability was assessed over time using the Tadpoling method. Relative viability was assessed by dividing the number of CFUs on day 0, 1, 2, or 5 by the number of CFUs on day 0. The average of three independent trials is shown.

One potential mechanism of FPP toxicity is to affect the mitochondria. Phenylpyruvate acts as an inhibitor of pyruvate transport into the mitochondria where pyruvate would normally be decarboxylated to acetate [14]. Indeed, accumulation of phenylpyruvate in phenylketonuria patients becomes toxic as it may disrupt oxidative metabolism by preventing normal pyruvate movement into the mitochondria. To test if phenylpyruvate alone may exert a toxic effect on yeast, exponential phase cells were inoculated to 500 mg L^{-1} phenylpyruvate and culture viability was measured over several days using the Tadpoling method. The ratio of CFUs of cells exposed to phenylpyruvate with control cells was calculated over the course of several days (Figure 8). There were no significant ($p < 0.05$) deviations from the ideal 1 ratio for all strains tested, indicating that phenylpyruvate does not significantly affect culture viability. Nonetheless, there were notable deviations in decreased viability of S288c cells exposed to phenylpyruvate and an increase in viability of AW052 cells exposed to phenylpyruvate. Thus, it is likely that fluorine is required for fluorophenylalanine toxicity.

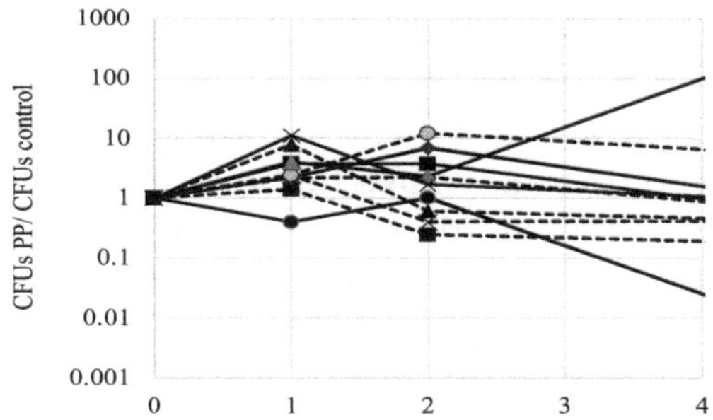

Figure 8. Phenylpyruvate does not affect culture viability. Yeast cells were inoculated to Synthetic Complete medium overnight, then inoculated to MM2 medium with 500 mg L^{-1} phenylpyruvate, or no drug, and incubated at 37 °C in an immobile 96-well plate. Culture viability was assessed over time using the Tadpoling method. To calculate the CFUs PP/CFUs control, the number of CFUs in phenylpyruvate treated cultures was divided by the number of CFUs in the untreated culture for each day. The average of three independent trials is shown. Markers are the same as used in Figure 7.

In *S. cerevisiae*, one method of increasing the production of the valuable chemical phenylethanol is to select for resistance to FPA [2]. We wanted to test whether FPA resistant strains obtained herein would produce high amounts of phenylethanol. To do this, we inoculated strains into medium containing phenylalanine as the source of nitrogen and measured the phenylethanol and phenylalanine concentration in the medium after three days of growth using GC-MS (Figure 9A).

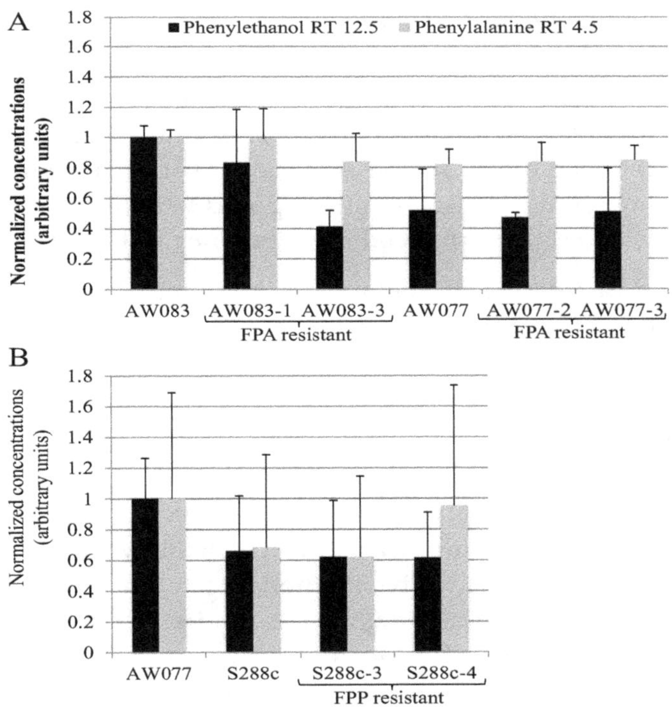

Figure 9. FPP or FPA resistance is not sufficient for increased phenylethanol production. Strains were incubated in medium containing phenylalanine as the source of nitrogen for three days, and then phenylethanol and phenylalanine concentration in the medium were measured using GC-MS. Each strain's production of phenylethanol or phenylalanine was normalized to 1 using (**A**) AW083 strain or (**B**) AW077 strain. The average and standard deviation of three independent trials is shown.

Resistance to FPA did not induce higher phenylethanol production in the four strains that were tested. This is not surprising, as previously, the frequency of FPA-resistant mutants with high levels of phenylethanol production was approximately 20% [15]. Thus, if we had tested five or more isolates we may have expected to obtain at least one. However, because of the low frequency of obtaining FPA-resistant strains we did not pursue this further. Because we had seen that FPP resistance was more difficult to attain than FPA, we predicted that these mutants may produce phenylalanine at a greater rate than FPA-resistant mutants and thus may have higher phenylethanol production. However, upon testing FPP-resistant strains, no increase in phenylethanol production was noted (Figure 9B). Thus, FPA and FPP resistance is not sufficient to induce high levels of phenylethanol production.

4. Discussion

Here, we report that FPP can act as a toxic agent in diverse strains of yeast, and this effect is at least partially mediated by *ARO8* and *ARO9* conversion to FPA. Whether FPP exerts its toxic effect solely by conversion to FPA or via another route rquires further investigation. FPP exhibited higher toxicity than FPA, suggesting a separate mechanism of toxicity, however this could be due to enhanced adsorption to cell wall, decreased cell uptake, or other enzymatic activities that decrease the effective concentration of FPA.

The mechanism of FPP toxicity still remains somewhat of a mystery. It is possible that there is preferred transport of phenylpyruvate over phenylalanine although evidence for this is absent. Another hypothesis is that FPP is converted into some other compound that FPA is not, such as fluoro-phenylethanol. Phenylalanine degradation follows this pathway: phenylalanine > phenylpyruvate > phenylacetaldehyde > phenylethanol. Phenylpyruvate is one enzymatic step closer to phenylethanol, and it is possible the fluorine moiety decreases its ability to convert to fluoro-phenylalanine, but instead allows it to become fluoro-phenylethanol. Phenylethanol has been shown to be toxic to *C. albicans* at 20 mg L^{-1} [16].

We observed differences in the sensitivity of strains to FPP and FPA depending on the method of delivery—solid or liquid medium. Phenylpyruvate shows increased solubility relative to phenylalanine (112 g L^{-1} versus 27 g L^{-1}) that may explain its increased toxicity in liquid medium [17]. This observation may be explained by the increased exposure of cells to all toxic molecules in a liquid, as opposed to a solid medium in which cells are only exposed to toxins within close physical proximity. Previous groups have reported similar enhanced drug potency in liquid media [18].

The effect of fluorinated phenylalanine metabolites has been observed previously in other fungal species. Fluorophenylalanine at concentrations as low as 2.5 mM can inhibit growth of fungal species *Cladosporium cucumerinum* and *Colletotrichum lagenarium* [19]. We found that FPP was toxic to a variety of yeast species at 2.74 mM indicating that it is a more potent fungal inhibitor than FPA.

Interestingly, we also found that *PHA2* was not required for slow growth. This is in contrast to several other works that indicate that *PHA2* causes an absolute requirement for phenylalanine in the medium [10,20]. However, strain growth is typically assayed over a period of up to three days on solid medium, while in our studies, liquid medium was used which allows for faster growth. *PHA2* converts prephenate to phenylpyruvate, but an alternative method to converting prephenate to phenylalanine is via the arogenate dehydratase enzyme. This enzyme has not been reported in *S. cerevisiae*. The closest species having a putative arogenate dehydratase is *Neurospora crassa*, which shares the same phylum with *S. cerevisiae*. Further investigation is needed to determine the mechanism of slow phenylalanine generation in strains lacking *PHA2*.

These observations should be helpful in better understanding phenylalanine metabolism and may be of interest in identifying potential secondary targets of FPP.

Author Contributions: Conceptualization, A.Z.W.; Methodology, A.Z.W.; Validation, A.Z.W. S.P., and I.M.; Formal Analysis, A.Z.W.; Investigation, A.W. S.P., and I.M.; Writing—Original Draft Preparation, A.Z.W.; Writing—Review

& Editing, A.Z.W. S.P., and I.M.; Visualization, A.Z.W.; Supervision, A.Z.W.; Project Administration, A.Z.W.; Funding Acquisition, A.Z.W.

Acknowledgments: The authors would like to thank Antonio Urrestarazu for the kind gift of strain 26704c. The authors would like to thank Darlene Miller for *Candida albicans* and *Candida tropicalis* strains. The authors would like to thank Doug Koshland for the kind gifts of *Saccharomyces bayanus* and *Saccharomyces paradoxus* strains. The authors would like to thank Stephanie How for her research contributions.

References

1. Etschmann, M.; Bluemke, W.; Sell, D.; Schrader, J. Biotechnological production of 2-phenylethanol. *Appl. Microbiol. Biotechnol.* **2002**, *59*, 1–8. [PubMed]
2. Akita, O.; Ida, T.; Obata, T.; Hara, S. Mutants of Saccharomyces cerevisiae producing a large quantity of β-phenethyl alcohol and β-phenethyl acetate. *J. Ferment. Bioeng.* **1990**, *69*, 125–128. [CrossRef]
3. Fukuda, K.; Watanabe, M.; Asano, K. Altered regulation of aromatic amino acid biosynthesis in β-phenylethyl-alcohol-overproducing mutants of sake yeast Saccharomyces cerevisiae. *Agric. Biol. Chem.* **1990**, *54*, 3151–3156. [CrossRef]
4. Forsburg, S.L. The art and design of genetic screens: Yeast. *Nat. Rev. Genet.* **2001**, *2*, 659–668. [CrossRef] [PubMed]
5. Toyn, J.H.; Gunyuzlu, P.L.; Hunter White, W.; Thompson, L.A.; Hollis, G.F. A counterselection for the tryptophan pathway in yeast: 5-fluoroanthranilic acid resistance. *Yeast* **2000**, *16*, 553–560. [CrossRef]
6. Boeke, J.D.; La Croute, F.; Fink, G.R. A positive selection for mutants lacking orotidine-5′-phosphate decarboxylase activity in yeast: 5-fluoro-orotic acid resistance. *Mol. Gen. Genet. MGG* **1984**, *197*, 345–346. [CrossRef] [PubMed]
7. Iraqui, I.; Vissers, S.; Cartiaux, M.; Urrestarazu, A. Characterisation of Saccharomyces cerevisiae ARO8 and ARO9 genes encoding aromatic aminotransferases I and II reveals a new aminotransferase subfamily. *Mol. Gen. Genet. MGG* **1998**, *257*, 238–248. [CrossRef] [PubMed]
8. Welch, A.Z.; Koshland, D.E. A simple colony-formation assay in liquid medium, termed tadpoling, provides a sensitive measure of Saccharomyces cerevisiae culture viability. *Yeast* **2013**, *30*, 501–509. [CrossRef] [PubMed]
9. Dittmer, K. The Synthesis and Microbiological Properties of β-3-Thienylalanine, a New anti-Phenylalanine. *J. Am. Chem. Soc.* **1949**, *71*, 1205–1207. [CrossRef] [PubMed]
10. Bross, C.D.; Corea, O.R.A.; Kaldis, A.; Menassa, R.; Bernards, M.A.; Kohalmi, S.E. Complementation of the pha2 yeast mutant suggests functional differences for arogenate dehydratases from Arabidopsis thaliana. *Plant Physiol. Biochem.* **2011**, *49*, 882–890. [CrossRef] [PubMed]
11. SRI International Yeast Pathways Database Website Home. Available online: https://pathway.yeastgenome.org/ (accessed on 1 May 2018).
12. Chattoo, B.B.; Sherman, F.; Azubalis, D.A.; Fjellstedt, T.A.; Mehnert, D.; Ogur, M. Selection of lys2 mutants of the yeast Saccharomyces cerevisiae by the utilization of α-aminoadipate. *Genetics* **1979**, *93*, 51–65. [PubMed]
13. Whelan, W.L.; Gocke, E.; Manney, T.R. The CAN1 locus of Saccharomyces cerevisiae: Fine-structure analysis and forward mutation rates. *Genetics* **1979**, *91*, 35–51. [PubMed]
14. Halestrap, A.P.; Brand, M.D.; Denton, R.M. Inhibition of mitochondrial pyruvate transport by phenylpyruvate and α-ketoisocaproate. *Biochim. Biophys. Acta (BBA) Biomembr.* **1974**, *367*, 102–108. [CrossRef]
15. Fukuda, K.; Watanabe, M.; Asano, K.; Ouchi, K.; Takasawa, S. Isolation and genetic study of p-fluoro-dl-phenylalanine-resistant mutants overproducing β-phenethyl-alcohol in Saccharomyces cerevisiae. *Curr. Genet.* **1991**, *20*, 449–452. [CrossRef] [PubMed]
16. Lingappa, B.T.; Prasad, M.; Lingappa, Y.; Hunt, D.F.; Biemann, K. Phenethyl alcohol and tryptophol: Autoantibiotics produced by the fungus Candida albicans. *Science* **1969**, *163*, 192–194. [CrossRef] [PubMed]
17. Wishart, D.S.; Feunang, Y.D.; Marcu, A.; Guo, A.C.; Liang, K.; Vázquez-Fresno, R.; Sajed, T.; Johnson, D.; Li, C.; et al. HMDB 4.0—The Human Metabolome Database for 2018. *Nucleic Acids Res.* **2018**, 608–617. [CrossRef] [PubMed]
18. Smith, A.M.; Ammar, R.; Nislow, C.; Giaever, G. A survey of yeast genomic assays for drug and target discovery. *Pharmacol. Ther.* **2010**, *127*, 156–164. [CrossRef] [PubMed]

19. Van Andel, O.M. Fluorophenylalanine as a systemic fungicide. *Nature* **1962**, *194*. [CrossRef]
20. Ma, Y.; Sugiura, R.; Saito, M.; Koike, A.; Sio, S.O.; Fujita, Y.; Takegawa, K.; Kuno, T. Six new amino acid-auxotrophic markers for targeted gene integration and disruption in fission yeast. *Curr. Genet.* **2007**, *52*, 97–105. [CrossRef] [PubMed]

15

Effect of Co-Inoculation with *Pichia fermentans* and *Pediococcus acidilactici* on Metabolite Produced during Fermentation and Volatile Composition of Coffee Beans

Alexander da Silva Vale [1], Gilberto Vinícius de Melo Pereira [1,*], Dão Pedro de Carvalho Neto [1], Cristine Rodrigues [1], Maria Giovana B. Pagnoncelli [2] and Carlos Ricardo Soccol [1,*]

1 Department of Bioprocess Engineering and Biotechnology, Federal University of Paraná (UFPR), 19011Curitiba, Paraná 81531-980, Brazil

2 Departament of Chemistry and Biology, Federal University of Technology—Paraná (UTFPR), Curitiba PR 80230-901, Brazil

* Correspondence: gilbertovinicius@gmail.com (G.V.d.M.P.); soccol@ufpr.br (C.R.S.)

Abstract: Removal of the mucilage layer of coffee fruits by a fermentation process has became an interesting strategy to improve coffee quality, which is able to assist the formation of flavored molecules. In this study, four sets of inoculation protocols were evaluated using ripe and immature coffee fruits, respectively, including (i) pure culture fermentation with *Pichia fermentans*, (ii) pure culture fermentation with *Pediococcus acidilactici*, (ii) combined fermentation with *P. fermentans* and *P. acidilactici*, and (iv) spontaneous, non-inoculated control. The initial pulp sugar concentration of ripe coffee fruits (0.57 and 1.13 g/L glucose and fructose content, respectively) was significantly higher than immature coffee pulp (0.13 and 0.26 g/L glucose and fructose content, respectively). Combined inoculation with *P. fermentans* and *P. acidilactici* of ripe coffee beans increased pulp sugar consumption and production of metabolites (lactic acid, ethanol, and ethyl acetate), evidencing a positive synergic interaction between these two microbial groups. On the other hand, when immature coffee fruits were used, only pure culture inoculation with *P. fermentans* was able to improve metabolite formation during fermentation, while combined treatment showed no significant effect. Altogether, 30 volatile compounds were identified and semi-quantified with HS- solid phase microextraction (SPME)-gas chromatography coupled to mass spectrophotometry (GC/MS) in fermented coffee beans. In comparison with pure cultures and spontaneous process, combined treatment prominently enhanced the aroma complexity of ripe coffee beans, with a sharp increase in benzeneacetaldehyde, 2-heptanol, and benzylalcohol. Consistent with the monitoring of the fermentation process, only *P. fermentans* treatment was able to impact the volatile composition of immature coffee beans. The major impacted compounds were 2-hexanol, nonanal, and D-limonene. In summary, this study demonstrated the great potential of the combined use of yeast and lactic acid bacteria to improve fermentation efficiency and to positively influence the chemical composition of coffee beans. Further studies are still required to investigate the mechanisms of synergism between these two microbial groups during the fermentation process and influence the sensory properties of coffee products.

Keywords: coffee processing; coffee fermentation; starter culture; coffee beverage; yeast

---

1. Introduction

Coffee plants are cultivated in more than 80 countries around the world, providing raw materials for a global industry valued at an excess of 10 billion US$ [1]. Production conditions and post-harvest operations, such as fruit harvesting, depulping, drying, and storage, have a direct impact on the quality of coffee products. Fruit harvesting is the first step in postharvest coffee processing. The heterogeneous development of coffee fruits leads to a simultaneous presence of different maturation stages in the same coffee tree, namely: (i) Green (immature) coffee fruits, presenting incomplete endosperm development and low reducing sugars content in the mucilaginous layer; (ii) cherry (ripe) fruits, presenting mucilage rich in reducing sugars, complete development of the endosperm and red or yellow exocarp color; and (iii) raisin (overripe), which are fruits showing the initial characteristics of the senescence cycle with metabolic pathway deviation for the catabolism of the nutrients accumulated in the beans [2]. In Brazil, the largest coffee producer in the world, it is estimated that 31% of the coffee fruits are harvested in the immature stage [3]. Immature coffee beans have a high content of chlorogenic acids (caffeine, trigonelline) and lower sugar content due to incomplete cycle of maturation, attributing astringency and depreciating the quality of coffee products [4,5].

After harvesting, coffee processing must begin as quickly as possible to prevent fruit spoilage by unfavorable fermentation or mold formation [6,7]. The outer layers of the coffee fruit (skin and pulp) are easily removed, while the mucilage, parchment, and silver skin are firmly attached to the beans [8]. The way that coffee growers use to remove the mucilage layer attached to the fruits classifies coffee in the international market: 'Natural coffee', where mucilage is removed by a simple method of sun-drying, known as dry processing; 'washed coffee', produced from coffee beans that undergo a relatively complex series of steps, including depulping, fermentation, and sun-drying known as wet processing; and "pulped natural coffee", which the fruits are mechanically husked and the mucilage is removed by a sun-drying process, known as semi-dry processing [9].

In the wet processing, coffee beans are submitted to underwater tank fermentation for mucilage breakdown and removal. The sugars present in the mucilage support microbial, especially yeasts and lactic acid bacteria [10]. Recent studies have been dedicated to the use of yeast and lactic acid bacteria (LAB) as pure starter cultures in post-harvest processing, in order to reduce the time required for fermentation and modulate the chemical and sensory characteristics of coffee beans [11–14]. Among the selected microorganisms, it is possible to highlight the yeast *Pichia fermentans* YC5.2 and the lactic acid bacteria *Pediococcus acidilactici* LPBC161, which are cultures with characteristics of efficient consumption of coffee pulp-sugars and adaptability to the stress factors of coffee processing [15,16]. Despite that the use of pure cultures offers advantages, recent studies in wine, meat, and dairy fermentations demonstrate that mixed starters are able to improve the sensorial and safety proprieties of the final product [17,18]. In this regard, the aim of this study was to evaluate the effects of co-inoculation with *Pichia fermentans* YC5.2 and *Pediococcus acidilactici* LPBC161 on metabolites produced during fermentation and the volatile composition of coffee beans.

2. Material and Methods

2.1. Microorganism and Inoculum Preparation

The selected yeast (*Pichia fermentans* YC5.2) and lactic acid bacteria (*Pediococcus acidilactici* LPBC161) strains used in this study were previously isolated and selected from spontaneous coffee fermentations, as detailed in Muynarsk et al. [15] and Pereira et al. [16]. The *P. fermentans* YC5.2 and *P. acidilactici* LPBC161 were reactivated in MRS (Merck Millipore, Burlington, MA) and YEPG broth (Himedia, Marg, India), respectively, at 28 °C during 24 h. Each microorganism was then grown up to a concentration of 10^9 CFU/mL. To reach this concentration, *P. acidilactici* LPBC161 was cultivated in Erlenmeyer containing 4 L of sugar cane molasses 3% (w/v) medium, enriched with yeast extract 0.5% (w/v), ammonium citrate 0.5% (w/v), ammonium phosphate 0.5% (w/v), sodium acetate 0.5% (w/v), Tween 80 0.1% (v/v), and manganese sulfate 0.005% (w/v) [19], and *P. fermentans* YC5.2 was grown in Erlenmeyer containing 4 L of sugar cane molasses 3% (w/v) medium enriched with yeast extract 0.5% (w/v). After incubation, the yeast and lactic acid bacteria (LAB) cells were separated from the medium by centrifugation at 5000 × *g* during 5 min, washed twice with sterile saline-peptone solution (0.1% [w/v] bacteriological peptone (Himedia), 0.8% (w/v; NaCl (Merck)), and resuspended in sterile saline solution (0.9% (w/v) NaCl).

2.2. *Farm Experiments*

The field experiments were conducted at the Fazenda Baobá (21°42′42.8′′ S, 46°49′42.2′′ W; 1400 m above sea level) situated in São Sebastião da Grama, São Paulo state, Brazil. Ripe and immature coffee fruits (10 kg) were, respectively, deposited in 20-L plastic buckets with 5 L of water. Four sets of inoculation protocols were performed in triplicate: (i) Pure culture fermentation with *P. fermentans*, (ii) pure culture fermentation with *P. acidilactici*, (ii) combined fermentation with *P. fermentans* and *P. acidilactici*, and (iv) spontaneous, non-inoculated control. Prior to inoculation, yeast and LAB cells were counted by a Thoma hemocytometer chamber using methylene blue dye as a marker of cell viability. Then, appropriate amounts of inoculum were used to reach an initial cell population of about 7 log CFU/mL. At the end of fermentation, coffee beans were sun dried until the value of 12% of moisture was reached.

2.3. *Sampling and pH Measurement*

Samples (50 mL) of the liquid fraction of the fermenting coffee pulp were collected in triplicate at intervals of 12 h to monitor sugars consumption and organic acids, ethanol, and volatile compounds production. At each sampling point, the pH was measured using a digital pH meter (Requipal, Curitiba, Brazil).

2.4. *HPLC Analysis of Fermenting Coffee Pulp*

The concentration of reducing sugars (glucose and fructose), organic acids (citric, succinic, lactic, acetic, and propionic acids), and ethanol in coffee pulp (liquid fraction) was determined in intervals of 12 h. Aliquots of 2 mL were centrifuged at 6000 × g for 15 min and filtered through 0.22 μm pore size filter (Millipore Corp., Billerica, MA, USA). Analysis parameters were performed according to Carvalho Neto et al. [20]. The filtered samples were injected into high-performance liquid chromatograph (HPLC) system equipped with an Aminex HPX 87 H column (300 × 7.8 mm; Bio-Rad, Richmond, CA, USA) and a refractive index (RI) detector (HPG1362A; Hewlett–Packard Company, Palo Alto, CA, USA). The column was eluted in an isocratic mode with a mobile phase of 5 mM H2SO4 at 60 °C and a flow rate of 0.6 mL/min.

2.5. *GC Analysis of Fermenting Coffee Pulp*

The formation of major volatile compounds was determined in intervals of 12 h by gas chromatography. For sample preparation, aliquots (4 mL) from the liquid fraction of the fermenting coffee pulp were placed in 20 mL hermetically sealed flasks containing NaCl 5% (w/v), followed by heating during 10 min at 60 °C. The headspace was then collected using a glass syringe (Hamilton, Bonaduz, Switzerland) and injected into a gas chromatograph (model 17A; Shimadzu, Kyoto, Japan) equipped with a flame ionization detector at 230 °C. The operation conditions were as follows: A 30 m × 0.32 mm HP-5 capillary column, column temperature of 40 to 150 °C at a rate of 20 °C/min [13]. A standard curve was constructed using authentic analytical standards purchased from Sigma and concentration of the compounds was expressed as μmol/L of headspace

2.6. *GC/MS Analysis of Fermented Coffee Beans*

The volatile aroma compound composition of spontaneous and inoculated coffee beans was determined by gas chromatography coupled to mass spectrophotometry (GC-MS) according to Carvalho Neto et al. [20]. The extraction of volatile compounds from the beans samples (2 g) was performed using a headspace vial coupled to a solid phase microextraction (SPME) fiber DVB/CAR/PDMS fiber (Supelco Co., Bellefonte, PA, USA). The flasks were heated at 70 °C for 10 min without agitation, followed by 15 min of exposition of the fiber in a COMBI-PAL system. The compounds were desorbed into the gas chromatograph injection system gas phase (CGMS-gun TQ Series 8040 and 2010 Plus GC-MS; Shimadzu, Tokyo, Japan) at 260 °C. The column oven temperature was maintained at 60 °C during 10 min, followed by two heating ramps of 4 and 10 °C /min until reaching the temperatures of 100 and 200 °C, respectively. The compounds were separated on a column 95% PDMS/5% PHENYL (30 m × 0.25 mm × 0.25 mm film thickness). The GC was equipped with an HP 5972 mass selective detector (Hewlett Packard, Palo Alto, CA, USA). Helium was used as carrier gas at a rate of 1.0 mL/min. Mass spectra were obtained by electron impact at 70 eV and a start and end mass-to-charge ratio (m/z) of 30 and 200, respectively. The compounds were identified by comparison to the mass spectra from library databases (Nist'98 and Wiley7N).

2.7. *Statistical Analysis*

The data obtained of target metabolite analysis were analyzed by post-hoc comparison of means by Duncan's test and a principal component analysis (PCA). Statistical analyses were performed using the SAS program (Statistical Analysis System Cary, NC, USA). Level of significance was established in a two-sided p-value <0.05.

3. Results and Discussion

3.1. *Field Experiment*

The use of mixed fermentation instead a single culture is a practice widely applied in winemaking in order to improve the aroma complexity or mouth-feel of wines [21,22]. It offers a number of advantages over conventional single-culture fermentations, including higher microbial growth rate and metabolite yield, better utilization of the substrate, and complex formation of aromatic compounds [23]. This work represents the first study on a mixed culture in the coffee beans fermentation process. We experimentally tested the impact of the combination of two selected cultures (*P. fermentans* YC5.2 and *P. acidilactici* LPBC161) in terms of the fermentation efficiency and volatile composition of coffee beans. The experiments were performed with individual and combined inoculations in ripe and immature coffee beans, compared to a spontaneous process. The changes in major non-volatiles (sugars, organic acids, and ethanol) and volatiles metabolites were quantified in the course of the fermentation time. The initial pulp sugar concentration of ripe coffee fruits (0.57 and 1.13 g/L glucose and fructose content, respectively) was significantly higher than immature coffee pulp (0.13 and 0.26 g/L glucose and fructose content, respectively; Figure 1). The low levels of sugars in the case of immature coffee are the consequence of incomplete maturation cycle fruits [24]. In all fermentation processes, the sugar concentration showed an increase during the initial 12 h. This phenomenon can be associated with the hydrolysis of pectin, cellulose, sucrose, and other coffee pulp complexes carbohydrates, into monomers of glucose and fructose [25,26]. After this increase, both glucose and fructose were partially consumed, resulting in a residual concentration of around 0.37 and 1.51 g/L (ripe coffee pulp) and 0.19 and 0.74 g/L (immature coffee pulp) of glucose and fructose, respectively. Residual pulp sugars are generally observed after the coffee fermentation process, mainly associated with the short fermentation cycle [14,27,28]. However, fructose consumption was more efficient in the treatments that the yeast was used (i.e., *P. fermentans*-pure culture and combined treatment with *P. fermentans* and *P. acidilactici*) than *P. acidilactici* pure culture and spontaneous assay (Figure 1). The yeast's ability to withstand stress tolerance factors and the production of pectinolytic enzymes confer advantages in comparison to lactic acid bacteria [16,29]. In addition, the low availability of initially willing nutrients in immature coffee pulp [24] may have restricted the development of *P. acidilactici*, showing poor sugar consumption (Figure 1B). Lactic acid bacteria are further distinguished by their limited biosynthetic abilities, being unable to synthesize multiple cofactors, vitamins, purines, pyrimidines, and other nutrients [30].

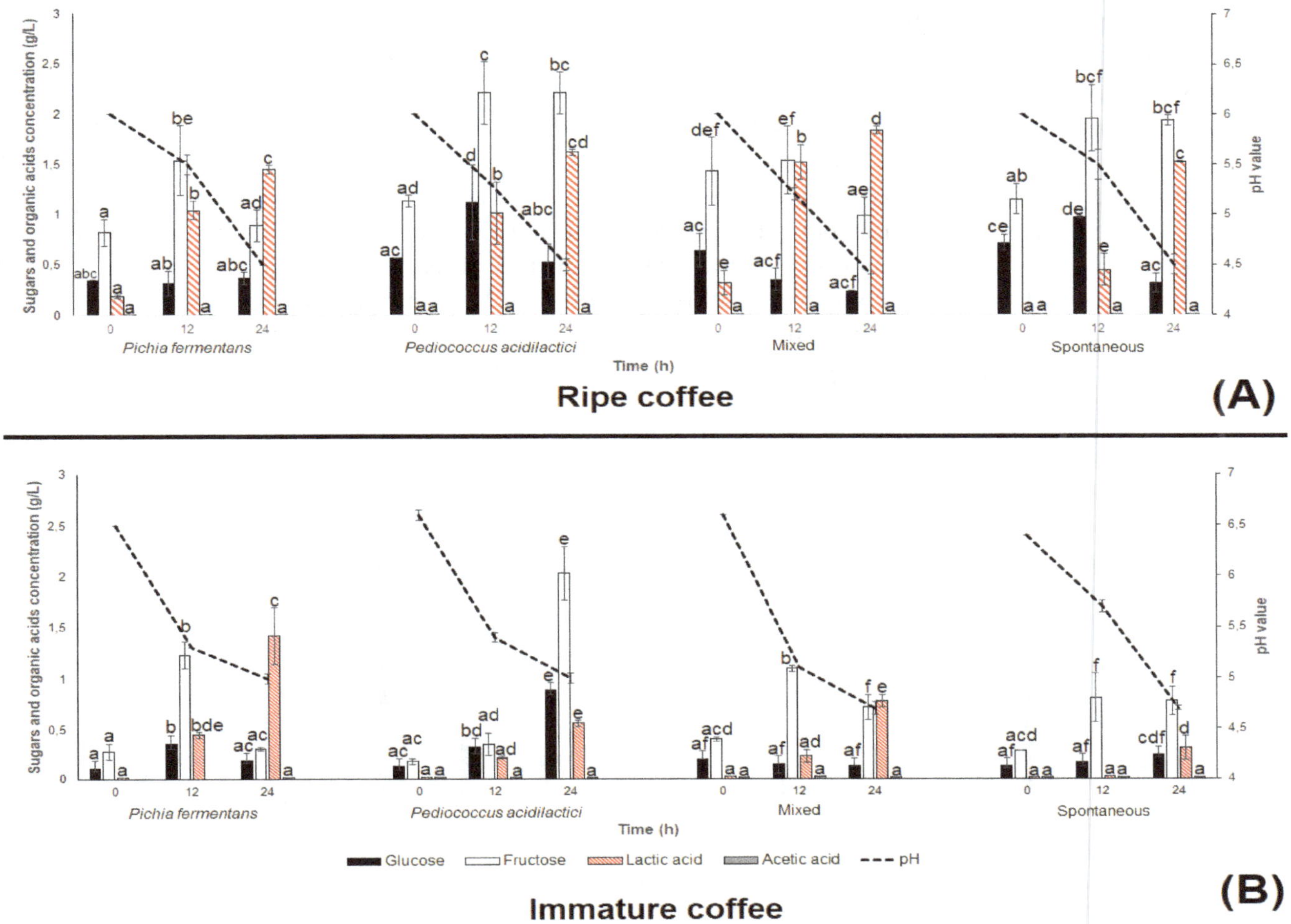

Figure 1. Pulp sugar consumption, organic acid production, and pH monitoring of inoculated (pure culture with *Pichia fermentans*, pure culture with *Pediococcus acidilactici*, and combined fermentation with *P. fermentans* and *P. acidilactici*) and spontaneous fermentation of ripe (**A**) and immature (**B**) coffee fruits. The significance of the results was assessed using an ANOVA with Duncan's post-hoc test at $p < 0.05$. Different lowercase letters (a, b, c, d, e, f) indicate significant differences within the same process (ripe and immature coffee beans) over fermentation time.

Lactic acid showed a significant increase through fermentative processes, reaching maximum concentration of ≥1.45 and ≥0.53 g/L in the ripe and immature treatments, respectively (Figure 1). Basal concentrations of acetic acid (≤0.01 g/L) were detected in all the processes, which can be associated with yeast metabolism or the heterofermentative nature of *P. acidilactici* [31,32]. Overfermentation acids, such as propionic and butyric acids, were not detected in both ripe and immature treatments. In this sense, the acidification of fermenting coffee pulp can be attributed mainly to lactic acid content. As expected, while the lactic acid concentration increased, the pH decreased progressively during all fermentation processes (Figure 1). Lactic acid is an important end-metabolite associated with coffee fermentation, which assists in the coffee-pulp acidification process without interference in the final product (Figure 1). The pH monitoring is a crucial parameter, since pH values below 4.5 are used as to indicate the end of the coffee fermentation process [33–35]. In immature coffee treatments, a pH higher than 4.5 was reported, which may be attributed to the insufficient development of *P. acidilactici*.

Ethanol and ethyl acetate were the major volatile compounds detected during fermentation processes (Figure 2). Interestingly, combined inoculations with *P. acidilactici* and *P. fermentans* resulted in a significant increase in the production of these metabolites when compared to pure cultures and a spontaneous process. The higher values of ethanol (27.04 and 14.8 μmol/L) and ethyl acetate (1.63 and 1.21 μmol/L) were reached after 24 h of ripe and immature combined fermentations, respectively. This agrees with the findings of Sun et al. [36] and Cañas et al. [37] that demonstrated an increase of ethyl- and acetate esters as a result of the co-inoculation of LAB and yeasts in wine fermentation. The significant sugar consumption and lactic acid, ethyl acetate, and ethanol production in treatments with combined inoculations indicated an ecological interaction between these two microbial groups. The complex nature of this interaction is highlighted by the observations that (i)the autolysis of yeasts release nutrients, such as amino acids, polysaccharides and riboflavin, favorable for bacterial growth, and that (ii) the acidification of the fermentation media by LAB creates a prone environment for yeast development [9,38,39]. These positive interactions have been shown to promote desired sensory attributes in wine, sourdough, and yogurt. However, information about these mechanisms in coffee fermentation is scarce [40].

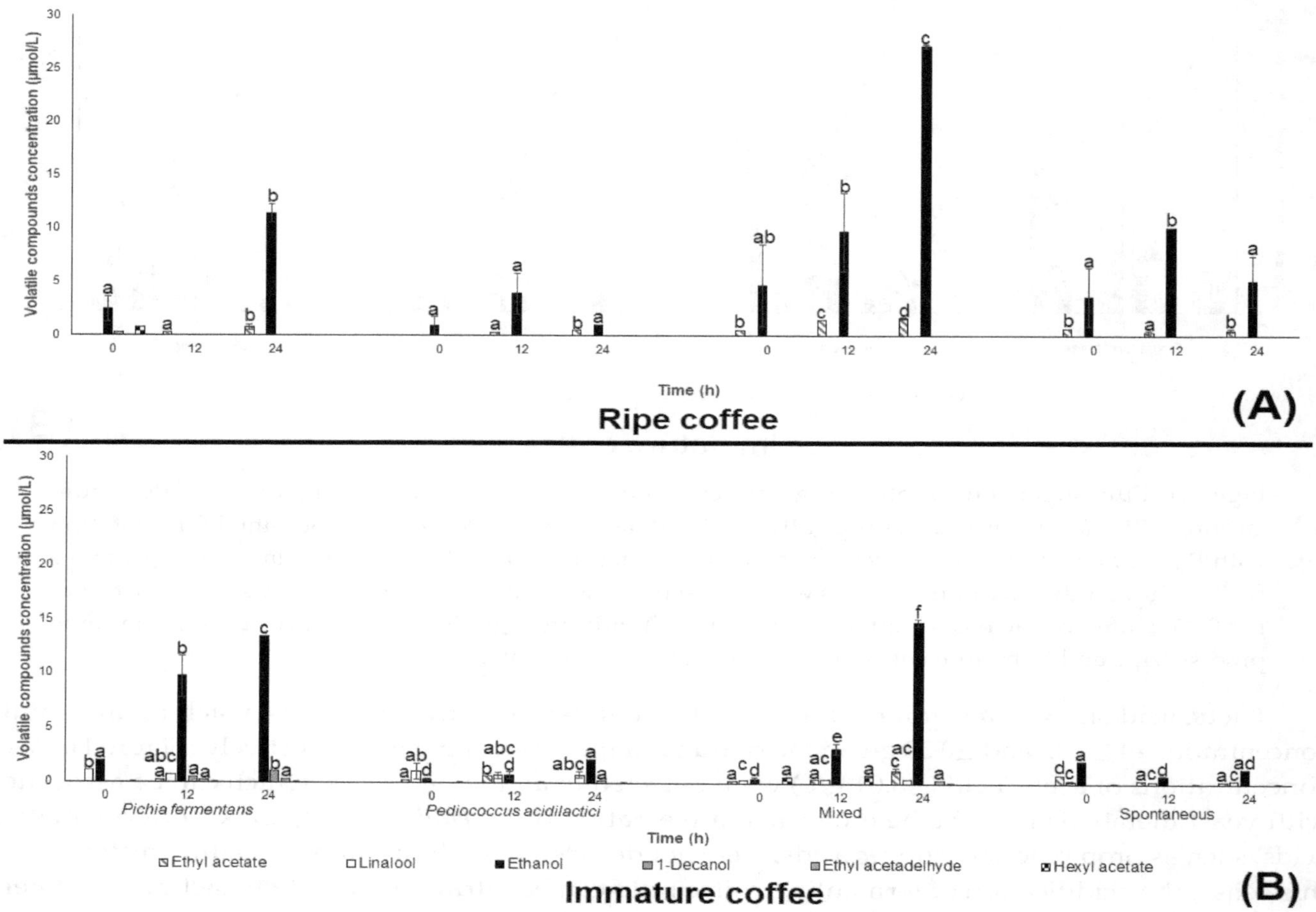

Figure 2. Concentration of volatile compounds produced in the course inoculated (pure culture with *P. fermentans*, pure culture with *P. acidilactici*, and combined fermentation with *P. fermentans* and *P. acidilactici*) and spontaneous fermentation of ripe (**A**) and immature (**B**) coffee fruits. The significance of the results was assessed using an ANOVA with Duncan's post-hoc test at $p < 0.05$. Different lowercase letters (a, b, c, d, e, f) indicate significant differences within the same process (ripe and immature coffee beans) over fermentation time.

Other minor volatile compounds that increased during the fermentation processes were 1-decanol, ethyl-acetaldehyde, and hexyl acetate (Figure 2). Yeast and lactic acid bacteria generate ethyl-acetaldehyde by a condensation reaction between fatty acids and an alcohol molecule [41], while 1-decanol can be derived from amino acid catabolism via the Ehrlich pathway [16,42]. The presence of higher concentrations of linalool using immature fruits can be associated with the inferior maturation stage of the coffee beans, since this compound has been considered a volatile marker of coffee beverages produced from immature fruits [43].

3.2. *Coffee Beans Chemical Composition*

Over 900 volatile compounds have already been identified in green and roasted coffee beans [27,44]. Among the major volatiles found, pyrazines, furans, ketones, aldehydes, higher alcohols, esters, and sulphur compounds can be highlighted [45–47]. Although some of these flavor-active compounds originate from the beans itself, recent studies have revealed that microbial-derived metabolites can also diffuse into the beans [11,14,27,33,48–50]. Upon characterization of the volatile composition of fermented coffee beans, it was observed that inoculation of *P. fermentans* and *P. acidilactici*, both in pure and combined treatments, resulted in the modulation of the volatile constitution of coffee beans. A total of 30 compounds were identified in fermented ripe coffee beans, including higher alcohols (seven compounds), aldehydes (six compounds), and terpenes (three compounds; Table 1). Among these, 1-hexanol, 2-heptanol, phenylethyl alcohol, and benzeneacetaldehyde were the major volatiles found. Single inoculation of *P. fermentans* and *P. acidilactici* resulted in the formation and diffusion of some volatile compounds, such as 3-octanol, 2-heptenal, benzaldehyde, dodecanal, and D-limonene, that were not detected in a spontaneous process. These compounds are strictly related to both yeast and LAB metabolism, such as aldehydes and higher alcohols formed from the catabolism of coffee pulp amino acids, and terpenes through mevalonic acid pathway or released from glycoside precursors during fermentation [10,51,52].

Table 1. Concentration of volatile compounds (Area*10^5) in ripe coffee beans after single cultures, a combined treatment, and a spontaneous assay. Means of triplicate in each row bearing the same lowercase letters (a, b) are not significantly different ($p > 0.05$) from one another using Duncan's Test (mean ± standard variation).

Compounds	Aroma Perception	Fermented Ripe Coffee Beans			
		Spontaneous	*P. fermentans*	*P. acidilactici*	*P. fermentans* + *P. acidilactici*
Organic acids (3)					
Butanoic acid, 3-methyl	-	7.02 ± 1.00 ^{a}	6.88 ± 1.48 ^{a}	6.10 ± 1.74 ^{a}	6.91 ± 0.18 ^{a}
Butanoic acid, 2-methyl	Fruity, dirty, acidic with a dairy buttery	1.77 ± 0.35 ^{a}	1.98 ± 0.28 ^{a}	ND	1.79 ± 0.28 ^{a}
Hexanoic acid	Sour, fatty, sweat, cheesy	ND	1.34 ± 0.61 ^{a}	0.99 ± 0.23 ^{a}	1.67 ± 0.88 ^{a}
Higher alcohols (7)					
2-Heptanol	Fresh, lemon grass, herbal	10.21 ± 0.00 ^{a}	ND	ND	12.12 ± 0.10 ^{b}
5-Methyl-2-Hexanol	-	6.09 ± 2.66	ND	ND	ND
1-Hexanol	Green, fruity, apple-skin and oily	15.90± 2.36 ^{a}	16.95 ± 2.23 ^{a}	17.10 ± 2.36 ^{a}	18.24 ± 0.17 ^{a}
1-Octen-3-ol	Earthy, green, oily, umami sensation	0.82 ± 0.00 ^{a}	0.80 ± 0.17 ^{a}	ND	0.86 ± 0.03 ^{a}
3-Octanol	Musty, mushroom, earthy, creamy dairy	ND	0.50 ±0.14 ^{a}	0.58 ± 0.00 ^{a}	0.67 ± 0.00 ^{a}
Benzylalcohol	Sweet, fruity with balsamic nuances	2.58 ± 0.05 ^{a}	3.26 ± 0.36 ^{a}	3.20 ± 0.68 ^{a}	4.31 ± 0.17 ^{b}
Phenylethyl alcohol	Floral, sweet and bready	9.52 ± 0.73 ^{a}	10.34 ± 0.41 ^{a}	12.17 ± 0.25 ^{b}	12.60 ± 1.31 ^{b}
Esters (3)					
Butanoic acid, 2-methyl, ethyl ester	-	0.58 ± 0.26 ^{a}	0.34 ± 0.00 ^{a}	ND	0.54 ± 0.20 ^{a}
Butanoic acid, 3-methyl- ethyl ester	-	4.62 ± 0.59 ^{a}	3.66 ± 1.71 ^{a}	ND	4.70 ± 0.57 ^{a}
Methyl salicylate	Wintergreen, mint-like	ND	0.28 ± 0.01	ND	ND
Aldehydes (6)					
2-Heptenal	Sweet, fresh fruity apple skin nuances	ND	0.80 ± 0.00 ^{a}	ND	0.80 ± 0.00 ^{a}
Benzaldehyde	Fruity, cherry	ND	0.32 ± 0.00 ^{a}	2.15 ± 0.84 ^{a}	2.56 ± 0.69 ^{a}
Dodecanal	Soapy, waxy, citrus, orange mandarin	ND	ND	0.23 ± 0.14	ND
Nonanal	With a fresh green lemon peel-like nuance	0.95 ± 0.05 ^{a}	0.93 ± 0.35 ^{a}	0.81 ± 0.06 ^{a}	0.84 ± 0.27 ^{a}
Benzeneacetaldehyde	Almond, fruity, powdery, nutty	1.83 ± 0.29 ^{a}	2.80 ± 0.73 ^{a}	2.28 ± 0.04 ^{a}	6.94 ± 0.00 ^{b}
Decanal	Sweet, aldehydic, orange, waxy and citrus rind	ND	0.37 ± 0.11 ^{a}	0.37 ± 0.09 ^{a}	0.34 ± 0.07 ^{a}
Ketone (1)					
2-Heptanone	Fruity, spice, herbal	3.49 ± 0.69 ^{a}	2.92 ± 0.28 ^{a}	2.29 ± 0.90 ^{a}	4.38 ± 1.75 ^{a}

Table 1. *Cont.*

Compounds	Aroma Perception	Fermented Ripe Coffee Beans			
		Spontaneous	***P. fermentans***	***P. acidilactici***	***P. fermentans + P. acidilactici***
Pyridine (2)					
Pyridine, 2,3-dimethyl	-	ND	1.55 ± 0.94^{a}	1.44 ± 0.26^{a}	1.25 ± 0.63^{a}
Pyridine, 2,6-Lutidine	Nutty, amine, woody, bready and vegetable-like	1.85 ± 0.19	ND	ND	ND
Lactone (1)					
Butyrolactone	Creamy, oily, fatty, caramellic	5.28 ± 0.10^{a}	6.17 ± 1.04^{a}	4.86 ± 1.13^{a}	6.09 ± 0.13^{a}
Terpenes (3)					
Linalool	Citrus, orange, lemon	2.36 ± 0.29^{a}	2.27 ± 0.21^{a}	2.22 ± 0.82^{a}	2.92 ± 0.17^{b}
D-Limonene	Sweet, orange, citrus	ND	1.29 ± 0.60^{a}	1.37 ± 0.49^{a}	1.42 ± 0.35^{a}
Anethole	-	4.10 ± 1.00^{b}	1.87 ± 0.26^{a}	1.96 ± 0.66^{a}	$3.13 \pm 0.20^{a,b}$
Hydrocarbons (2)					
Styrene	Sweet, balsamic, floral	ND	2.89 ± 0.00^{a}	ND	3.15 ± 0.00^{a}
Tetradecane	Waxy	0.91 ± 0.07^{a}	0.82 ± 0.06^{a}	0.88 ± 0.16^{a}	0.80 ± 0.06^{a}
Pyrzine (1)					
Pyrazine, 2-methoxy-3-(2-methylpropyl)	Roasted almond hazelnut peanut	0.92 ± 0.04^{a}	ND	0.93 ± 0.08^{a}	ND
Furan (1)					
Furan, 2-pentyl	Waxy, with musty, cooked caramellic nuances	1.16 ± 0.04	ND	ND	ND

Interestingly, coffee beans generated from combined treatments showed significantly increased ($p < 0.05$) of specific volatile compounds, such as benzeneacetaldehyde, 2-heptanol, and benzylalcohol. These findings are in accordance with Englezos et al. [18] and Plessas et al. [53], which evidences that mixed treatments of yeast and LAB starter cultures enable higher production of esters, aldehydes, and higher alcohols in sourdough and wine fermentations when compared to single inoculations. However, further studies are required to evidence the metabolic pathways associated with the positive interaction between these microorganisms in coffee products.

Chemical analysis of immature coffee beans revealed a composition with lower diversity and concentration of volatile compounds (Table 2). A total of 19 compounds were detected, including higher alcohols (four compounds), organic acids (three compounds), and aldehydes (three compounds). *P. fermentans*-single inoculation resulted in coffee beans with significantly higher concentrations ($p < 0.05$) of 2-hexanol, nonanal, and D-limonene when compared to the spontaneous process. These compounds are commonly attributed to *Pichia* metabolism [10,54,55]. This corroborates with results from fermentation process monitoring, which demonstrated intense microbial activity of *P. fermentans* in immature coffee pulp. On the other hand, coffee beans derived from *P. acidilactici*-pure culture and combined treatment showed no significant increase ($p > 0.05$) in the volatile constituents when compared to the control (spontaneous process). This fact can be correlated to the insufficient growth of the LAB starter culture in the nutrient-scarce environment from the pulp of immature coffee beans. The auxotrophism of several amino acids turns LAB directly dependent on a rich growth medium for its full development [31].

Table 2. Concentration of volatile compounds (Area*10^5) in immature coffee beans after single cultures, a combined treatment, and a spontaneous assay. Means of triplicate in each row bearing the same lowercase letters (a, b, c) are not significantly different ($p > 0.05$) from one another using Duncan's Test (mean ± standard variation).

Compounds	Aroma Perception	Fermented Immature Coffee Beans			
		Spontaneous	*P. fermentans*	*P. acidilactici*	*P. fermentans* + *P. acidilactici*
Organic acids (3)					
Butanoic acid, 3-methyl	-	7.77 ± 1.05^{a}	5.99 ± 2.31^{a}	$8.56 \pm 1.57^{a,b}$	$5.15 \pm 0.98^{a,c}$
Butanoic acid, 2-methyl	Fruity, acidic with a dairy buttery	1.32 ± 0.36^{a}	1.24 ± 0.86^{a}	$1.65 \pm 0.13^{a,b}$	$0.69 \pm 0.15^{a,c}$
Hexanoic acid	Sour, fatty, sweat, cheesy	ND	0.63 ± 0.15^{a}	0.50 ± 0.11^{a}	ND
Higher alcohols (4)					
1-Hexanol	Green, fruity, apple-skin and oily	10.02 ± 0.26^{a}	10.57 ± 1.33^{a}	9.14 ± 1.51^{a}	9.85 ± 1.17^{a}
2-Hexanol	-	0.36 ± 0.05^{a}	0.75 ± 0.01^{b}	ND	0.45 ± 0.10^{a}
Benzylalcohol	Sweet, fruity with balsamic nuances	0.54 ± 0.15^{a}	0.62 ± 0.08^{a}	0.51 ± 0.20^{a}	0.61 ± 0.04^{a}
Phenylethyl alcohol	Floral, sweet and bready	2.54 ± 0.38^{a}	2.78 ± 0.17^{b}	2.24 ± 0.12^{a}	2.20 ± 0.18^{a}
Aldehydes (3)					
Nonanal	Citrus, with a fresh green lemon peel-like nuance	0.30 ± 0.08^{a}	0.74 ± 0.04^{b}	$0.46 \pm 0.07^{c,d}$	0.48 ± 0.05^{d}
Benzeneacetaldehyde	Almond, fruity, powdery, nutty	0.21 ± 0.00^{a}	0.28 ± 0.01^{a}	0.27 ± 0.13^{a}	0.38 ± 0.00^{a}
Decanal	Sweet, orange, waxy and citrus rind	0.24 ± 0.09^{a}	0.43 ± 0.06^{a}	0.30 ± 0.10^{a}	0.42 ± 0.08^{a}
Pyridines (2)					
Pyridine, 2,3-dimethyl	-	ND	1.45 ± 0.34^{a}	0.62 ± 0.14^{a}	1.10 ± 0.36^{a}
Pyridine, 2,6-Lutidine	Nutty, woody, bready and vegetable-like	0.95 ± 0.35	ND	ND	ND
Lactone (1)					
Butyrolactone	Creamy, oily, fatty, caramellic	0.92 ± 0.07^{a}	0.46 ± 0.20^{a}	0.66 ± 0.26^{a}	0.49 ± 0.17^{a}
Terpenes (1)					
D-Limonene	Sweet, orange, citrus	0.36 ± 0.06^{a}	0.72 ± 0.16^{b}	0.16 ± 0.09^{c}	0.41 ± 0.19^{a}

Table 2. *Cont.*

Compounds	Aroma Perception	Fermented Immature Coffee Beans			
		Spontaneous	***P. fermentans***	***P. acidilactici***	***P. fermentans* + *P. acidilactici***
Furans (2)					
Furan-2-pentyl	Waxy, cooked caramellic nuances	0.57 ± 0.17^{a}	0.74 ± 0.12^{a}	0.58 ± 0.15^{a}	0.69 ± 0.13^{a}
2(3)-Furanone, dihydro-5-methyl	Creamy, waxy with a citrus fruity nuance	0.55 ± 0.14^{a}	0.38 ± 0.16^{a}	ND	0.45 ± 0.15^{a}
Hydrocarbons (2)					
Hexadecane	-	ND	0.55 ± 0.01	ND	ND
Tetradecane	Waxy	0.62 ± 0.02^{a}	0.24 ± 0.07^{b}	0.30 ± 0.06^{c}	$0.76 \pm 0.10^{a,c}$
Pyrzine (1)					
Pyrazine, 2-methoxy-3-(2-methylpropyl)	Roasted almond hazelnut peanut	0.78 ± 0.07^{a}	0.80 ± 0.01^{a}	0.85 ± 0.12^{a}	0.92 ± 0.08^{a}

In order to explain the chemical characteristics and grouping of the samples, the parameters in Tables 1 and 2 were analyzed by a PCA (Figure 3). The first and second principal components explained, together, 73.66% of the total variability within the data. The samples were categorized into two clusters, *viz.*, ripe and immature coffee beans. This distinction was mainly related to the richer constitution of volatiles of ripe coffee beans relative to immature treatments. In addition, the presence of specific compounds (benzyl alcohol, phenylethyl alcohol, benzeneacetaldehyde, decanal, and D-limonene in ripe coffee beans, and furan-2-pentyl, 2-methyl-butanoic acid, and pyrazine, 2-methoxy-3-(2-methylpropyl) in immature coffee beans) also contributed to the separation of ripe and immature coffee beans in the PCA analysis. Interestingly, only the treatment with *P. fermentans* grouped immature coffee beans in the positive axis, which corroborates with the better yeast' adaptation and generation of volatiles in immature coffee pulp.

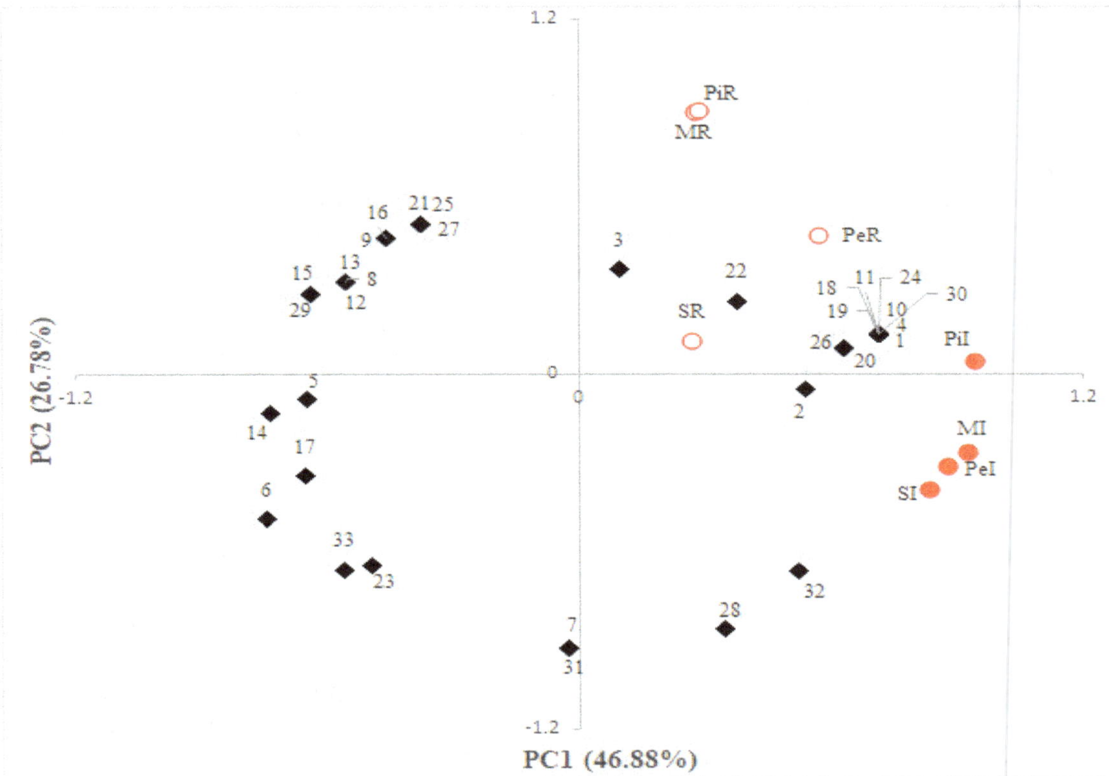

Figure 3. Principal component analysis (PCA) of volatile compounds (lozenges) identified in the different treatments of ripe (open circles) and immature (closed circles) coffee beans. Abbreviations: *SR*—spontaneous, ripe control; *PiR*—*Pichia* inoculation in ripe coffee beans; *PeR*—*Pediococcus* inoculation in ripe coffee beans; *MR*—mixed (*Pichia* plus *Pediococcus*) inoculation in ripe coffee beans; *SI*—spontaneous, immature control; *PiI*—*Pichia* inoculation in immature coffee beans; *PeI*—*Pediococcus* inoculation in immature coffee beans; *MI*—mixed (*Pichia* plus *Pediococcus*) inoculation in immature coffee beans. 1—butanoic acid, 3-methyl; 2—butanoic acid, 2-methyl; 3—hexanoic acid; 4—1-hexanol; 5—2-heptanol; 6—5-methyl-2-hexanol; 7—2-hexanol; 8—1-octen-3-ol; 9—3-octanol; 10—benzylalcohol; 11—phenylethyl alcohol; 12—butanoic acid, 2-methyl, ethyl ester; 13 - butanoic acid, 2-methyl, ethyl ester; 14—methyl salicylate; 15—2-heptenal; 16—benzaldehyde; 17—dodecanal; 18—nonanal; 19—benzeneacetaldehyde; 20—decanal; 21—2-heptanone; 22—pyridine, 2,3-dimethyl; 23—pyridine, 2,6-lutidine; 24—butyrolactone; 25—linalool; 26—D-limonene; 27—anethole; 28—furan-2-pentyl; 29—styrene; 30—tetradecane; 31—2(3)-furanone, dihydro-5-methyl; 32—pyrazine, 2-methoxy-3-(2-methylpropyl); 33—hexadecane.

4. Conclusions

This is the first study investigating the impact of co-inoculation with yeast and LAB on the fermentation of ripe and immature coffee fruits. Among the different treatments, combined inoculations with *Pichia fermentans* YC5.2 and *Pediococcus acidilactici* LPBC161 in ripe coffee fruits showed interesting features. It was possible to reach increased coffee pulp-sugar consumption and production of metabolites (lactic acid, ethanol, and ethyl acetate), evidencing a positive synergic interaction between these two microbial groups. On the other hand, when using immature coffee fruits, only *Pichia fermentans* was able to improve metabolite formation during fermentation and impact volatile composition of resulting coffee beans. This may be due to the high nutritional requirement of LAB species and poor adaptability in immature coffee pulp. Howsoever, since immature coffee beans usually have a low quality because the formation of flavored precursors is incomplete, yeast metabolism has great potential to add flavor quality to these beans.

Chemical analysis revealed a more complex volatile composition of fermented coffee beans from combined treatment in relation to pure inoculations and spontaneous process. The major compounds impacted were benzeneacetaldehyde, 1-hexanol, benzylalcohol, 2-heptanol, and phenylethyl alcohol, which are reported as important aroma-impacting compounds. Thus, this study shows the great potential of combined inoculation for the formation of desirable aroma compounds and production of specialty coffees. Further studies are still required to investigate the mechanisms of synergism between yeast and LAB and influence on sensory properties of coffee products.

Author Contributions: C.R.S. and G.V.d.M.P. designed the experiments. C.R. conducted the physicochemical characterization of fermenting coffee pulp. A.d.S.V. and D.P.d.C.N. performed the experiments and wrote the manuscript. C.R.S., G.V.d.M.P., and M.G.B.P., reviewed and edited the manuscript. All authors read and approved the final manuscript.

References

1. Huch, M.; Franz, C.M.A.P. Coffee: Fermentation and microbiota. In *Advances in Fermented Foods and Beverages: Improving Quality, Technologies and Health Benefits*, 1st ed.; Holzapfel, W., Ed.; Woodhead Publishing: Sawston, UK, 2015; Volume 4, pp. 501–513.
2. DaMatta, F.M.; Ronchi, C.P.; Maestri, M.; Barros, R.S. Ecophysiology of coffee growth and production. *Braz. J. Plant Physiol.* **2007**, *19*, 485–510. [CrossRef]
3. Mesquita, C.M.; Rezende, J.E.; De Carvalho, J.S.; Junior, M.A.F.; Moraes, N.C.; Dias, P.T.; Carvalho, R.M.; de Araújo, W.G. *Manual do café: Colheita e preparo*; Emater: Belo Horizonte, Brazil, 2016; pp. 6–21.
4. Franca, A.S.; Oliveira, L.S.; Mendonça, J.C.F.; Silva, X.A. Physical and chemical attributes of defective crude and roasted coffee beans. *Food Chem.* **2005**, *90*, 89–94. [CrossRef]
5. Oliveira, L.S.; Franca, A.S.; Mendonça, J.C.F.; Barros-Júnior, M.C. Proximate composition and fatty acids profile of green and roasted defective coffee beans. *LWT Food Sci. Technol.* **2006**, *39*, 235–239. [CrossRef]
6. Bee, S.; Brando, C.H.J.; Brumen, G.; Carvalhaes, N.; Kölling-Speer, I.; Speer, K.; Liverani, F.S.; Teixeira, A.A.; Thomaziello, R.A.; Viani, R.; et al. The raw coffee. In *Expresso Coffee: The Science of Quality*; Illy, A., Vinai, R., Eds.; Elsevier Academic Press: London, UK, 2005; pp. 87–178.
7. Illy, E. The complexity of coffee. *Sci. Am.* **2002**, *286*, 86–91. [CrossRef] [PubMed]
8. De Bruyn, F.; Zhang, S.J.; Pothakos, V.; Torres, J.; Lambot, C.; Moroni, A.V.; Callanan, M.; Sybesma, W.; Weckx, S.; De Vuyst, L. Exploring the impacts of postharvest processing on the microbiota and metabolite profiles during green coffee bean production. *Appl. Environ. Microbiol.* **2017**, *83*, e02398-16. [CrossRef] [PubMed]
9. Pereira, G.V.M.; Soccol, V.T.; Brar, S.K.; Neto, E.; Soccol, C.R. Microbial ecology and starter culture technology in coffee processing. *Crit. Rev. Food Sci. Nutr.* **2017**, *57*, 2775–2788. [CrossRef] [PubMed]
10. Pereira, G.V.M.; Carvalho Neto, D.P.; Júnior, A.I.M.; Vásquez, Z.S.; Medeiros, A.B.P.; Vandenberghe, L.P.S.; Soccol, C.R. Exploring the impacts of postharvest processing on the aroma formation of coffee beans—A review. *Food Chem.* **2019**, *272*, 441–452. [CrossRef] [PubMed]

11. Wang, C.; Sun, J.; Lassabliere, B.; Yu, B.; Zhao, F.; Zhao, F.; Chen, Y.; Liu, S.Q. Potential of lactic acid bacteria to modulate coffee volatiles and effect of glucose supplementation: Fermentation of green coffee beans and impact of coffee roasting. *J. Sci. Food Agric.* **2019**, *99*, 409–420. [CrossRef]
12. Lee, L.W.; Cheong, M.W.; Curran, P.; Yu, B.; Liu, S.Q. Modulation of coffee aroma via the fermentation of green coffee beans with *Rhizopus oligosporus*: I. Green coffee. *Food Chem.* **2016**, *211*, 916–924. [CrossRef]
13. Pereira, G.V.M.; Carvalho Neto, D.P.; Medeiros, A.B.P.; Soccol, V.T.; Neto, E.; Woiciechowski, A.L.; Soccol, C.R. Potential of lactic acid bacteria to improve the fermentation and quality of coffee during on-farm processing. *Int. J. Food Sci. Technol.* **2016**, *51*, 1689–1695. [CrossRef]
14. Pereira, G.V.M.; Neto, E.; Soccol, V.T.; Medeiros, A.B.P.; Woiciechowski, A.L.; Soccol, C.R. Conducting starter culture-controlled fermentations of coffee beans during on-farm wet processing: Growth, metabolic analyses and sensorial effects. *Food Res. Int.* **2015**, *75*, 348–356. [CrossRef] [PubMed]
15. Muynarsk, E.S.M.; Pereira, G.V.M.; Mesa, D.; Thomaz-Soccol, V.; Carvalho, J.C.; Pagnoncelli, M.G.B.; Soccol, C.R. Draft genome sequence of *Pediococcus acidilactici* strain LPBC161, isolated from mature coffee cherries during natural fermentation. *Microbiol. Resour. Announc.* **2019**, *8*, e00332-19. [CrossRef] [PubMed]
16. Pereira, G.V.M.; Soccol, V.T.; Pandey, A.; Medeiros, A.B.P.; Lara, J.M.R.A.; Gollo, A.L.; Soccol, C.R. Isolation, selection and evaluation of yeasts for use in fermentation of coffee beans by the wet process. *Int. J. Food Microbiol.* **2014**, *188*, 60–66. [CrossRef] [PubMed]
17. Xia, X.; Luo, Y.; Zhang, Q.; Huang, Y.; Zhang, B. Mixed starter culture regulates biogenic amines formation via decarboxylation and transamination during Chinese rice wine fermentation. *J. Agric. Food Chem.* **2018**, *66*, 6348–6356. [CrossRef] [PubMed]
18. Englezos, V.; Torchio, F.; Cravero, F.; Marengo, F.; Giacosa, S.; Gerbi, V.; Rantsiou, K.; Rolle, L.; Cocolin, L. Aroma profile and composition of Barbera wines obtained by mixed fermentations of *Starmerella bacillaris* (synonym *Candida zemplinina*) and *Saccharomyces cerevisiae*. *LWT Food Sci. Technol.* **2016**, *73*, 567–575. [CrossRef]
19. Feltrin, V.P.; Sant'Anna, E.S.; Porto, A.C.S.; Torres, R.C.O. Produção de Lactobacillus plantarum em melaço de cana-de-açûcar. *Braz. Arch. Biol. Technol.* **2000**, *43*, 119–124. [CrossRef]
20. Carvalho Neto, D.P.; Pereira, G.V.M.; Tanobe, V.O.A.; Thomaz-Soccol, V.; da Silva, B.G.J.; Rodrigues, C.; Soccol, C.R. Yeast diversity and physicochemical characteristics associated with coffee bean fermentation from the Brazilian cerrado mineiro region. *Fermentation* **2017**, *3*, 11. [CrossRef]
21. Gobbi, M.; Comitini, F.; Domizio, P.; Romani, C.; Lencioni, L.; Mannazzu, I.; Ciani, M. *Lachancea thermotolerans* and *Saccharomyces cerevisiae* in simultaneous and sequential co-fermentation: A strategy to enhance acidity and improve the overall quality of wine. *Food Microbiol.* **2013**, *33*, 271–281. [CrossRef]
22. Cinai, M.; Comitini, F.; Mannazzu, I.; Domizio, P. Controlled mixed culture fermentation: A new perspective on the use of non-*Saccharomyces* yeasts in winemaking. *FEMS Yeast Res.* **2010**, *10*, 123–133. [CrossRef]
23. Gaden, E.L., Jr.; Bokanga, M.; Harlander, S.; Hesseltine, C.W.; Steinkraus, K.H. *Applications of Biotechnology to Traditional Fermented Foods*; National Academy Press: Washington, DC, USA, 1992.
24. Eira, M.T.S.; da Silva, E.A.A.; de Castro, R.D.; Dussert, S.; Walters, C.; Bewley, J.D.; Hilhorst, W.M. Coffee seed physiology. *Braz. J. Plant Physiol.* **2006**, *18*, 149–163. [CrossRef]
25. Marques, W.L.; Raghavendran, V.; Stambuk, B.U.; Gombert, A.K. Sucrose and *Saccharomyces cerevisiae*: A relationship most sweet. *FEMS Yeast Res.* **2016**, *16*, fov107. [CrossRef] [PubMed]
26. Murthy, P.S.; Naidu, M. Improvement of Robusta coffee fermentation with microbial enzymes. *Eur. J. Appl. Sci.* **2011**, *3*, 130–139.
27. Evangelista, S.R.; Miguel, M.G.C.P.; Cordeiro, C.S.; Silva, C.F.; Pinheiro, A.C.M.; Schwan, R.F. Inoculation of starter cultures in a semi-dry coffee (*Coffea arabica*) fermentation process. *Food Microbiol.* **2014**, *44*, 87–95. [CrossRef] [PubMed]
28. Evangelista, S.R.; Silva, C.F.; Miguel, M.G.P.C.; Cordeiro, C.S.; Pinheiro, A.C.M.; Duarte, W.F.; Schwan, R.F. Improvement of coffee beverage quality by using selected yeasts strains during the fermentation in dry process. *Food Res. Int.* **2014**, *61*, 183–195. [CrossRef]
29. Avallone, S.; Brillouet, J.M.; Guyot, B.; Olguin, E.; Guiraud, J.P. Involvement of pectolytic micro-organisms in coffee fermentation. *Int. J. Food Sci. Technol.* **2002**, *37*, 191–198. [CrossRef]
30. Muñoz, R.; Moreno-Arribas, M.V.; de las Rivas, B. Lactic acid bacteria. In *Molecular Wine Microbiology*, 1st ed.; Carrascosa, A.V., Muñoz, R., González, R., Eds.; Academic Press: Burlington, VT, USA, 2011; pp. 191–226.

31. Endo, A.; Dicks, L.M.T. Physiology of the LAB. In *Lactic Acid Bacteria: Biodiversity and Taxonomy*; Holzapfel, W.H., Wood, B.J.B., Eds.; Wiley Blackwell: Chichester, UK, 2014; pp. 13–30.
32. Rantsiou, K.; Dolci, P.; Giacosa, S.; Torchio, F.; Tofalo, R.; Torriani, S.; Suzzi, G.; Rolle, L.; Cocolin, L. *Candida zemplinina* can reduce acetic acid produced by *Saccharomyces cerevisiae* in sweet wine fermentations. *Appl. Environ. Microbiol.* **2012**, *78*, 1987–1994. [CrossRef] [PubMed]
33. Carvalho Neto, D.P.; Pereira, G.V.M.; Finco, A.M.O.; Letti, L.A.J.; Silva, B.J.G.; Vandenberghe, L.P.S.; Soccol, C.R. Efficient coffee beans mucilage layer removal using lactic acid fermentation in a stirred-tank bioreactor: Kinetic, metabolic and sensorial studies. *Food Biosci.* **2018**, *26*, 80–87. [CrossRef]
34. Velmourougane, K. Impact of natural fermentation on physicochemical, microbiological and cup quality characteristics of Arabica and Robusta coffee. *Proc. Natl. Acad. Sci. USA India Sect. B Biol. Sci.* **2013**, *83*, 233–239. [CrossRef]
35. Jackels, S.C.; Jackels, C.F. Characterization of the coffee mucilage fermentation process using chemical indicators: A field study in Nicaragua. *Food Chem. Toxicol.* **2005**, *70*, 321–325. [CrossRef]
36. Sun, S.Y.; Gong, H.S.; Zhao, K.; Wang, X.L.; Wang, X.; Zhao, X.H.; Yu, B.; Wang, H.X. Co-inoculation of yeast and lactic acid bacteria to improve cherry wines sensory quality. *Int. J. Food Sci. Technol.* **2013**, *48*, 1783–1790. [CrossRef]
37. Cañas, P.M.I.; Romero, E.G.; Pérez-Martín, F.; Seseña, S.; Palop, M.L. Sequential inoculation versus co-inoculation in Cabernet Franc wine fermentation. *Food Sci. Technol. Int.* **2015**, *21*, 203–212. [CrossRef] [PubMed]
38. Alexandre, H.; Guilloux-Benatier, M. Yeast autolysis in sparkling wine—A review. *Aust. J. Grape Wine Res.* **2006**, *12*, 119–127. [CrossRef]
39. Fleet, G.H. Yeast interactions and wine flavour. *Int. J. Food Microbiol.* **2003**, *86*, 11–22. [CrossRef]
40. Junqueira, A.C.O.; Pereira, G.V.M.; Medina, J.D.C.; Alvear, M.C.R.; Rosero, R.; Carvalho Neto, D.P.; Enríquez, H.G.; Soccol, C.R. First description of bacterial and fungal communities in Colombian coffee beans fermentation analysed using Illumina-based amplicon sequencing. *Sci. Rep.* **2019**, *9*, 8794. [CrossRef] [PubMed]
41. Saerens, S.M.G.; Delvaux, F.R.; Verstrepen, K.J.; Thevelein, J.M. Production and biological function of volatile esters in *Saccharomyces cerevisiae*. *Microb. Biotechnol.* **2010**, *3*, 165–177. [CrossRef] [PubMed]
42. Elías, L.G. Chemical composition of coffee-berry by-products. In *Coffee Pulp: Composition, Technology, and Utilization*; Braham, J.E., Bressani, R., Eds.; IDRC: Ottawa, ON, Canada, 1979; pp. 11–16.
43. Toci, A.T.; Farah, A. Volatile fingerprint of Brazilian defective coffee seeds: Corroboration of potential marker compounds and identification of new low quality indicators. *Food Chem.* **2014**, *153*, 298–314. [CrossRef] [PubMed]
44. Oestreich-Janzen, S. Chemistry of Coffee. In *Comprehensive Natural Products II*; Liu, H.-W., Mander, L., Eds.; Elsevier Science: Kindlington, UK, 2013; Volume 3, pp. 1085–1117.
45. Toledo, P.R.A.B.; Pezza, L.; Pezza, H.R.; Toci, A.T. Relationship between the different aspects related to coffee quality and their volatile compounds. *Compr. Rev. Food Sci. Food Saf.* **2016**, *15*, 705–719. [CrossRef]
46. Gonzalez-Rios, O.; Suarez-Quiroz, M.L.; Boulanger, R.; Barel, M.; Guyot, B.; Guiraud, J.P.; Schorr-Galindo, S. Impact of "ecological" post-harvest processing on the volatile fraction of coffee beans: I. Green coffee. *J. Food Compos. Anal.* **2007**, *20*, 289–296. [CrossRef]
47. Akiyama, M.; Murakami, K.; Hirano, Y.; Ikeda, M.; Iwatsuki, K.; Wada, A.; Tokuno, K.; Onishi, M.; Iwabuchi, H. Characterization of headspace aroma compounds of freshly brewed arabica coffees and studies on a characteristic aroma compound of Ethiopian coffee. *J. Food Sci.* **2008**, *73*, C335–C346. [CrossRef]
48. Lee, L.W.; Cheong, M.W.; Curran, P.; Yu, B.; Liu, S.Q. Modulation of coffee aroma via the fermentation of green coffee beans with *Rhizopus oligosporus*: II. Effects of different roast levels. *Food Chem.* **2016**, *211*, 925–936. [CrossRef]
49. Afriliana, A.; Pratiwi, D.; Giyarto; Belgis, M.; Harada, H.; Yushiharu, M.; Taizo, M. Volatile compounds change in unfermented Robusta coffee by re-fermentation using commercial kefir. *Nutr. Food Sci. Int. J.* **2019**, *8*, 555745. [CrossRef]
50. Peñuela-Martínez, A.E.; Zapata-Zapata, A.D.; Durango-Restrepo, D.L. Performance of different fermentation methods and the effect on coffee quality (*Coffea arabica* L.). *Coffee Sci.* **2018**, *13*, 465–476. [CrossRef]

51. Mendes-Ferreira, A.; Barbosa, C.; Falco, V.; Leão, C.; Mendes-Faia, A. The production of hydrogen sulphide and other aroma compounds by wine strains of *Saccharomyces cerevisiae* in synthetic media with different nitrogen concentrations. *J. Ind. Microbiol. Biotechnol.* **2009**, *36*, 571–583. [CrossRef] [PubMed]
52. Yvon, M.; Rijnen, L. Cheese flavour formation by amino acid catabolism. *Int. Dairy J.* **2001**, *11*, 185–201. [CrossRef]
53. Plessas, S.; Bekatorou, A.; Gallanagh, J.; Nigam, P.; Koutinas, A.A.; Psarianos, C. Evolution of aroma volatiles during storage of sourdough breads made by mixed cultures of *Kluyveromyces marxianus* and *Lactobacillus delbrueckii* ssp. *bulgaricus* or *Lactobacillus helveticus*. *Food Chem.* **2008**, *107*, 883–889. [CrossRef]
54. Patrignani, F.; Chinnici, F.; Serrazanetti, D.I.; Vernocchi, P.; Ndagijimana, M.; Riponi, C.; Lanciotti, R. Production of volatile and sulfur compounds by 10 *Saccharomyces cerevisiae* strains inoculated in trebbiano must. *Front. Microbiol.* **2016**, *7*, 1–11. [CrossRef] [PubMed]
55. Leclercq-Perlat, M.-N.; Corrieu, G.; Spinnler, H.-E. Comparison of volatile compounds produced in model cheese medium deacidified by *Debaryomyces hansenii* or *Kluyveromyces marxianus*. *J. Dairy Sci.* **2010**, *87*, 1545–1550. [CrossRef]

16

Purification and Properties of Yeast Proteases Secreted by *Wickerhamomyces anomalus* 227 and *Metschnikovia pulcherrima* 446 during Growth in a White Grape Juice

Martina Schlander [1], Ute Distler [2], Stefan Tenzer [2], Eckhard Thines [1] and Harald Claus [1,*]

1 Institute for Microbiology and Wine Research, Johannes Gutenberg University of Mainz, Becherweg 15, D-55099 Mainz, Germany; schlande@uni-mainz.de (M.S.); thines@uni-mainz.de (E.T.)

2 Institute of Immunology, University Medical Centre of the Johannes Gutenberg University Mainz, Langenbeckstr. 1, D-55131 Mainz, Germany; ute.distler@uni-mainz.de (U.D.); tenzer@uni-mainz.de (S.T.)

* Correspondence: hclaus@uni-mainz.de

Academic Editor: Ronnie G. Willaert

Abstract: Aspartic proteases are of significant importance for medicine and biotechnology. In spite of sufficient evidence that many non-*Saccharomyce*s yeasts produce extracellular proteases, previous research has focused on the enzymes of *Candida* species because of their role as virulence factors. Nowadays, there is also increasing interest for their applications in industrial processes, mainly because of their activities at low pH values. Here, we report the features of new acid proteases isolated from wine-relevant yeasts *Metschnikovia pulcherrima* and *Wickerhamomyces anomalus*. To our knowledge, this is the first detailed description of such an enzyme derived from strains of *W. anomalus*. Deviating to most former studies, we could demonstrate that the yeasts produce these enzymes in a natural substrate (grape juice) during the active growth phase. The enzymes were purified from concentrated grape juice by preparative isoelectric focusing. Biochemical data (maximum activity at $\approx$ pH 3.0, inhibition by pepstatin A) classify them as aspartic proteases. For *W. anomalus* 227, this assumption was confirmed by the protein sequence of WaAPR1 determined by LC-MS/MS. The sequence revealed a signal peptide for secretion, as well as a peptidase A1 domain with two aspartate residues in the active site. The enzyme has a calculated molecular mass of 47 kDa and an isolelectric point of 4.11.

Keywords: aspartic protease; *Wickerhamomyces*; *Metschnikovia*; grape juice; wine protein

---

1. Introduction

The use of mixed starters of selected non-*Saccharomyces* yeasts and *Saccharomyces cerevisiae* is of increasing interest for production of novel wines with more complex organoleptic characteristics [1] and/or lower ethanol contents [2]. Non-*Saccharomyces* wine yeasts, also called "wild" yeasts, can enhance the analytical composition and aroma profile of wine by production of secondary metabolites and secretion of enzymes [3–5]. Yeasts of the genera *Kloeckera, Candida, Debaryomyces, Rhodotorula, Pichia, Wickerhamomyces, Zygosaccharomyces, Hanseniaspora, Kluyveromyces* and *Metschnikowia* produce hydrolytic exoenzymes (esterases, lipases, glycosidases, glucanases, pectinases, amylases, and proteases) that interact with grape compounds [6]. Particularly, glycoside hydrolases can release aroma active compounds in grape must from their odourless glycosidic precursors [7]. Others produce pectinolytic enzymes that can promote grape must clarification and may substitute for fungal enzymes, which are currently used for winemaking [8].

Recent research in wine biotechnology has focused on acid proteases to prevent formation of wine haze. A thermotolerant fungal protease (aspergilloglutamic peptidase) has already been approved for winemaking in Australia [9].

Degradation of haze forming-proteins by enzymes is an attractive alternative to bentonite fining because it would minimize losses of wine volume and aroma. Appropriate proteases must be active under harsh winemaking conditions, i.e., low pH (~3.5), low temperature (~15 °C), presence of ethanol ($\geq$ 10% v/v), phenolic compounds and sulphites. Another problem is the intrinsic stability of haze forming proteins due to their high numbers of disulphide bonds as those present in lipid transfer proteins, chitinases and thaumatin-related proteins [9].

Yeasts producing acid proteases may offer an alternative or supplement to bentonite treatment for removal of undesirable wine proteins [10–13]. In contrast to the classical wine yeast *S. cerevisiae*, non-*Saccharomyces* yeasts are important sources of extracellular enzymes including proteases [6,14]. In the study of Fernández et al. [15], 53 of 141 isolates of "wild yeasts" hydrolyzed casein. The positive strains were identified as *Metschnikowa pulcherrima* and *Pichia membranifaciens*. In a similar study with 245 yeast isolates, 10 strains of *Candida stellata, C. pulcherrima, Kloeckera apiculata* and one strain of *Debaromyces hansenii* showed proteolytic activity [16]. Oenological isolates of *Hanseniaspora* [17], *Metchnikowia pulcherrima* and *Candida apicola* [18] produce extracellular proteases with potential applications in biotechnological processes.

Wine yeasts secreting proteolytic enzymes are of high biotechnological interest for protein haze prevention because they could be directly added as starter cultures to the grape must. Besides cost reductions, there are no administrative restrictions for their applications in must and wine, which should be considered with enzyme preparations.

In this study, we describe the purification and properties of extracellular acid proteases isolated from wine relevant yeasts strains of *Wickerhamomyces anomalus* and *Metschnikovia pulcherrima*, which are produced during cultivation in grape juice.

2. Materials and Methods

2.1. Yeast Strains and Cultures

Yeast strains investigated are deposited at the local culture collection of the Institute for Microbiology and Wine Research (IMW), Johannes Gutenberg University Mainz, Country. The identity of *Wickerhamomyces anomalus* strain 227 and *Metschnikowia pulcherrima* 446 have been verified by sequence analysis of the ITS (internal transcribed spacer region). The primers used for PCR amplification were ITS (F) GGAAGTAAAAGTCGTAACAAGG and ITS (R) TCCTCCGCTTATTGATATGC. Sequencing was performed by LCG Genomics (Berlin, Germany) and identification was accomplished by BLAST searches in public databases.

Yeasts were maintained on GYP medium (20 $g \cdot L^{-1}$ glucose, 10 $g \cdot L^{-1}$ yeast extract, meat peptone 20 $g \cdot L^{-1}$). Solid media were prepared with 15 $g \cdot L^{-1}$ agar. For cultivation and protease production, a white grape juice (Lindavia®, Niehoffs-Vaihinger Fruchtsaft GmbH, Lauterecken, Germany) was diluted to 50% (v/v) with deionized water and steam-sterilized for 10 min at 100 °C. Volumes of 100 mL or 300 mL of this medium in Erlenmeyer flasks were inoculated with 1% (v/v) of a washed yeast preculture grown in GYP medium. The cultures were incubated on a rotary shaker (100 rpm) at 20 °C for 7 days. On each day, 2 mL samples were taken to monitor cell growth (OD_{600nm}) and proteolytic activities in the supernatants.

2.2. Buffer Solutions

Tri-sodium phosphate (20.0 $g \cdot L^{-1}$, pH 12); sodium phosphate (100 mM, pH 7.0: 17.8 $g \cdot L^{-1}$ $NaHPO_4 \times H_2O$ adjusted with 0.1 M HCl); sodium acetate (100 mM: 6.0 $g \cdot L^{-1}$ glacial acetic acid adjusted to pH 4.5 with 0.1 M NaOH); sodium tartrate (5.0 $g \cdot L^{-1}$ tartaric acid adjusted to pH 3.5 with

0.1 M NaOH); and universal buffer (tris 6.5 g·L^{-1}, maleic acid 2.32 g·L^{-1}, boric acid 1.26 g·L^{-1}; the pH desired was adjusted with 0.1 M NaOH).

2.3. *Endo-Protease Activity of Secreted Yeast Enzymes*

Endoprotease activities were measured with Megazyme® protease substrate AZCL-collagen® (Megazyme International, Wicklow, Ireland). This compound is prepared by dying and crosslinking collagen to the intensively blue-coloured copper protein azurine to produce a material, which hydrates in water but is still water insoluble. Hydrolysis by proteases produces water-soluble dyed fragments, and the rate of release of these (increase in absorbance at 590_{nm}) is directly related to enzyme activity. The standard test was performed as follows: suspensions of AZCL-collagen® were prepared in appropriate buffers depending on the individual experiment. Afterwards, 100 µL of these suspensions were incubated with 100 µL sample for 24 h at 40 °C under vigorous shaking (750 rpm) in a thermo mixer (Eppendorf, Hamburg, Germany). The reaction was stopped by adding 1.0 mL tri-sodium phosphate-buffer (pH 12) and the tubes centrifuged at 16,000 *g* for 30 min. The absorbance of the supernatants were measured at 590_{nm} in a spectrophotometer (Shimadzu UV-2450, Duisburg, Germany). The control solutions were prepared as above, but test samples were heat-treated (30 min, 100 °C) to inactivate any protease activity. Experiments were performed in triplicate.

2.4. *Protease Inhibition Studies*

Effects of selected inhibitors on proteolytic activities were tested in triplicate by incubation of the samples (1 h) with phenylmethylsulfonylfluoride (PMSF, 2 mM), pepstatin A (20 µM) or Na-EDTA (10 mM) before starting the reaction at 40 °C.

2.5. *pH Optimum*

The pH optimum of yeast proteases was determined in triplicate with the AZCL-collagen® substrate suspended in universal buffer between pH 3.0 and 8.0. Data presented are mean values of triplicate determinations.

2.6. *Temperature Optimum*

The temperature optimum of yeast proteases was determined in triplicate with the standard AZCL-collagen® assay at pH 3.5 between 20 and 60 °C.

2.7. *Real-Time Monitoring of Proteolytic Yeast Enzymes*

A test introduced by Chasseriaud et al. [13] was used with some modifications to monitor proteolytic activity by yeasts directly during growth in grape must. In brief, white grape juice (50% v/v) was steam-sterilized for 10 min at 100 °C. A stock solution (20 mg·mL^{-1}) of the chromogenic protease substrate azocasein (Sigma-Aldrich, Munich, Germany) was prepared in 0.1 M NaOH and added to the grape juice at a final concentration of 2.0 mg·mL^{-1}.

At regular intervals, 600 µL samples of the yeast cultures were withdrawn and mixed with 90 µL trichloroacetic acid (20% w/v). Samples were centrifuged at 16,000 *g* for 30 min. In addition, 500 µL NaOH (1 M) were added to the supernatants and colour release from azocasein was measured in a spectrophotometer (Shimadzu UV-2450, Duisburg, Germany) at 500_{nm}. Non-inoculated culture media served as controls. Experiments were performed in duplicate.

2.8. *Preparative Isoelectric Focusing (pIEF)*

Purification of the extracellular proteases was achieved by preparative IEF using the Rotofor® Preparative IEF-Cell (BioRad, Munich, Germany). Yeasts were grown for 6 days in 100 mL grape juice on a shaker (100 rpm) at 20 °C. After centrifugation (10,000 *g* for 10 min), supernatants were dialyzed (cut-off: 10 kDa) and lyophilized. Concentrated culture supernatants (3.0–5.0 mL) were filled up to 50 mL with double-deionized water and mixed with 1.0 mL of 40% (*w*/*v*) ampholyte solution pH 3.0 to 5.0 (Rotilyte®, Roth, Karlsruhe, Germany). The electrode buffers were 0.5 M acetic acid (anode) and 0.25 M HEPES (cathode), respectively. The separation was performed under a preset power of 15 W until constant voltage was reached after 4 to 5 h at 10 °C. The pH of the 20 fractions obtained was measured, their protein content checked by SDS-PAGE, and protease activities tested with AZCL-collagen®. The protease positive fractions were concentrated in 50 mL spin columns (cut-off: 10 kDa, Vivaspin®, Sartorius AG, Goettingen, Germany), washed three times with deionized water and stored at −20 °C.

2.9. Gel Electrophoretic Methods

Protein compositions in the culture filtrates and fractions of pIEF were checked by SDS-PAGE. Samples were separated in 12.5% (*w*/*v*) SDS gels (10 cm × 10 cm) with a 4% (*w*/*v*) stacking gel at room temperature for 1 h. A prestained protein ladder (Pink Color Protein Standard II®; Serva, Heidelberg, Germany) served as the molecular mass standard. Protein staining was performed with Quick Coomassie Stain® (Serva, Heidelberg, Germany).

2.10. Identification of Yeast Exoenzymes

The concentrated and dialyzed culture filtrates were separated by SDS-PAGE and gels treated with Quick Coomassie Stain®. *W. anomalus* 227 delivered two well separated protein bands after 24 h incubation, which were excised and sliced into small pieces. After destaining and drying, proteins were reduced with 2 mM DTT at 55 °C and alkylated with 15 mM iodacetamide at room temperature in the dark for 1 h each. After washing and drying, trypsin digests were done at 37 °C overnight (0.5 µg of trypsin per gel slice). Tryptic peptides were transferred into an autosampler vial for peptide analysis via LC-MS/MS. Nanoscale liquid chromatography of tryptic peptides was performed with a Waters NanoAcquity UPLC system (Eschborn, Germany) equipped with a 75 µm × 250 mm HSS-T3 reversed phase column and a 2.6 µL PEEKSIL-sample loop (SGE, Darmstadt, Germany) as described before (PMID: 23265486). Online mass spectrometry analysis of tryptic peptides was performed using a Waters Synapt G2-S QTOF mass spectrometer, operated at a resolving power of $R = 20.000$. All analyses were performed using positive mode ESI using a NanoLockSpray source as described (PMID: 23265486). Resulting liquid chromatography tandem MS (LC-MS/MS) data were processed and searched by using PROTEINLYNX GLOBAL SERVER, Ver. 3.0.2. (Waters, Eschborn, Germany). Protein identifications were assigned by searching a custom compiled database containing open reading frames obtained from transcriptome sequencing of *W. anomalus* [19] for gel bands derived from *W. anomalus* and all available (212.809) RefSeq database entries from the order of *Saccharomycetales* for gel bands derived from *Metschnikowia pulcherrima*. Sequence databases were supplemented by known possible contaminants (trypsin, human keratins) based on the precursor and fragmentation data afforded by the LC-MS/MS acquisition method as described before [20]. The false discovery rate (FDR) for peptide and protein identification was assessed searching a reverse database generated automatically in PLGS. FDR was set to 0.01 for database search.

2.11. In Silico Analysis

Protein sequence was analysed by public databases and tools (BLAST; ScanPROSITE; PROTPARAM; SIGNAL P, NetNGlyc) offered by the ExPASY Bioinformatics Portal.

2.12. Degradation Experiments

Proteins from a German Riesling wine (vintage 2012) were prepared by dialysis and lyophilisation as recently described by Jaeckels et al. [20]. The proteolytic activity against isolated wine proteins and bovine serum albumin (Sigma-Aldrich, Munich, Germany) were tested after incubation at 20 °C in sodium tartrate buffer (pH 3.5) and subsequently in an actual German Riesling wine (vintage 2014,

11.5% *v/v* alcohol). The specific reaction conditions are indicated in the corresponding results section. Degradation of the model proteins was checked by SDS-PAGE.

3. Results

In a previous screening study with 102 yeasts from the internal strain collection (IMW), *W. anomalus* 227 and *M. pulcherrima* 446 were identified as potential protease producers as evidenced by clearing zones on turbid milk powder agar plates (unpublished). In the present study, both strains grew equally well in 50% grape juice and reached a maximum cell density after six days (Figure 1).

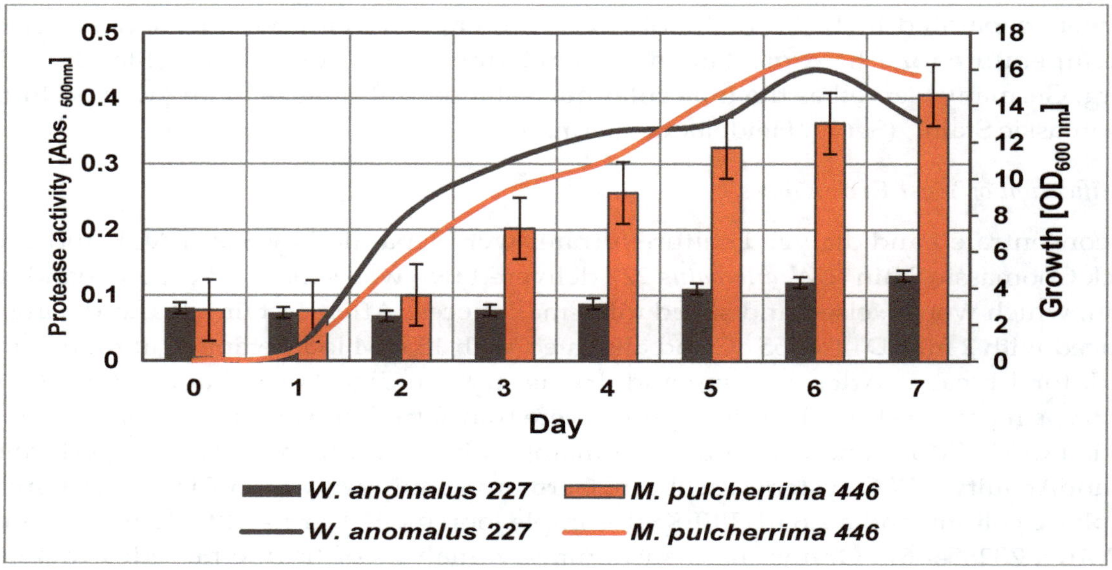

Figure 1. Cell densities (lines) and proteolytic activities (columns) of *Wickerhamoyces anomalus* 227 and *Metschnikovia pulcherrima* 446 during cultivation in a white grape juice at 20 °C. Proteolytic activity was directly detected in the cultures by hydrolysis of supplemented azocasein.

In the course of cultivation, yeasts exhibited proteolytic activities as indicated by cleavage of azocasein supplemented to the medium, although at quite different levels. In order to characterize these activities in more detail, the cultures were harvested after six days and the supernatants were dialyzed and subsequently concentrated by lyophilisation. As detected by SDS-PAGE (Figure 2), only a few distinct proteins were present in the culture concentrates but not in the juice controls (not shown). The protein bands were cut out of the gel and tryptic fragment patterns were analysed by LC-MS/MS. By database research, two extracellular proteins of *W. anomalus* 227 could be identified. The band at ca. 30 kDa corresponded with 100% identity to an exo-β-1,3-glucanase of *W. anomalus* AS1 described in our recent study [21].

The higher molecular mass protein delivered 100% identity with a hypothetic protein (NCBI Acc. No. XP_019036036) annotated in the genome of *W. anomalus* NRRL Y-366-8. It could be attributed to peptidase family A1 (Figure 3). The sequence contains two aspartyl residues (D_{79} and D_{291}) as typically found in the active site of acid proteases. In addition, a putative signal peptide for secretion, possible cysteine bridges and a single N-glycosylation site were detected. Due to the lack of reference sequence data, the extracellular proteins of *M. pulcherrima* 446 were not identified by database searches.

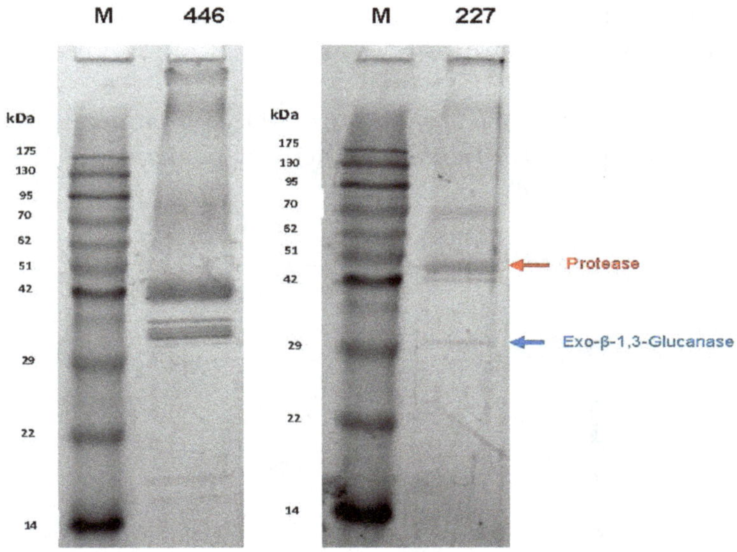

Figure 2. SDS-PAGE (12.5%) of dialyzed and lyophilized culture supernatants after six days of yeast growth in 50% (v/v) grape juice. Strains: *M. pulcherrima* 446 and *W. anomalus* 227. M: Molecular mass standard. Arrows indicate proteins identified by LC-MS/MS.

```
1     MKFIQALSIS   ALITTALAID   NVVKLPVTKT   KTNKLSKAVH   SGLVKRAQDH
51    QESLADDGVG   YFATIEVGSP   AQTIRVQIDT   GSSDLWFPGT   NNPQCPQGTA
101   TPIDQGSPYQ   DDFDYCLRSA   QYDPGASSSW   KVDSSAPAFH   IEYVDYSFAT
151   GSWGTDNVQW   NDVTIDNFFL   ASAQNSNATS   VFGIALTTDE   SSNEGSDAFQ
201   YPNFPQRLKS   DGYISKIVYS   LYPSEDPYTT   SGDIDITLLF   GGVDTAKYSG
251   TLEVFPLDSN   YDLAITLTGI   STDISGQTSV   ASTQSLSAVL   DSGTTLQALP
301   YNIVEQLAYN   LGGSGETDSY   GYFIVPCNYG   SDDHITYTFG   TKNINVPVEA
351   VVSSDGNGNC   ALAIEPTNGL   TILGDTFLIN   AYVVYDLEDR   EIAIAQAKYT
401   STENIQPVIS   SIPGAVRAAA   VAAPSVAASA   TGGFVTSIID   SSATSSA
```

Matched to a peptide

Matched to a partial peptide

Matched to a modified peptide

Matched to a partial modified peptide

MKFIQALSISALITTALAIDNVVKLPVTKTKTNKLSKAVHSGLVKRAQDHQESLADDGVGYFATIEV
GSPAQTIRVQI**D**TGSSDLWFPGTNNPQCPQGTATPIDQGSPYQDDFDYCLRSAQYDPGASSSWKVDS
SAPAFHIEYVDYSFATGSWGTDNVQWNDVTIDNFFLASAQNS**NATS**VFGIALTTDESSNEGSDAFQY
PNFPQRLKSDGYISKIVYSLYPSEDPYTTSGDIDITLLFGGVDTAKYSGTLEVFPLDSNYDLAITLTGIS
TDISGQTSVASTQSLSAVL**D**SGTTLQALPYNIVEQLAYNLGGSGETDSYGYFIVP**C**NYGSDDHITYTF
GTKNINVPVEAVVSSDGNGN**C**ALAIEPTNGLTILGDTFLINAYVVYDLEDREIAIAQAKYTSTENIQPV
ISSIPGAVRAAAVAAPSVAASATGGFVTSIIDSSATSSA

Figure 3. Sequence coverage (peptides identified by LC-MS/MS) and annotated putative protein sequence of the extracellular protease WaAPR1 of *W. anomalus* 227. Amino acids are given in the short code.

Above: Peptides identified by LC-MS/MS are highlighted according to their type (see legend). Total sequence coverage was 49.4%.

Below: Annotated putative protein sequence obtained from the *W. anomalus* transcriptome database. Number of amino acids: 447; Molecular mass: 47328 Da (PROTPARAM); Theoretical pI: 4.11 (PROTPARAM). A possible signal peptide of 19 amino acids (SIGNAL P) is shown in red and underlined. The sequence contains a peptidase family A1 domain (amino acids$_{61\text{-}395}$) with aspartyl residues D_{79} and D_{291} (green) in the active site (ScanPROSITE) and a single N-glycoyslation signature (blue; NetGlyc 1.0 server). Cysteine residues C_{329} and C_{260} (yellow) may build a disulfide bridge (ScanPROSITE).

The extracellular proteases of both yeasts were purified by preparative IEF. The proteolytic activity of *W. anomalus* 227 focused at ca. pH 3.4, whereas two activity peaks at pH 3.96 and 4.61 were found in fractions of *M. pulcherrima* 446 (Figure 4).

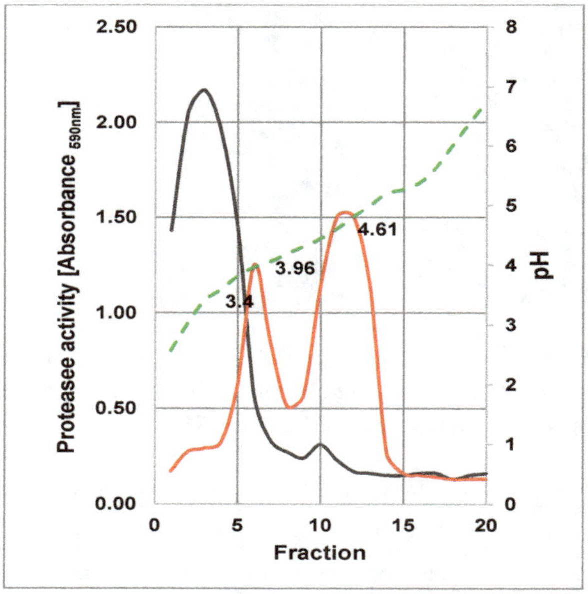

Figure 4. Preparative IEF of concentrated culture supernatants of *W. anomalus* 227 (**black** line) and *M. pulcherrima* 446 (**red** line). Collected fractions were examined for protease activities under the standard assay conditions. The pH gradient is illustrated as the broken **green** line.

The protease-active fractions were pooled, concentrated and subjected to SDS-PAGE and sensitive protein staining (Figure 5). One single band appeared in the fractions collected from the main activity peak of *W. anomalus* 227 (at $\approx$ pH 3.4) and in the second peak of *M. pulcherrima* 446 (at $\approx$ pH 4.61).

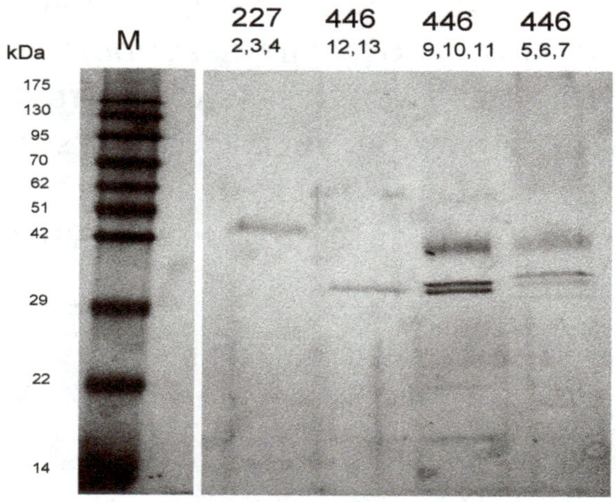

Figure 5. SDS-PAGE (12.5%) of protease-active fractions from *W. anomalus* 227 and *M. pulcherrima* 446 as obtained by preparative IEF.

The isoelectric point and the apparent molecular mass of the purified protease WaAPR1 from *W. anomalus* 227 correspond fairly well with the theoretical values (pI 4.11, MW 47 kDa) derived from the amino acid sequence (Figure 3). The protein purified from the culture supernatant (in fractions 12 and 13) of *M. pulcherrima* 446 is probably one of several protease isoenzymes secreted by this strain. This can be concluded from the appearance of at least five protein bands in SDS-PAGE (Figure 2) and by two activity peaks displayed in preparative IEF. The protein concentrations and enzymatic activities at different stages of purification are listed in Table 1.

Table 1. Protein concentrations and proteolytic activities at different stages of purification.

Species	Sample	Protein Concentration (μg/mL)	Activity * (Absorbance $_{590\ nm}$)
W. anomalus 227	Original culture supernatant (7 days)	4.87	0.85
	After dialysation and lyophilisation	20.21	1.75
	Concentrated pIEF fractions (2,3,4)	5.85	1.63
M. pulcherrima 446	Original culture supernatant (7 days)	5.27	0.59
	After dialysation and lyophilisation	23.18	2.09
	Concentrated pIEF fractions (12,13)	8.43	1.72

* Determined with Protazyme OL® tablets.

In accordance with the sequence data, inhibitor studies confirmed that both yeasts secrete acid proteases: activities were most significantly impaired by low concentrations of pepstatin A, which specifically inhibits aspartic proteases (Table 2).

Table 2. Effect of inhibitors on protease activities of *W. anomalus* 227 and *M. pulcherrima* 446.

Inhibitor	Concentration	Activity (%)*	
		227	446
Pepstatin A	20 μM	18.7	7.4
PMFS	2 mM	58.7	47.1
EDTA	10 mM	49.6	44.0

* Relative to the control without inhibitor. Activity was determined with Protazyme OL® tablets.

Inhibitors for serine proteases (PMFS) and metalloproteinase (EDTA) were far less effective even at high concentrations. In addition, enzymatic activity on azurine collagen was highest at pH 3–4 and dropped at higher pH values (Figure 6).

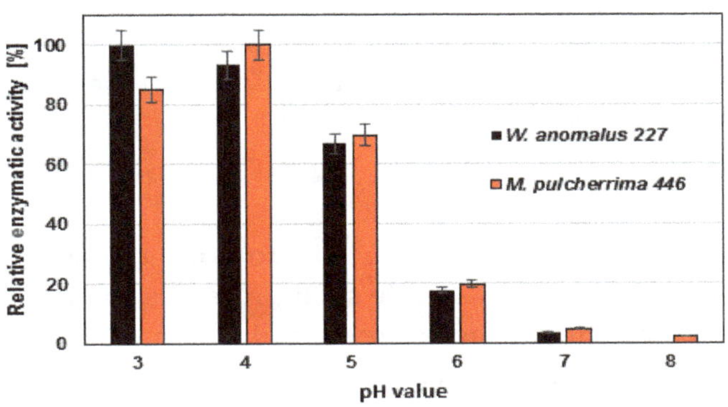

Figure 6. Effect of pH on protease activities of *W. anomalus* 227 and *M. pulcherrima* 446.

The temperature maximum of proteolytic activity for both yeast strains was found at 40 °C, although significant activities could be determined at lower and higher temperatures (Figure 7).

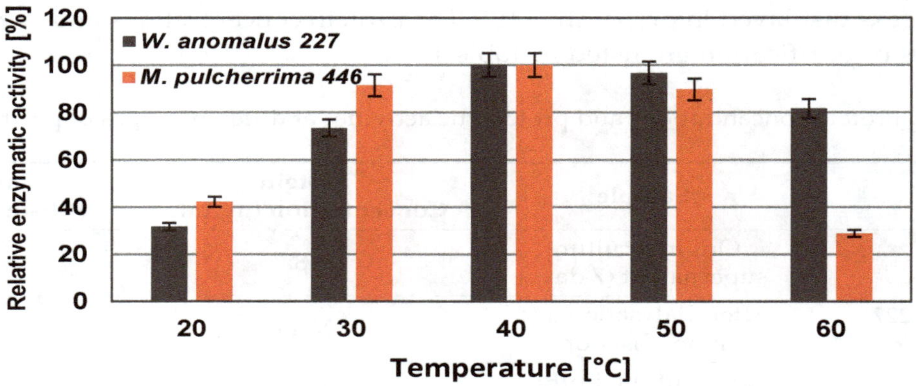

Figure 7. Effect of temperature on protease activities of *W. anomalus* 227 and *M. pulcherrima* 446.

With respect to a potential application for protein haze reduction, we finally assessed the activity at wine-relevant conditions. Bovine serum albumin was completely degraded after 24 h in a white wine. However, thaumatin-like wine proteins were resistant to enzymatic hydrolysis (Figure 8).

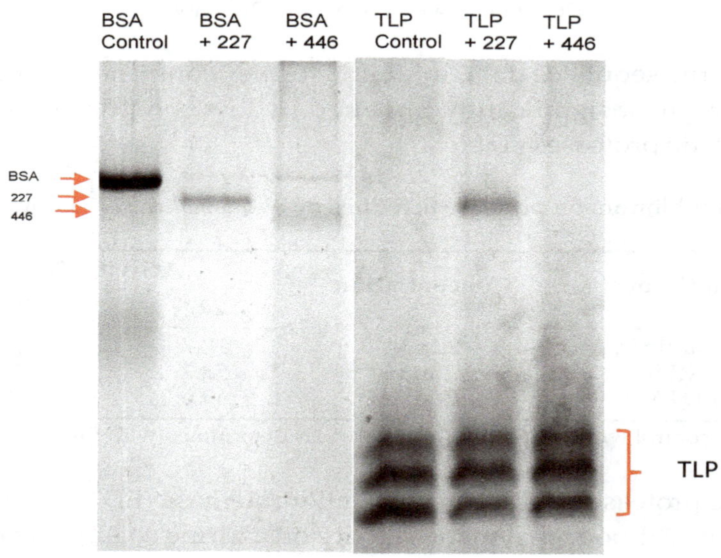

Figure 8. Detection of proteolytic activity of yeasts by SDS-PAGE (12.5%). Bovine serum albumine (BSA, **left**) and thaumatin-like wine proteins (TLP, **right**) were added to a white wine at final concentrations of 100 μg·mL^{-1} and incubated for 24 h at 20 °C with concentrated culture supernatants of *W. anomalus* 227 and *M. pulcherrima* 448.

4. Discussion

Aspartic proteases are ubiquitously distributed proteolytic enzymes that are active in acidic environments [22]. Several microorganisms secrete such proteases as virulence factors and/or in order to break down proteins, thereby deliberating sources of nitrogen.

Extracellular proteinases secreted by yeasts have been isolated and examined by several researchers, especially with respect to their enzymatic properties and the physiology of their induction and secretion. Most of these studies focus on the acid aspartic proteases of *Candida albicans* because of their involvement in human diseases. Nowadays, there is growing interest on yeast acid proteases for industrial applications [10].

Although simple screening procedures with solid casein agar plates have repeatedly demonstrated proteolytic activities in various *non-Saccharomyces* species, only limited knowledge exists on their cellular location, substrate preferences, catalytic properties or molecular structure [11,15,16]. In contrast to some previous studies using synthetic culture media and proteins as inducers [18,23], we demonstrated the secretion of proteases by two wine associated yeasts directly in a natural substrate, i.e., grape juice. Remarkably, they were the only dominant extracellular enzymes produced during the active growth phase. Recent studies have already demonstrated protease activities in *Wickerhamomyces* isolates from enological ecosystems [24]. However, the activities detected were relatively low and not expressed in all strains examined [25]. The extracellular protease WaAPR1 of *W. anomalus* 227 with a molecular mass of 47 kDa and an isoelectric point of 4.11 is in the typical range of microbial aspartic proteases [10,22]. As only a single N-glycosylation site is present in the protein sequence, any covalent glycosylation would be difficult to demonstrate. A similar situation has been reported for the aspartic protease from *M. pulcherrima* IWBT Y1123 [18]. The bioinformatic tool ScanPROSITE predicts the position of the two essential aspartyl residues in the active site at D_{79} and D_{291}. Results of a BLAST search reveals 100% identity with a hypothetic protein (NCBI Acc. No. XP_019036036) in the genome of *W. anomalus* NRRL Y-366-8 and ca. 35% identity with putative acid proteases of other yeast species belonging to the genera *Wickerhamomyces*, *Saccharomyces*, *Zygosaccharomyces*, *Metschnikovia*, *Yarrowia*, *Torulaspora* and *Candida*.

The amino acid sequence of the *M. pulcherrima* 446 protease could not be identified in this study. This is attributed to a lack of reference sequence data. In addition, the identification attempts reported in our study were performed with a non-purified concentrate (Figure 2), which probably contained a mixture of several protease isoenzymes. The MALDI analysis will be repeated with a purified sample. Nevertheless, biochemical data (pH optimum, inhibition by pepstatin A) and results of other researchers give strong hints for an aspartic protease. Reid et al. [18] reported on genetic and biochemical levels of the existence of an extracellular aspartic protease in *M. pulcherrima* strain IWBT Y1123. Gotoh et al. [23] purified a protease from the culture supernatant from strain KSY 188-5. The latter had an apparent molecular mass of 37 kDa, an isolelectric point of 4.7, and an optimum activity around 45 °C at pH 3.0. These parameters are very close to the data obtained in the present study for the protease of *M. pulcherrima* 446.

Constitutive production in the natural substrate grape juice without need of external inducers combined with maximum activity at low pH suggests an essential physiological role of the secreted enzyme. The protease activity might help to gain nitrogen from grape proteins necessary for cell growth, especially considering the surplus of sugar carbon in nitrogen-limited milieus [26,27]. The secreted proteases may also act as a survival tool by degrading cell wall proteins of competitive microorganisms, analogous to its role as a virulence factor in *C. albicans*. The membrane-bound aspartic protease of *S. cerevisiae*, a member of the yapsin protease family, is involved in cell wall growth and maintenance. It is a challenge to get more insights in the natural role of secreted yeast aspartic proteases. On the other hand, their biotechnological potential needs to be further examined in future experiments. The use of aspartic proteases as alternatives or supplements to clarifying agents in various beverage industries is under intensive investigation, and the potential applications in the wine industry are thoroughly discussed [10]. Under wine-relevant conditions (pH 3.5, 20 °C), the proteases of both yeast strains were able to degrade bovine serum albumin but not wine specific thaumatin-like proteins (TLP). Nevertheless, it is well known that TLP are very inert to proteolytic hydrolysis and among the few protein classes that finally reach the bottled wine. One should keep in mind that even the actual used commercial fungal protease preparation is only effective at elevated (not wine-relevant) temperatures and cannot completely eliminate wine proteins. Other grape must proteins than TLP may be more susceptible to proteolytic hydrolysis by yeast enzymes. The fact that the proteolytic enzymes were produced during growth in grape juice offers the chance to lower the protein content by using our yeast strains directly for must fermentations. Recent studies have already shown that strains of *Wickerhamomyces anomalus* and *Metschnikovia pulcherrima* positively influence wine sensory properties [28]. Any reduction of grape must protein content would mean an advantage for wine-makers by lowering the effective bentonite dosages. These aspects are topics of our current research.

Acknowledgments: The authors would like to thank undergraduate student Hendrik Leibhan for conducting some of the experiments.

Author Contributions: Martina Schlander performed most of the experiments; Ute Distler performed HPLC-MS analysis; Stefan Tenzer supervised and evaluated the HPLC-MS/MS analysis; Eckhard Thines provided financial, laboratory and personnel capacities; Harald Claus conceived and designed the experiments, analyzed the data and wrote the manuscript.

Abbreviations

TLP	Thaumatin-like Proteins
pIEF	Preparative Isoelectric Focusing

References

1. Padilla, B.; Gil, J.V.; Manzanares, P. Past and future of Non-*Saccharomyces* yeasts: From spoilage microorganisms to biotechnological tools for improving wine aroma complexity. *Front. Microbiol.* **2016**. [CrossRef] [PubMed]
2. Ciani, M.; Morales, P.; Comitini, F.; Tronchoni, J.; Canonico, L.; Curiel, J.A.; Oro, L.; Rodrigues, A.J.; Gonzalez, R. Non-conventional yeast species for lowering ethanol content of wines. *Front. Microbiol.* **2016**. [CrossRef] [PubMed]
3. Maturano, Y.P.; Assof, M.; Fabani, M.P.; Nally, M.C.; Jofré, V.; Rodríguez Assaf, L.A.; Toro, M.E.; Castellanos de Figueroa, L.I.; Vazquez, F. Enzymatic activities produced by mixed *Saccharomyces* and non-*Saccharomyces* cultures: Relationship with wine voltile composition. *Antonie van Leeuwenhoek* **2015**, *108*, 1239–1256. [CrossRef] [PubMed]
4. Suzzi, G.; Schirone, M.; Sergi, M.; Marianella, R.M.; Fasoli, G.; Aguzzi, I.; Tofalo, R. Multistarter from organic viticulture for red wine Montepulciano d´Abruzzo production. *Front. Microbiol.* **2012**. [CrossRef]
5. Tofalo, R.; Patrignani, F.; Lanciotti, R.; Perpetuini, G.; Schirone, M.; Di Gianvito, P.; Pizzoni, D.; Arfelli, G.; Suzzi, G. Aroma profile of Montepulciano d´Abruzzo wine fermented by single and co-culture starters of autochthonous *Saccharomyces* and non-*Saccharomyces* yeasts. **2016**. [CrossRef]
6. Claus, H. Exoenzymes of wine microoorganisms. In *Biology of Microorganisms on Grapes, in Must and Wine*; König, H., Unden, G., Fröhlich, J., Eds.; Springer: Berlin/Heidelberg, Germany, 2009; pp. 259–271.
7. Mateo, J.J.; Maicas, S. Application of Non-*Saccharomyces* yeasts to wine-making process. *Fermentation* **2016**, *2*, 14. [CrossRef]
8. Ugliano, M. Enzymes in winemaking. In *Wine Chemistry and Biochemistry*; Moreana-Arribas, M.V., Polo, C., Eds.; Springer Science-Business Media: Adelaide, Australia, 2009; pp. 103–126.
9. Van Sluyter, S.C.; McRae, J.M.; Falconer, R.J.; Smith, P.A.; Bacic, A.; Waters, E.J.; Marangon, M. Wine protein haze: Mechanisms of formation and advances in prevention. *J. Agric. Food. Chem.* **2015**, *63*, 4020–4030. [CrossRef] [PubMed]
10. Theron, L.W.; Divol, B. Microbial aspartic proteases: Current and potential applications in industry. *Appl. Microbiol. Biotechnol.* **2014**, *98*, 8853–8863. [CrossRef] [PubMed]
11. Rosi, I.; Costamagna, L.; Bertuccioli, M. Screening for extracellular protease(s) production by wine yeasts. *J. Inst. Brew.* **1987**, *93*, 322–324. [CrossRef]
12. Lagace, L.S.; Bisson, L.F. Survey of yeast acid proteases for effectiveness of wine haze reduction. *Am. J. Enol. Vitic.* **1990**, *41*, 147–162.
13. Chasseriaud, L.; Miot-Sertier, C.; Coulon, J.; Iturmendi, N.; Moine, V.; Albertin, W.; Bely, M. A new method for monitoring the extracellular proteolytic activity of wine yeasts during alcoholic fermentation of grape must. *J. Microbiol. Methods* **2015**, *119*, 176–179. [CrossRef] [PubMed]

14. Molnárova, J.; Vadkertiová, R.; Stratilová, E. Extracellular enzymatic activities and physiological profiles of yeasts colonizing fruit trees. *J. Basic Microbiol.* **2014**, *51*, S74–S84. [CrossRef] [PubMed]
15. Fernández, M.; Úbeda, J.F.; Briones, A.I. Typing of non-*Saccharomyces* yeasts with enzymatic activities of interest in wine-making. *Int. J. Food. Microbiol.* **2000**, *59*, 29–36. [CrossRef]
16. Strauss, M.L.A.; Jolly, N.P.; Lambrechts, M.G.; van Rensburg, P. Screening for the production of extracellular hydrolytic enzymes by non-*Saccharomyces* wine yeasts. *J. Appl. Microbiol.* **2001**, *91*, 182–190. [CrossRef] [PubMed]
17. Mateo, J.J.; Maicas, S.; Thießen, C. Biotechnological characterisation of extracellular proteases produced by enological *Hanseniaspora* isolates. *Int. J. Food Sci. Technol.* **2015**, *50*, 218–225. [CrossRef]
18. Reid, V.J.; Theron, L.W.; du Toit, M.; Divol, B. Identification and partial characterization of extracellular aspartic protease genes from *Metschnikowia pulcherrima* IWBT Y1123 and *Candida apicola* IWBT Y1384. *Appl. Environm. Microbiol.* **2012**, *78*, 6838–6849. [CrossRef] [PubMed]
19. Schneider, J.; Fupp, O.; Trost, S.; Jaenicke, S.; Passoth, V.; Goesmann, A.; Tauch, A.; Brinkrolf, K. Genome sequence of *Wickerhamomyces anomalus* DSM 6766 reveals genetic basis of biotechnologically important antimicrobial activities. *FEMS Yeast Res.* **2012**, *12*, 382–386. [CrossRef] [PubMed]
20. Jaeckels, N.; Tenzer, S.; Rosfa, S.; Schild, H.; Decker, H.; Wigand, P. Purification and structural characterization of lipid transfer protein from red wine and grapes. *Food Chem.* **2013**, *138*, 263–269. [CrossRef] [PubMed]
21. Schwentke, J.; Sabel, A.; Petri, A.; König, H.; Claus, H. The wine yeast *Wickerhamomyces anomalus* AS1 secretes a multifunctional exo-β-1,3 glucanase with implications for winemaking. *Yeast* **2014**, *31*, 349–359. [CrossRef] [PubMed]
22. Ogrydziak, D.M. Yeast extracellular proteases. *Crit. Rev. Biotechnol.* **1993**, *13*, 1–55. [CrossRef] [PubMed]
23. Gotoh, T.; Kikuchi, K.; Kodama, K.; Konno, H.; Kakuta, T.; Koizumi, T.; Nojiro, K. Purification and properties of extracellular carboxyl proteinase secreted by *Candida pulcherrima*. *Biosci. Biotech. Biochem.* **1995**, *59*, 367–371. [CrossRef]
24. Madrigal, T.; Maicas, S.; Mateo Tolosa, J.J. Glucose and ethanol tolerant enzymes produced by *Pichia* (*Wickerhamomyces*) isolates from enological ecosystems. *Am. J. Enol. Vitic.* **2013**, *64*, 126–133. [CrossRef]
25. López, S.; Mateo, J.J.; Maicas, S. Screening of *Hanseniaspora* strains for production of enzymes with potential interest for winemaking. *Fermentation* **2016**. [CrossRef]
26. Christ, E.; Kowalczyk, M.; Zuchowska, M.; Claus, H.; Löwenstein, R.; Szopinska-Morawska, A.; Renaut, J.; König, H. An exemplary model study for overcoming stuck fermentation during spontaneous fermentation with the aid of a *Saccharomyces* triple hybrid. *J. Agric. Sci.* **2015**, *7*, 18–34. [CrossRef]
27. Guo, Z.P.; Zhang, L.; Ding, Z.Y.; Wang, Z.X.; Shi, G.Y. Improving the performance of industrial ethanol-producing yeast by expressing the aspartyl protease on the cell surface. *Yeast* **2010**, *27*, 1017–1027. [CrossRef] [PubMed]
28. Varela, C. The impact of non-*Saccharomyces* yeasts in the production of alcoholic beverages. *Appl. Microbiol. Biotechnol.* **2016**, *100*, 9861–9874. [CrossRef] [PubMed]

17

Adaptive Evolution of Industrial Brewer's Yeast Strains towards a Snowflake Phenotype

Yeseren Kayacan [1,2,†], Thijs Van Mieghem [1,2,†], Filip Delvaux [3], Freddy R. Delvaux [3] and Ronnie Willaert [1,2,4,*]

1 Alliance Research Group VUB-UGent NanoMicrobiology (NAMI), Research Group Structural Biology, 1050 Brussels, Belgium; Yeseren.Kayacan@vub.be (Y.K.); thijs.van.mieghem@hotmail.com (T.V.M.)
2 International Joint Research Group VUB-EPFL NanoBiotechnology & NanoMedicine (NANO), Vrije Universiteit Brussel, 1050 Brussels, Belgium
3 Biercentrum Delvaux, 3040 Neerijse, Belgium; filip@biercentrum.be (F.D.); Freddy.Delvaux@biw.kuleuven.be (F.R.D.)
4 Department Bioscience Engineering, University Antwerp, 2020 Antwerp, Belgium
* Correspondence: Ronnie.Willaert@vub.be
† These authors contributed equally to this work.

Abstract: Flocculation or cell aggregation is a well-appreciated characteristic of industrial brewer's strains, since it allows removal of the cells from the beer in a cost-efficient and environmentally-friendly manner. However, many industrial strains are non-flocculent and genetic interference to increase the flocculation characteristics are not appreciated by the consumers. We applied adaptive laboratory evolution (ALE) to three non-flocculent, industrial *Saccharomyces cerevisiae* brewer's strains using small continuous bioreactors (ministats) to obtain an aggregative phenotype, i.e., the "snowflake" phenotype. These aggregates could increase yeast sedimentation considerably. We evaluated the performance of these evolved strains and their produced flavor during lab scale beer fermentations. The small aggregates did not result in a premature sedimentation during the fermentation and did not result in major flavor changes of the produced beer. These results show that ALE could be used to increase the sedimentation behavior of non-flocculent brewer's strains.

Keywords: *Saccharomyces cerevisiae*; industrial brewer's strains; adaptive laboratory evolution (ALE); snowflake phenotype; beer fermentation

---

1. Introduction

Bulk sedimentation of yeast cells during fermentation is a crucial part of the brewing process. At the end of the fermentation, single yeast cells aggregate and form macroscopic "flocs" [1,2]. These clumps of cells then rapidly sediment from the beer and can be harvested from the bottom (lager fermentation) or float and can be harvested from the top (open ale fermentation) at the end of the primary fermentation. This phenomenon allows the brewer to separate the yeast from the beer in an effective, cost-efficient, and environmentally-friendly way, leaving only the clear and almost cell-free product. The neatly harvested yeast can also be "repitched" into the next fermentation. The timing of sedimentation is of considerable importance to the process. Sedimentation should not take place prematurely and cause stuck fermentation leading to beers with low quality flavor profiles. Complete sedimentation at the end of fermentation is preferred by the brewer, which provides the opportunity for a neat separation of the yeast cells from the beer [3].

Flocculation is the reversible, asexual self-adhesion of yeast cells which leads to their sedimentation [4]. Although single yeast cells do sediment, the large clumps formed by flocculating cells sediment at a much higher rate. The flocculation capacity of yeast is highly strain-dependent, influenced mainly by the genetic background since expressed cell wall flocculins that are lectins, effectuate cell-cell binding [5–7]. However, several environmental factors can affect flocculation too. Calcium availability, pH, temperature, ethanol concentration and oxygen concentration are some of the physiological factors that influence flocculation while physical factors such as cell surface hydrophobicity and favorable hydrodynamic conditions can also affect formation of flocs [8–10]. The sedimentation rate is dependent on the size, shape and density of these flocs.

Optimization of the brewer yeast towards a more flocculating phenotype can lead to a more efficient beer production and a higher final beer quality. Recent advances in DNA sequencing, high-throughput technologies and genetic manipulation methods have led to the molecular and genomic characterization of the brewer's yeast. However, the exponential increase in knowledge generated in the field of functional genomics of yeast can only facilitate strain improvement efforts to some degree. Procedures to obtain approval for modified GMO yeasts are complicated and consumer acceptance for a GMO-produced beer is lacking [11,12]. These hurdles have guided researchers to look elsewhere to generate strains with desired properties.

One attractive more "natural" approach to enhance the attributes of microorganisms is the adaptive laboratory evolution (ALE) approach [13]. In ALE, microorganisms are cultivated under clearly defined conditions for long periods of time, allowing metabolic engineering of microorganisms utilizing genetic variation and selection for beneficial mutations [14,15]. Already a more-and-more used tool in microbial strain improvement, ALE has been applied for improving yeast strains such as for the utilization of alternative sugars by *S. cerevisiae* [16], increasing tolerance of *S. cerevisiae* to environmental conditions [17], for increasing the fermentation capacity of a lager *S. pastorianus* brewing strain under hyperosmotic conditions [18], modifying the production of flavor compounds by *S. pastorianus* for alcohol-free beer production [19], the adaptation of lager strains to very high-gravity brewing conditions [20,21], and for enhancing the fermentation rate with decreased formation of acetate and greater production of fermentative aroma of *S. cerevisiae* wine strains [22,23].

Microbial cells can be cultivated in parallel serial cultures for ALE but varying population densities, fluctuating growth rate, nutrient supply, and environmental conditions characterize this batch cultivation. Continuous (chemostat) cultures, however, ensure more stable conditions such as constant growth rate, tightly controlled nutrient supply and stable pH and oxygen availability [24–26].

Previously, we performed ALE of a *S. cerevisiae* strain in a 3D-printed continuous mini tower fermentor using gravity as a selective pressure to obtain a snowflake phenotype [27]. In this work, we've used the ALE approach for the continuous cultivation of three non-flocculating industrial *S. cerevisiae* brewing strains in miniature chemostats (ministats) [28]. This simple and low-cost setup was used to carry out adaptive evolution experiments where gravity is also the selective pressure on the planktonic cells, which are continuously removed while aggregating cells' sediment are retained. Stable aggregating cells were observed during continuous cultivation, showing a "snowflake" phenotype of unseparated daughter and mother cells. Finally, we have also demonstrated the beer fermentation performance of these evolved strains in lab scale tall tubes fermentors.

2. Materials and Methods

2.1. Yeast Strains and Media

The industrial *Saccharomyces cerevisiae* brewer's strains BCD1, BCD2, BCD3, and BCD4 were provided by Biercentrum Delvaux (Neerijse, Belgium). The lab strains BY4742 [29], BY4742 [*FLO1*], and BY4742::*FLO8* [5] and the strong flocculating industrial strain BCD4 were used as control strains in the flocculation assay. All strains were precultured in YPD (Yeast extract–peptone–dextrose) medium (1% *m*/*v* yeast extract, 2% *m*/*v* peptone, 4% *m*/*v* glucose) overnight at 30 °C. For the continuous ALE in ministats, a high-glucose medium (100 g/L D-glucose, 4 g/L $(NH_4)SO_4$, 1.5 g/L KH_2PO_4, 1 g/L $MgSO_4{\cdot}7H_2O$, and 5 g/L yeast extract) was used. The yeast cells and aggregates were visualized by microscopy (Nikon Eclipse Ti2, Tokyo, Japan).

2.2. Flocculation Assay

The assay described by D'Hautcourt and Smart [30] was used with minor modifications. Cells were cultivated for 24 h in YPD, harvested by centrifugation and resuspended in EDTA buffer (50 mM EDTA, pH 7) to reach an OD_{600nm} value of 10. A sample of 50 µL was taken at 0.5 mL below the meniscus, and the sample was diluted 20 times in a 1.5 mL cuvette with EDTA buffer (50 mM EDTA, pH 7). The tubes were centrifuged (4000 rpm, 3 min), and the supernatant was discarded. The cells were resuspended in 1 mL flocculation buffer A (3 mM $CaSO_4$). The last step was repeated, but the cells were resuspended in flocculation buffer B (3 mM $CaSO_4$, 83 mM CH_3COONa, 4% *v*/*v* ethanol, pH 4.5). The tubes were shaken at 100 rpm for 10 min. Prior to taking 50 µL samples 0.5 mL below the meniscus, 3 min of sedimentation in a vertical position took place. The sample was diluted 20 times with EDTA buffer in a 1.5 mL cuvette. The absorbance of both suspensions in the cuvettes was determined, and the related flocculation percentage was calculated:

$$\text{Flocculation percentage (\%)} = \frac{OD_{EDTA} - OD_{Flocculation\ buffer}}{OD_{EDTA}} \times 100$$

For the evolved strains, OD_{EDTA} corresponded to the OD_{600nm} value of the non-evolved reference strain.

2.3. Experimental Setup with Ministats

The continuous fermentation of industrial strains was carried out in a ministat set-up (Figure S1) [28]. Briefly, 15 mL test tubes were kept in an analog heat block (VWR®, Bridgeport, NJ, USA) at 30 °C and fed with high-glucose growth medium using a peristaltic pump (Type ISM833A, Ismatec®, Zurich, Switzerland). This high-glucose growth medium was previously also used in the adaptive evolution of *S. cerevisiae* strains towards a snowflake phenotype using a mini tower fermentor [27]. Medium was supplied at a flow rate of 30 µl/min. Air from a 4-port aquarium pump was fed to the medium in the test tube through an air filter (0.20 µm) and the needle was pushed to the bottom of the tube, in order to agitate the solution. During the experiment, a volume of 7–10 mL was maintained. The ministats were inoculated with 1 mL of an overnight culture. The pH, cell concentration (OD at 600 nm) and the glucose concentration (estimated using a refractometer (Brouwland, Belgium)) were measured during the ALE experiments.

2.4. Wort Fermentations in Tall Tubes

Laboratory-scale tall tubes, made from glass (75 cm high and 8-cm diameter), were used to assess beer fermentation with the evolved strains (Figure S2a). The tall tubes were filled with 2 L of wort with a density of 11 °P, which was provided by Biercentrum Delvaux (Neerijse, Belgium), and autoclaved before inoculation. The evolved strains from the ALE experiment and the original brewer's strains BCD1, BCD2, and BCD3 were added to the tall tubes at a cell concentration of 10×10^6 cells/mL. Fermentations were carried out in duplicates and sampled daily.

Alcolyzer Plus Beer Analyzing System (Anton Paar®, Graz, Austria) and headspace gas chromatography (GC) (Autosystem XL, Perkin Elmer®, Waltham, MA, USA) were used for the analyses of the fermentation process. For the GC analysis, the samples were filtered through a filter paper (Grade MN 713 $\frac{1}{4}$, Macherey-Nagel®, Düren, Germany). During the experiment, the apparent extract (% *m*/*m*) and the ethanol content (% *v*/*v*) were measured. The apparent extract (Ea) is a direct measurement of the dissolved solids in brewer's wort, gauged according to specific gravity. During fermentation, the fermentable carbohydrates (glucose, maltose, and maltotriose) are consumed by the yeast and the progress of the fermentation is monitored by measuring the disappearance of these solids [31].

1-butanol), and ethyl caproate) in the beer samples were determined by headspace gas chromatography (HS-GC FID/ECD) as previously described [32,33]. Shortly, collected samples were cooled on ice and after centrifugation, 5 mL of the cooled supernatant was transferred to a vial. The vials were analyzed with a calibrated Autosystem XL gas chromatograph with a headspace autosampler (HS40; Perkin Elmer, Wellesley, MA, USA), equipped with a Chrompack-Wax 52 CB column (length 50 m, 0.32 mm internal diameter, 1.2 µm layer thickness; Varian, Palo Alto, CA, USA). Samples were heated for 16 min at 60 °C in the headspace autosampler before injection (needle temperature 70 °C). Helium was used as the carrier gas. The oven temperature was kept at 50 °C for 7.5 min, increased to 110 °C at 25 °C/min, and was held at that temperature for 3.5 min. Detection of esters, and higher alcohols was established with a flame ionization detector (FID); diacetyl was detected with an electron capture detector (ECD). The FID and ECD temperatures were kept constant at 250 °C and 200 °C, respectively.

3. Results

3.1. Adaptive Evolution in Ministats

Adaptive evolution experiments were performed with the three selected non-flocculating industrial strains using continuous cultivation in the ministats. The flocculation behavior before evolution was assessed and compared to the non-flocculating haploid lab strain BY4742, the strongly flocculating BY4742 [*FLO1*] (constitutively overexpressed *FLO1*), the naturally flocculating BY4742::*FLO8* (functional Flo8p) and the strongly flocculating industrial strain BCD4 (Figure 1a). The flocculation percentages of the three industrial strains were low (<21% ± 2%), ranking them below the natural flocculating BY4742::*FLO8* reference lab strain (46% ± 4%). The strongly flocculating industrial brewer's strain BCD4 shows the same flocculation capacity as the BY4742 lab strain with constitutively overexpressed *FLO1*.

For the ALE experiments, cultivation with the three strains was initiated with a dilution rate of approximately 0.2 h^{-1} using a medium feeding flow rate of 30 µL/min. The cultivation was monitored daily by measuring the glucose concentration, pH, and cell density (Figure 2). Between days 5–7, steady-state was reached and over time, and the dilution rate was increased gradually to 0.35 h^{-1} (BCD1) and 0.45 h^{-1} (BCD2, BCD3) to avoid wash-out and to select for larger aggregates. The continuous cultivation was stopped after 45 days, and the yeast populations were examined by microscopy (Figure 3). The evolved BCD2 aggregates were smaller than the BCD1 and BCD3 aggregates. The nature of the cell clusters was determined by resuspending the cells in EDTA-buffer, which chelates Ca^{+2} ions and disrupts yeast cell clusters if they are formed via flocculin-dependent adhesion. The aggregates persisted for all three strains. This indicated that the clusters are not the result of flocculin interactions and are likely due to failure in separation of the mother and daughter cells, described previously as the "snowflake" phenotype [27,34–36]. The evolved strains were subsequently cultivated in batch cultures and the aggregating phenotype was found to be stable.

The flocculation assay was repeated with the evolved strains to estimate and compare their sedimentation velocity to that of the BCD1, BCD2, and BCD3 strains before evolution (Figure 1). All three evolved strains showed an increase in "flocculation" percentage. Even though none of the strains evolved towards a real flocculating phenotype, their multicellular aggregates contributed to a significant larger sedimentation velocity.

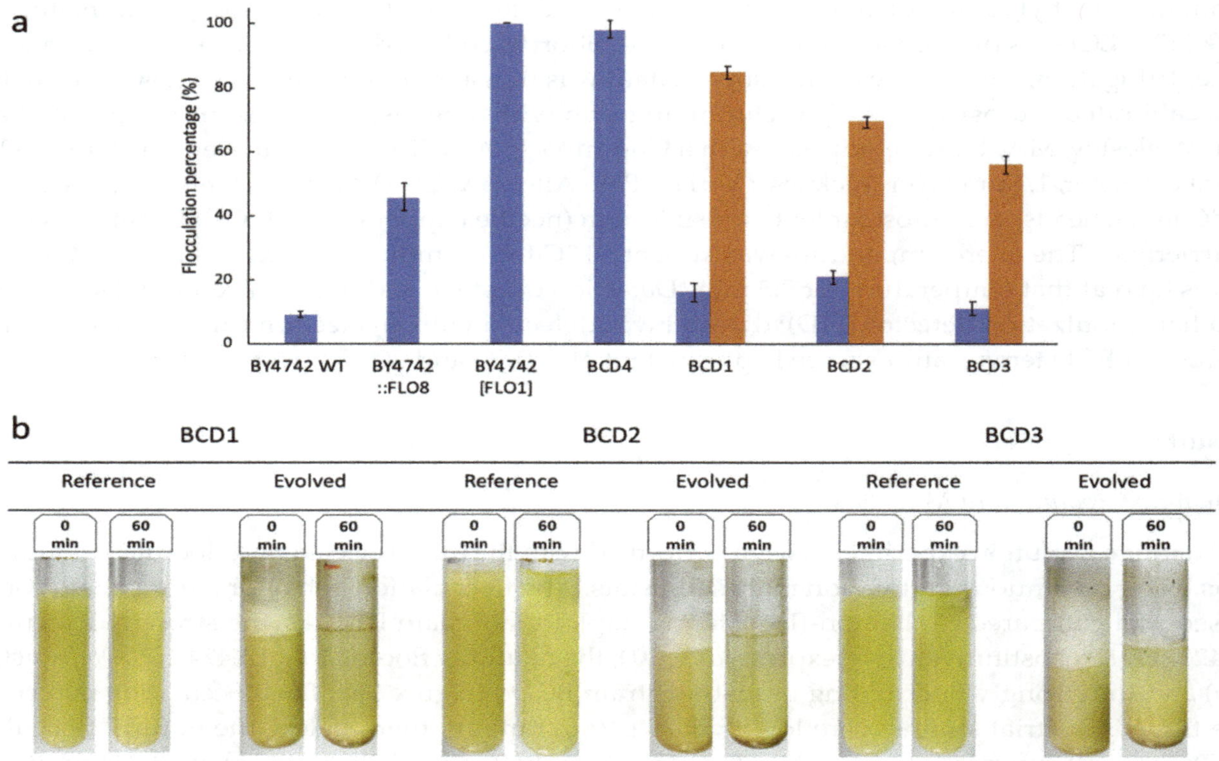

Figure 1. Flocculation and sedimentation behavior of laboratory and industrial strains. (**a**) Flocculation percentages determined for the control strains (BY4742, BY4742::*FLO8*, BY4742 (*FLO1*), the industrial BCD4 strain) and for the three industrial non-flocculating strains BCD1, BCD2, and BCD3 before (■) and after (■) ALE in ministats. All measurements were performed in triplicates. (**b**) The sedimentation behavior of the three industrial brewer's strains before and after adaptive laboratory evolution (ALE) in the ministats.

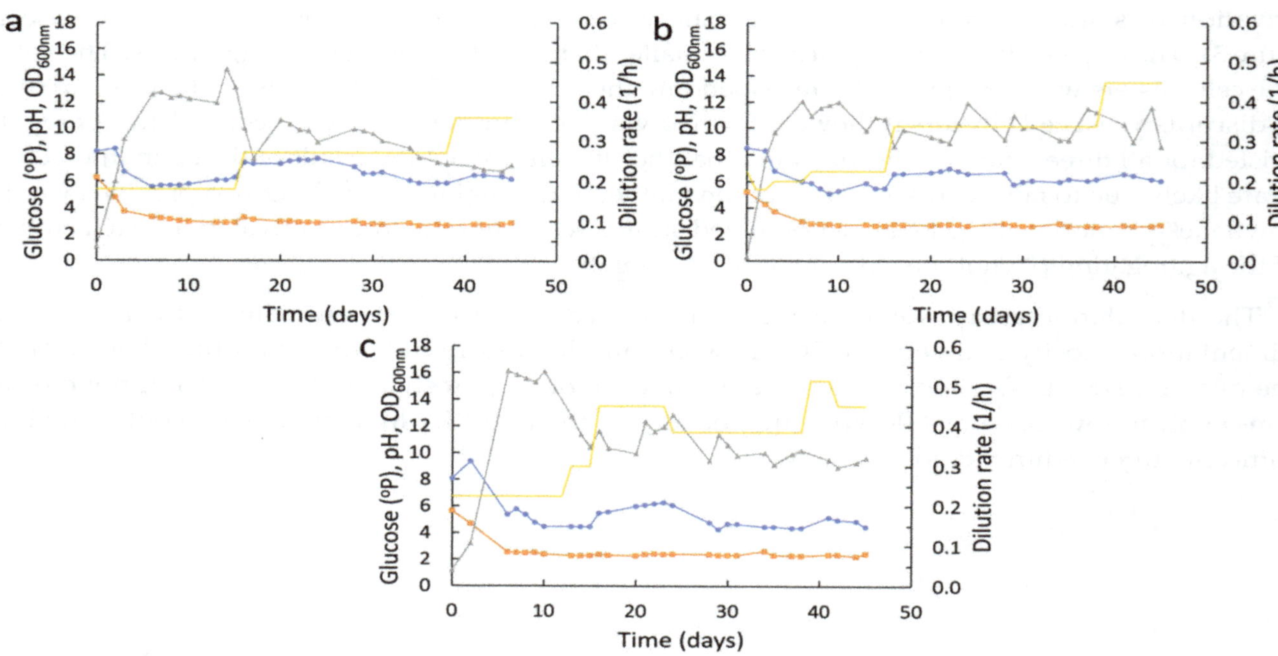

Figure 2. Cultivation of the industrial strains in the ministats: (**a**) BCD1, (**b**) BCD2, and (**c**) BCD3. Glucose content (•), pH (■), $OD_{600\ nm}$ (▲), and the dilution rate (1/h) (—).

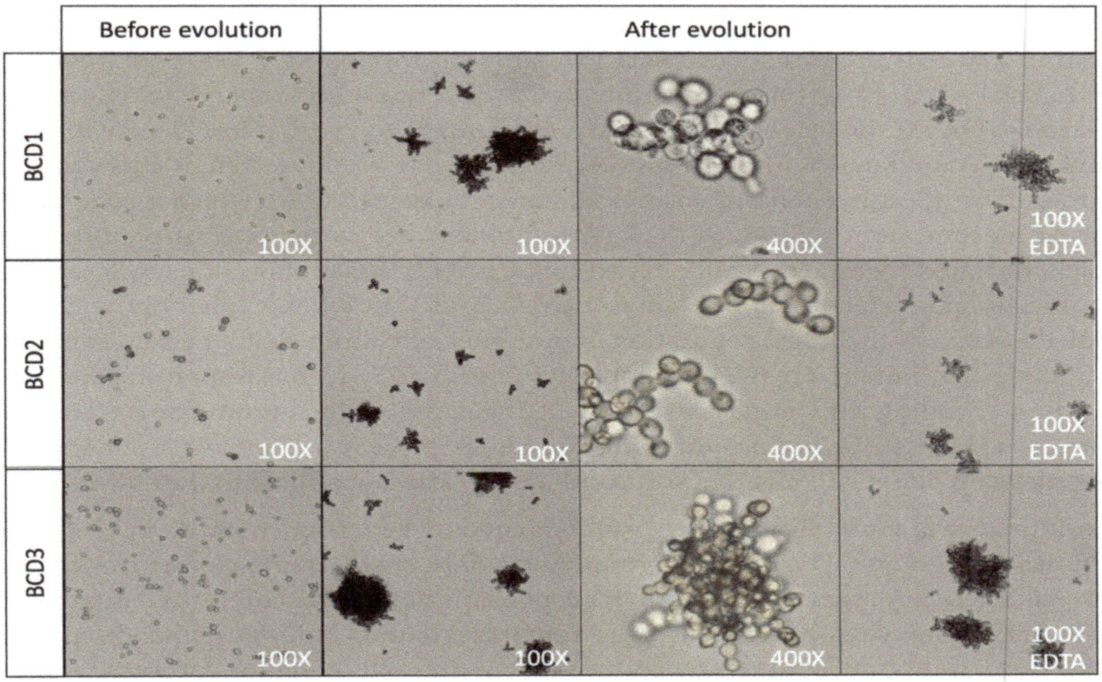

Figure 3. Microscopic observations of the industrial strains before and after the evolution in the ministats. Snowflake clumps are observed for each strain and could not be disrupted by treatment with 50 mM EDTA.

3.2. Performance of Evolved Strains During Beer Fermentation

The performance and behavior of the evolved yeast strains, compared to their reference strains, were evaluated during wort fermentations in tall tubes (Figure S2a). The fermentations were monitored by sampling and measuring cell concentration and beer characteristics such as apparent extract (E_a), ethanol content (Table S1), and the concentration of several flavor compounds (Figure S3 and S4).

The reference strain BCD1 showed a faster fermentation capacity than the evolved strain: The fermentation was almost completed after only 1 day as observed from the evolution of the apparent extract and ethanol concentration (Table S1) as well as from the suspended cell concentration (Figure S2b). The fermentation capacity of the evolved and reference BCD2 strains were similar (Table S1). In contrast, the evolved BCD3 strain showed a faster fermentation than the reference strain (Table S1). The evolution of the suspended cell concentrations during the fermentations for the evolved and reference strains is shown in Figure S1b. There are no large differences observable in the number of suspended cells between the evolved and the reference strains, except for the evolved BCD1 at the second day of fermentation where a much lower cell concentration of the evolved strain was present.

Flavor compounds in the beer were quantified by headspace gas chromatography (Figure S3 and S4). In general, no major influence of the evolved yeast strains on the development of the flavor profile was observed. Some remarkable observations include an increased content of vicinal diketones diacetyl (up to 2.1 ppm) and 2,3-pentanedione (up to 0.9 ppm) during the initial stages of the fermentation for the evolved BCD1 strain, which was decreased by the third day for both compounds to 0.1 ppm. Also, the aliphatic higher alcohol isobutanol concentration increased to 178 ppm (compared to 53 ppm for the reference strain) at the third day of the fermentation. The isobutanol concentration of the evolved BCD2 strain fermentation was doubled at the second day compared to the reference strain, but was still below the isobutanol flavor threshold of 100–200 ppm [37]. By the third day, the evolved BCD3 gave a lower concentration of acetaldehyde than its reference, but a higher concentration of the higher alcohol propanol and isobutanol, and the esters ethylacetate and isoamylacetate.

4. Discussion

Miniature, low-cost chemostats (ministats) were used for adaptive laboratory evolution (ALE) of three industrial *S. cerevisiae* brewer's strains towards a more favorable, aggregating phenotype. The three strains—BCD1, BCD2, and BCD3—were characterized with a low flocculation ability and were continuously cultivated with high-glucose medium for 45 days. Small clusters of cells were observed around day 15, corresponding to approximately 110 generations, which is comparable to other *S. cerevisiae* ALE experiments [27,38,39]. The yeast cell clusters were not disrupted by treatment by EDTA, which indicated that the cell-cell interactions are not based on flocculins. Microscopy showed that the multicellular clusters look like "snowflakes". This "snowflake" phenotype was previously described as the result of the failed separation of daughter cells from mother cells. This phenotype is caused by a frameshift mutation in the transcription factor *ACE2*, which is responsible for the activation of the *CTS1* gene encoding the chitinase necessary to break down the septum between the mother and the daughter cells [34,36,40]. The clusters of cells formed in this way are unlike flocs in that they consist entirely of genetically identical cells and surrounding cells can not adhere to the cluster [36].

The performance of the evolved strains was compared to the reference strains in beer fermentations using tall tubes fermentors. During the fermentation only small aggregates of the evolved strains were observed. These aggregates were kept in suspension during the convective mixing by the CO_2 release by the fermenting yeast cells as was clear from the evolution of the suspended cell concentration (Figure S2b). Apparently, shear stress by convective mixing will break up large aggregates and the presence of these small aggregates will not lead to premature sedimentation during fermentation. The CO_2 production stops at the end of the fermentation and convective currents are reduced significantly. At this moment, the sedimentation of the snowflake aggregates will be significantly faster (Figure 1b) than the reference strains.

To assess the effect of adapted evolution of the 3 strains on the beer flavor, the evolution of a few flavor compounds was determined during the first 3 days of the fermentation. The evolved BCD1 strain showed an increased production of the vicinal diketones and the higher alcohol isobutanol. The synthesis of these compounds is linked to the isoleucine–leucine–valine (ILV) pathway. Although the flavor threshold of diacetyl (0.1–0.15 ppm [41]) was exceeded and the 2,3-pentanedione concentration was close to the flavor threshold (1.0–1.5 ppm [41]) during the first 2 days of fermentation, these concentrations were reduced significantly below the flavor threshold at the third day. Also, the isobutanol content in the beer fermented by the evolved BCD1 strain was much higher than for the reference strain, but did not exceed the flavor threshold of 100–200 ppm [37]. After 7 days, the green beer from the evolved and the reference BCD1 strain both tasted fruity (isoamyl acetate and acetaldehyde). Although the isobutanol concentration was larger in the beer from the evolved BCD2 strain, no difference was tasted. In both beers, the apple flavor (acetaldehyde) could be recognized. The beer from the evolved BCD3 strain contained a higher concentration of the ester isoamyl acetate (banana aroma) and ethyl acetate ester (fruity, solvent-like aroma), which presence could be tasted in the green beer.

Author Contributions: Conceptualization, R.W. and F.R.D.; methodology, R.W., T.V.M.; formal analysis, T.V.M.; investigation, T.V.M., Y.K., F.D., F.R.D., and R.W..; resources, R.W.; writing—original draft preparation, Y.K., T.V.M.; writing—review and editing, Y.K., R.W., F.R., and F.R.D.; supervision, R.W., F.R.D., and F.D.; project administration, R.W.; funding acquisition, R.W. All authors have read and agree to the published version of the manuscript.

Acknowledgments: The Belgian Federal Science Policy Office (Belspo) and the European Space Agency (ESA) PRODEX program supported this work. The Research Council of the Vrije Universiteit Brussel (Belgium) and the University of Ghent (Belgium) are acknowledged to support the Alliance Research Group VUB-UGent NanoMicrobiology (NAMI), and the International Joint Research Group (IJRG) VUB-EPFL BioNanotechnology & NanoMedicine (NANO).

References

1. Bowden, C.P.; Leaver, G.; Melling, J.; Norton, M.G.; Whittington, P.N. Recent and novel developments in the recovery of cells from fermentation broths. In *Separations for Biotechnology*; Verral, M.S., Hudson, M.J., Eds.; Ellis Horwood: Chichester, UK, 1987; pp. 49–61.

2. Stratford, M. Yeast flocculation: A new perspective. *Adv. Microb. Physiol.* **1992**, *33*, 1–71.
3. Vidgren, V.; Londesborough, J. 125th anniversary review: Yeast flocculation and sedimentation in brewing. *J. Inst. Brew.* **2011**, *117*, 475–487. [CrossRef]
4. Goossens, K.; Willaert, R. Flocculation protein structure and cell-cell adhesion mechanism in *Saccharomyces cerevisiae*. *Biotechnol. Lett.* **2010**, *32*, 1571–1585. [CrossRef]
5. Van Mulders, S.E.; Christianen, E.; Saerens, S.M.G.; Daenen, L.; Verbelen, P.J.; Willaert, R.; Verstrepen, K.J.; Delvaux, F.R. Phenotypic diversity of Flo protein family-mediated adhesion in *Saccharomyces cerevisiae*. *FEMS Yeast Res.* **2009**, *9*, 178–190. [CrossRef] [PubMed]
6. Willaert, R.G. Adhesins of yeasts: Protein structure and interactions. *J. Fungi* **2018**, *4*, 119. [CrossRef] [PubMed]
7. Brückner, S.; Mösch, H.-U. Choosing the right lifestyle: Adhesion and development in *Saccharomyces cerevisiae*. *FEMS Microbiol. Rev.* **2011**, *36*, 25–58. [CrossRef] [PubMed]
8. Verstrepen, K.J.; Derdelinckx, G.; Verachtert, H.; Delvaux, F.R. Yeast flocculation: What brewers should know. *Appl. Microbiol. Biotechnol.* **2003**, *61*, 197–205. [CrossRef] [PubMed]
9. Gibson, B.R.; Lawrence, S.J.; Leclaire, J.P.R.; Powell, C.D.; Smart, K.A. Yeast Responses to Stresses Associated with Industrial Brewery Handling. *FEMS Microbiol. Rev.* **2007**, *31*, 535–569. [CrossRef]
10. Soares, E.V. Flocculation in *Saccharomyces cerevisiae*: A review. *J. Appl. Microbiol.* **2010**, *110*, 1–18. [CrossRef]
11. Dequin, S. The potential of genetic engineering for improving brewing, wine-making and baking yeasts. *Appl. Microbiol. Biotechnol.* **2001**, *56*, 577–588. [CrossRef]
12. Saerens, S.M.G.; Duong, C.T.; Nevoigt, E. Genetic improvement of brewer's yeast: Current state, perspectives and limits. *Appl. Microbiol. Biotechnol.* **2010**, *86*, 1195–1212. [CrossRef] [PubMed]
13. Conrad, T.M.; Lewis, N.E.; Palsson, B.Ø. Microbial laboratory evolution in the era of genome-scale science. *Mol. Syst. Biol.* **2011**, *7*, 509. [CrossRef] [PubMed]
14. Portnoy, V.A.; Bezdan, D.; Zengler, K. Adaptive laboratory evolution-harnessing the power of biology for metabolic engineering. *Curr. Opin. Biotechnol.* **2011**, *22*, 590–594. [CrossRef] [PubMed]
15. Dragosits, M.; Mattanovich, D. Adaptive laboratory evolution—Principles and applications for biotechnology. *Microb. Cell Fact.* **2013**, *12*, 64. [CrossRef]
16. Sonderegger, M.; Sauer, U. Evolutionary engineering of *Saccharomyces cerevisiae* for anaerobic growth on xylose. *Appl. Environ. Microbiol.* **2003**, *69*, 1990–1998. [CrossRef]
17. Dhar, R.; Sägesser, R.; Weikert, C.; Wagner, A. Yeast adapts to a changing stressful environment by evolving cross-protection and anticipatory gene regulation. *Mol. Biol. Evol.* **2013**, *30*, 573–588. [CrossRef]
18. Ekberg, J.; Rautio, J.; Mattinen, L.; Vidgren, V.; Londesborough, J.; Gibson, B.R. Adaptive evolution of the lager brewing yeast *Saccharomyces pastorianus* for improved growth under hyperosmotic conditions and its influence on fermentation performance. *FEMS Yeast Res.* **2013**, *13*, 335–349. [CrossRef]
19. Strejc, J.; Siříšťová, L.; Karabín, M.; Almeida e Silva, J.B.; Brányik, T. Production of alcohol-free beer with elevated amounts of flavouring compounds using lager yeast mutants. *J. Inst. Brew.* **2013**, *119*, 149–155. [CrossRef]
20. Huuskonen, A.; Markkula, T.; Vidgren, V.; Lima, L.; Mulder, L.; Geurts, W.; Walsh, M.; Londesborough, J. Selection from Industrial Lager Yeast Strains of Variants with Improved Fermentation Performance in Very-High-Gravity Worts. *Appl. Environ. Microbiol.* **2010**, *76*, 1563–1573. [CrossRef]
21. Yu, Z.; Zhao, H.; Li, H.; Zhang, Q.; Lei, H.; Zhao, M. Selection of *Saccharomyces pastorianus* variants with improved fermentation performance under very high gravity wort conditions. *Biotechnol. Lett.* **2012**, *34*, 365–370. [CrossRef]
22. Cadière, A.; Ortiz-Julien, A.; Camarasa, C.; Dequin, S. Evolutionary engineered *Saccharomyces cerevisiae* wine yeast strains with increased in vivo flux through the pentose phosphate pathway. *Metab. Eng.* **2011**, *13*, 263–271. [CrossRef] [PubMed]
23. Cadière, A.; Aguera, E.; Caillé, S.; Ortiz-Julien, A.; Dequin, S. Pilot-scale evaluation the enological traits of a novel, aromatic wine yeast strain obtained by adaptive evolution. *Food Microbiol.* **2012**, *32*, 332–337. [CrossRef] [PubMed]
24. Weikert, C.; Sauer, U.; Bailey, J.E. Use of a glycerol-limited, long-term chemostat for isolation of Escherichia coli mutants with improved physiological properties. *Microbiology* **1997**, *143*, 1567–1574. [CrossRef] [PubMed]
25. Koppram, R.; Albers, E.; Olsson, L. Evolutionary engineering strategies to enhance tolerance of xylose utilizing recombinant yeast to inhibitors derived from spruce biomass. *Biotechnol. Biofuels* **2012**, *5*, 32. [CrossRef] [PubMed]

26. Wang, L.; Spira, B.; Zhou, Z.; Feng, L.; Maharjan, R.P.; Li, X.; Li, F.; McKenzie, C.; Reeves, P.R.; Ferenci, T. Divergence involving global regulatory gene mutations in an Escherichia coli population evolving under phosphate limitation. *Genome Biol. Evol.* **2010**, *2*, 478–487. [CrossRef]
27. Conjaerts, A.; Willaert, R. Gravity-Driven Adaptive Evolution of an Industrial Brewer's Yeast Strain towards a Snowflake Phenotype in a 3D-Printed Mini Tower Fermentor. *Fermentation* **2017**, *3*, 4. [CrossRef]
28. Miller, A.W.; Befort, C.; Kerr, E.O.; Dunham, M.J. Design and use of multiplexed chemostat arrays. *J. Vis. Exp.* **2013**, e50262. [CrossRef]
29. Brachmann, C.B.; Davies, A.; Cost, G.J.; Caputo, E.; Li, J.; Hieter, P.; Boeke, J.D. Designer Deletion Strains derived from *Saccharomyces cerevisiae* S288C: A Useful set of Strains and Plasmids for PCR-mediated Gene Disruption and Other Applications. *Yeast* **1998**, *14*, 115–132. [CrossRef]
30. D'Hautcourt, O.; Smart, K.A. Measurement of brewing yeast flocculation. *J. Am. Soc. Brew. Chem.* **1999**, *57*, 123–128. [CrossRef]
31. Shimizu, H.; Mizuno, S.; Hiroshima, T.; Shioya, S. Effect of carbon and nitrogen additions on consumption activity of apparent extract of yeast cells in a brewing process. *J. Am. Soc. Brew. Chem.* **2002**, *60*, 163–169. [CrossRef]
32. Verbelen, P.J.; Dekoninck, T.M.L.; Saerens, S.M.G.; Van Mulders, S.E.; Thevelein, J.M.; Delvaux, F.R. Impact of pitching rate on yeast fermentation performance and beer flavour. *Appl. Microbiol. Biotechnol.* **2009**, *82*, 155–167. [CrossRef] [PubMed]
33. Dekoninck, T.M.L.; Mertens, T.; Delvaux, F.; Delvaux, F.R. Influence of beer characteristics on yeast refermentation performance during bottle conditioning of belgian beers. *J. Am. Soc. Brew. Chem.* **2013**, *71*, 23–34. [CrossRef]
34. Ratcliff, W.C.; Fankhauser, J.D.; Rogers, D.W.; Greig, D.; Travisano, M. Origins of multicellular evolvability in snowflake yeast. *Nat. Commun.* **2015**, *6*, 6102. [CrossRef] [PubMed]
35. Ratcliff, W.C.; Denison, R.F.; Borrello, M.; Travisano, M. Experimental evolution of multicellularity. *Proc. Natl. Acad. Sci. USA* **2012**, *109*, 1595–1600. [CrossRef]
36. Oud, B.; Guadalupe-Medina, V.; Nijkamp, J.F.; De Ridder, D.; Pronk, J.T.; Van Maris, A.J.A.; Daran, J.M. Genome duplication and mutations in ACE2 cause multicellular, fast-sedimenting phenotypes in evolved *Saccharomyces cerevisiae*. *Proc. Natl. Acad. Sci. USA* **2013**, *110*, E4223–E4231. [CrossRef]
37. Djordjević, V.; Willaert, R.; Gibson, B.; Nedović, V. Immobilized Yeast Cells and Secondary Metabolites. In *Fungal Metabolites*; Springer International Publishing: Berlin/Heidelberg, Germany, 2016; pp. 1–40.
38. Ratcliff, W.C.; Pentz, J.T.; Travisano, M. Tempo and mode of multicellular adaptation in experimentally evolved *Saccharomyces cerevisiae*. *Evolution* **2013**, *67*, 1573–1581. [CrossRef]
39. Hope, E.A.; Amorosi, C.J.; Miller, A.W.; Dang, K.; Heil, C.S.; Dunham, M.J. Experimental evolution reveals favored adaptive routes to cell aggregation in yeast. *Genetics* **2017**, *206*, 1153–1167. [CrossRef]
40. King, L.; Butler, G. Ace2p, a regulator of CTS1 (chitinase) expression, affects pseudohyphal production in *Saccharomyces cerevisiae*. *Curr. Genet.* **1998**, *34*, 183–191. [CrossRef]
41. Meilgaard, M.C. Flavor chemistry of beer: Part ii: Flavour and threshold of 239 aroma volatiles. *MBAA Tech. Q.* **1975**, *12*, 151–168.

PERMISSIONS

All chapters in this book were first published in MDPI; hereby published with permission under the Creative Commons Attribution License or equivalent. Every chapter published in this book has been scrutinized by our experts. Their significance has been extensively debated. The topics covered herein carry significant findings which will fuel the growth of the discipline. They may even be implemented as practical applications or may be referred to as a beginning point for another development.

The contributors of this book come from diverse backgrounds, making this book a truly international effort. This book will bring forth new frontiers with its revolutionizing research information and detailed analysis of the nascent developments around the world.

We would like to thank all the contributing authors for lending their expertise to make the book truly unique. They have played a crucial role in the development of this book. Without their invaluable contributions this book wouldn't have been possible. They have made vital efforts to compile up to date information on the varied aspects of this subject to make this book a valuable addition to the collection of many professionals and students.

This book was conceptualized with the vision of imparting up-to-date information and advanced data in this field. To ensure the same, a matchless editorial board was set up. Every individual on the board went through rigorous rounds of assessment to prove their worth. After which they invested a large part of their time researching and compiling the most relevant data for our readers.

The editorial board has been involved in producing this book since its inception. They have spent rigorous hours researching and exploring the diverse topics which have resulted in the successful publishing of this book. They have passed on their knowledge of decades through this book. To expedite this challenging task, the publisher supported the team at every step. A small team of assistant editors was also appointed to further simplify the editing procedure and attain best results for the readers.

Apart from the editorial board, the designing team has also invested a significant amount of their time in understanding the subject and creating the most relevant covers. They scrutinized every image to scout for the most suitable representation of the subject and create an appropriate cover for the book.

The publishing team has been an ardent support to the editorial, designing and production team. Their endless efforts to recruit the best for this project, has resulted in the accomplishment of this book. They are a veteran in the field of academics and their pool of knowledge is as vast as their experience in printing. Their expertise and guidance has proved useful at every step. Their uncompromising quality standards have made this book an exceptional effort. Their encouragement from time to time has been an inspiration for everyone.

The publisher and the editorial board hope that this book will prove to be a valuable piece of knowledge for researchers, students, practitioners and scholars across the globe.

LIST OF CONTRIBUTORS

Bartłomiej Zieniuk, Patrycja Mazurczak-Zieniuk and Agata Fabiszewska
Department of Chemistry, Institute of Food Sciences, Warsaw University of Life Sciences-SGGW, 159c Nowoursynowska St, 02-776 Warsaw, Poland

Ricardo Vejarano
Dirección de Investigación y Desarrollo, Universidad Privada del Norte (UPN), Trujillo 13001, Peru

Friederike Rex and Adeline Hirschler
Institute for Viticulture and Oenology, Dienstleistungszentrum Ländlicher Raum (DLR) Rheinpfalz, Breitenweg 71, D-67435 Neustadt, Germany

Maren Scharfenberger-Schmeer
Institute for Viticulture and Oenology, Dienstleistungszentrum Ländlicher Raum (DLR) Rheinpfalz, Breitenweg 71, D-67435 Neustadt, Germany
Hochschule Kaiserslautern, Weincampus Neustadt, Breitenweg 71, D-67435 Neustadt, Germany

Jozef Sabo
Department of Fruit Sciences, Viticulture and Enology, Faculty of Horticulture and Landscape Engineering, Slovak University of Agriculture, Tr. A. Hlinku 2, 94976 Nitra, Slovakia

Miroslava Kačániová
Department of Fruit Sciences, Viticulture and Enology, Faculty of Horticulture and Landscape Engineering, Slovak University of Agriculture, Tr. A. Hlinku 2, 94976 Nitra, Slovakia
Department of Bioenergy, Food Technology and Microbiology, Institute of Food Technology and Nutrition, University of Rzeszow, Zelwerowicza St. 4, 35601 Rzeszow, Poland

Simona Kunová
Department of Food Hygiene and Safety, Faculty of Biotechnology and Food Sciences, Slovak University of Agriculture, Tr. A. Hlinku 2, 94976 Nitra, Slovakia

Eva Ivanišová
Department of Technology and Quality of Plant Products, Faculty of Biotechnology and Food Sciences, Slovak University of Agriculture, Tr. A. Hlinku 2, 94976 Nitra, Slovakia

Jana Žiarovská
Department of Plant Genetics and Breeding, Faculty of Agrobiology and Food Resources, Slovak University of Agriculture, Tr. A. Hlinku 2, 94976 Nitra, Slovakia

Soňa Felsöciová
Department of Microbiology, Faculty of Biotechnology and Food Sciences, Slovak University of Agriculture, Tr. A. Hlinku 2, 94976 Nitra, Slovakia

Margarita Terentjeva
Institute of Food and Environmental Hygiene, Faculty of Veterinary Medicine, Latvia University of Life Sciences and Technologies, K. Helmaņa iela 8, LV-3004 Jelgava, Latvia

Marina Ruíz-Muñoz, Maria del Carmen Bernal-Grande, Gustavo Cordero-Bueso, Mónica González, David Hughes-Herrera and Jesús Manuel Cantoral
Department of Biomedicine, Biotechnology and Public Health, University of Cádiz, Cádiz 11510, Spain

Mingyan Yang
School Environment Science Engineering, Chang'an University, Xi'an 710064, China
USDA Forest Service, Forest Products Laboratory, Madison, WI 53726, USA
Key Laboratory of Subsurface Hydrology and Ecological Effects in Arid Region, Ministry of Education, Chang'an University, Xi'an 710064, China

Hairui Ji
USDA Forest Service, Forest Products Laboratory, Madison, WI 53726, USA
College of Life Science and Technology, Beijing University of Chemical Technology, Beijing 100029, China

J.Y. Zhu
USDA Forest Service, Forest Products Laboratory, Madison, WI 53726, USA

Jared Muysson and Robert Allie
Centre for Biotechnology, Brock University, St. Catharines, ON L2S 3A1, Canada

Laurianne Miller
Department of Biological Sciences, Brock University, St. Catharines, ON L2S 3A1, Canada

Debra L. Inglis
Centre for Biotechnology, Brock University, St. Catharines, ON L2S 3A1, Canada
Department of Biological Sciences, Brock University, St. Catharines, ON L2S 3A1, Canada
Cool Climate Oenology and Viticulture Institute, Brock University, St. Catharines, ON L2S 3A1, Canada

Charlotte Yvanoff
Research Group Structural Biology Brussels, Vrije Universiteit Brussel, 1050 Brussels, Belgium
Alliance Research Group VUB-UGent NanoMicrobiology (NAMI), International Joint Research Group
VUB-EPFL NanoBiotechnology & NanoMedicine (NANO), Vrije Universiteit Brussel, 1050 Brussels, Belgium

Stefania Torino
Research Group Structural Biology Brussels, Vrije Universiteit Brussel, 1050 Brussels, Belgium
Center for Structural Biology, Vlaams Instituut voor Biotechnologie (VIB), 1050 Brussels, Belgium

Ronnie G. Willaert
Research Group Structural Biology Brussels, Vrije Universiteit Brussel, 1050 Brussels, Belgium
Alliance Research Group VUB-UGent NanoMicrobiology (NAMI), International Joint Research Group VUB-EPFL NanoBiotechnology & NanoMedicine (NANO), Vrije Universiteit Brussel, 1050 Brussels, Belgium
Department Bioscience Engineering, University Antwerp, 2000 Antwerp, Belgium

Davide Ravasio and Andrea Walther
Carlsberg Research Laboratory, Yeast & Fermentation, J. C. Jacobsens Gade 4, DK-1799 Copenhagen V, Denmark

Silvia Carlin and Urska Vrhovsek
Fondazione Edmund Mach, Research and Innovation Centre, Food Quality and Nutrition Department, Via E. Mach 1, I-38010 S. Michele all' Adige, Italy

Teun Boekhout
Westerdijk Fungal Biodiversity Institute, 3584 CT Utrecht, The Netherlands
Institute of Biodiversity and Ecosystem Dynamics (IBED), University of Amsterdam, 1000 GG Amsterdam, The Netherlands

Marizeth Groenewald
Westerdijk Fungal Biodiversity Institute, 3584 CT Utrecht, The Netherlands

Jürgen Wendland
Carlsberg Research Laboratory, Yeast & Fermentation, J. C. Jacobsens Gade 4, DK-1799 Copenhagen V, Denmark
Vrije Universiteit Brussel, Research Group of Microbiology (MICR)—Functional Yeast Genomics, BE-1050 Brussels, Belgium

Frida S. Gustafsson, Marissa Neuner, Chrystal M. Scholl, Sydney C. Morgan and Daniel M. Durall
The University of British Columbia (UBC), Okanagan, Biology Department, 1177 Research Rd, Kelowna, BC V1V 1V7, Canada

Vladimir Jiranek
Department of Wine and Food Science, The University of Adelaide, PMB1, Glen Osmond SA 5064, Australia

Laura Canonico, Edoardo Galli, Francesca Comitini and Maurizio Ciani
Department of Life and Environmental Sciences, Marche Polytechnic University, Via Brecce Bianche, 60131 Ancona, Italy

Enrico Ciani
Birra dell'Eremo, Microbrewery, Via Monte Peglia, 5, 06081 Assisi, Italy

Ali Abghari and Shulin Chen
Department of Biological Systems Engineering, Bioprocessing and Bioproducts Engineering Laboratory, Washington State University, Pullman, WA 99163, USA

Catherine Madzak
UMR 782 GMPA, CBAI, INRA—AgroParisTech-Université Paris-Saclay, F-78850 Thiverval Grignon, France

Katrin Matti, Silvia Brezina, Heike Semmler, Christian von Wallbrunn, Doris Rauhut and Jürgen Wendland
Department of Microbiology and Biochemistry, Hochschule Geisenheim University, Von-Lade-Strasse 1, D-65366 Geisenheim, Germany

Beatrice Bernardi
Department of Microbiology and Biochemistry, Hochschule Geisenheim University, Von-Lade-Strasse 1, D-65366 Geisenheim, Germany
Research Group of Microbiology (MICR)—Functional Yeast Genomics, Vrije Universiteit Brussel, Pleinlaan 2, BE-1050 Brussels, Belgium

Ian S. Murdoch, Samantha L. Powers and Aaron Z. Welch
Biomolecular Sciences Institute, Florida International University, Miami 33199, FL, USA

Alexander da Silva Vale, Gilberto Vinícius de Melo Pereira, Dão Pedro de Carvalho Neto, Cristine Rodrigues and Carlos Ricardo Soccol
Department of Bioprocess Engineering and Biotechnology, Federal University of Paraná (UFPR), 19011 Curitiba, Paraná 81531-980, Brazil

Maria Giovana B. Pagnoncelli
Departament of Chemistry and Biology, Federal University of Technology—Paraná (UTFPR), Curitiba PR 80230-901, Brazil

Martina Schlander, Eckhard Thines and Harald Claus
Institute for Microbiology and Wine Research, Johannes Gutenberg University of Mainz, Becherweg 15, D-55099 Mainz, Germany

Ute Distler and Stefan Tenzer
Institute of Immunology, University Medical Centre of the Johannes Gutenberg University Mainz, Langenbeckstr. 1, D-55131 Mainz, Germany

Yeseren Kayacan and Thijs Van Mieghem
Alliance Research Group VUB-UGent NanoMicrobiology (NAMI), Research Group Structural Biology, 1050 Brussels, Belgium
International Joint Research Group VUB-EPFL NanoBiotechnology & NanoMedicine (NANO), Vrije Universiteit Brussel, 1050 Brussels, Belgium

Filip Delvaux and Freddy R. Delvaux
Biercentrum Delvaux, 3040 Neerijse, Belgium

Ronnie Willaert
Alliance Research Group VUB-UGent NanoMicrobiology (NAMI), Research Group Structural Biology, 1050 Brussels, Belgium
International Joint Research Group VUB-EPFL NanoBiotechnology & NanoMedicine (NANO), Vrije Universiteit Brussel, 1050 Brussels, Belgium
Department Bioscience Engineering, University Antwerp, 2020 Antwerp, Belgium

Index

www.ingramcontent.com/pod-product-compliance
Lightning Source LLC
LaVergne TN
LVHW060514210926
848485LV00005B/1078